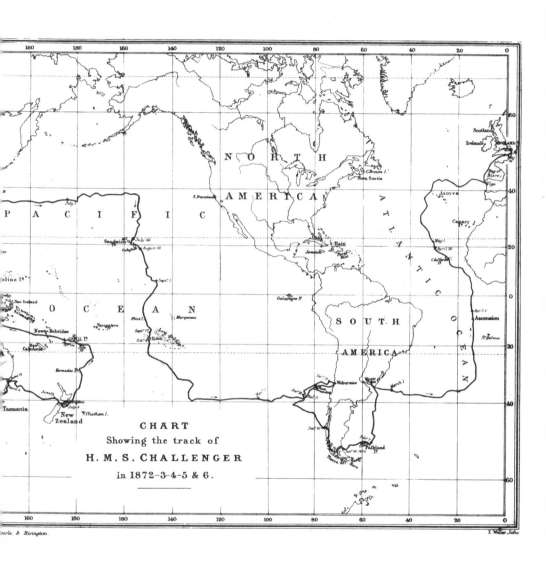

CHART
Showing the track of
H. M. S. CHALLENGER
in 1872-3-4-5 & 6.

The *Challenger* Letters of
Joseph Matkin
Edited by Philip F. Rehbock

At Sea with the Scientifics

University of Hawaii Press

Honolulu

© 1992 University of Hawaii Press
All rights reserved
Printed in the United States of America
92 93 94 95 96 97 5 4 3 2 1

Library of Congress Cataloging-in-Publication Data
Matkin, Joseph.
   At sea with the scientifics : the Challenger letters of Joseph
Matkin / edited by Philip F. Rehbock.
      p.   cm.
   Includes bibliographical references and index.
   ISBN 0-8248-1424-X (acid-free)
   1. Scientific expeditions.   2. Matkin, Joseph—
Correspondence.
3. Challenger (Ship)   4. Oceanographers—
Great Britain—Biography.
I. Rehbock, Philip F., 1942-   II. Title.
Q115.M42   1992
910.4'1'092—dc20
[B]   92-29881
                                                              CIP

Endpapers: Chart of *Challenger*'s track
(W. J. J. Spry, *The Cruise of her Majesty's Ship "Challenger"*
[London: Sampson Low, Marston, Searle, and Rivington,
1876], at p. xvi).

University of Hawaii Press books are printed on acid-free
paper and meet the guidelines for permanence and durability
of the Council on Library Resources

Designed by Cameron Poulter

To the descendants of
JOSEPH MATKIN
and to Karen

Contents

Preface

Few events in the life of the historian evoke more excitement than the appearance of a major new collection of original manuscripts bearing on the subject of his or her research specialty. Since my earliest days in graduate school, I had harbored a fascination, if not a reverence, for that famous Victorian scientific voyage, the oceanographic expedition of H.M.S. *Challenger*. Consequently, when in 1985 Scripps Institution of Oceanography announced its acquisition of twenty-three original letters written during the expedition by a member of *Challenger's* below-decks crew, I was understandably curious. The expedition had, of course, been well chronicled by both the ship's officers and its scientific staff. But letters by a member of the crew hinted of a new perspective on the expedition, a less formal, less official, more vernacular, and perhaps closer-to-the-raw-facts account of the events during those long three and a half years at sea.

My imagination certainly did not prepare me, however, for the detailed contents, and indeed the sheer volume, of the chronicle so carefully penned by Joseph Matkin, ship's steward's assistant. In addition to the original letters from the voyage, Scripps eventually acquired microfilm of an 846-page letterbook containing transcriptions of a further thirty-six letters, plus a small collection of letters Matkin wrote during the years immediately preceding the expedition. Moreover, by coincidence the British Museum (Natural History) had independently acquired at auction a collection of nine Matkin letters just months before the Scripps donation was received. In sum, there were no fewer than sixty-nine *Challenger* letters, some more than a dozen pages long.

From the outset, Scripps director Dr. William Nierenberg and Scripps archivist Deborah Day recognized the momentousness of their new resource and took immediate steps to ensure its proper development. A Matkin Publication Committee of prominent oceanographic scientists was formed, chaired by Nierenberg and including geologist Robert L. Fisher, geophysicist Walter Munk, ichthyologist Richard Rosenblatt, and the late Bill Menard, geologist, along with archivists Day and Eileen Brunton, the latter of the British Museum (Natural History). To be appointed editor of the Matkin letters by this prestigious group was both a joyous and a humbling event.

That was, however, only the beginning of my indebtedness. By the time I arrived at Scripps, the detailed archival work associated with the Matkin collection, including an extensive correspondence with Matkin descendants, had been all but concluded by Deborah Day. Deborah has been a delightful and devoted co-worker in every aspect of this project; in fact, I doubt very much that there is a more competent and energetic archivist at any scientific institution in this country.

At the British Museum (Natural History) Palaeontological Library I have had the great fortune to work closely with Eileen Brunton, who has for many years taken a personal interest in the museum's unexcelled *Challenger* collections; her knowledge of the *Challenger* photographic collection has made possible many of the illustrations in this book. Dr. Tony Rice of the Institute of Oceanographic Sciences, Wormley, has been exceptionally generous in providing both materials and encouragement for the project. Margaret Deacon and an anonymous referee gave very careful readings of the entire manuscript and made many useful suggestions, far beyond those specifically cited in the notes. And a special note of appreciation also goes to my good friend Peter James, whose native knowledge of the byways of Rutland made idyllic a day spent exploring the haunts of Joseph Matkin— "Matkinizing" as we called it.

Matkin descendants have contributed substantially to this project, most notably his granddaughter, who donated the original Matkin collection to Scripps and who produced the first typescript of the Matkin letters, a task of deciphering at which I continue to marvel.

Presentation of the Matkin story at several conferences and seminars has elicited some important insights from colleagues. George Simson and Anthony Friedson of the University of Hawaii Biography Brown Bag Seminar were especially useful in this respect.

Finally, it is to my wife Karen most of all that I owe the progress of this project. Long before I could begin the task of editing, she spent countless hours—mostly stolen from holidays—transcribing the Matkin letters to word-processor. Many of the insights I have explored here began in her imaginative questioning of Matkin passages. From our first acquaintance with the Matkin letters to the final reading of proofs, she has been an enthusiastic and dedicated conspirer.

Abbreviations

JM	Joseph Matkin
JTS	John Thomas Swann
M1, M2, etc.	Serial numbering of Matkin letters. *See* Appendix B for a complete calendar.
NHM	The Natural History Museum, London (formerly British Museum [Natural History])
PRO	Public Record Office, Kew
SIO	Archives, Scripps Institution of Oceanography, University of California, San Diego
Campbell, Log-letters	Lord George Campbell, *Log-letters from "The Challenger,"* 2d revised ed. (London: Macmillan and Co., 1877)
Challenger Reports	*Report on the Scientific Results of the Voyage of H.M.S. Challenger during the Years 1873-76,* 50 vols. (London: HMSO, 1880-1895)
Moseley, Notes	H. N. Moseley, *Notes by a Naturalist on the "Challenger," being an Account of Various Observations made during the Voyage of H.M.S. "Challenger" Round the World, in the Years 1872-1876* (London: Macmillan and Co., 1879)
Spry, Cruise	W. J. J. Spry, *The Cruise of her Majesty's Ship "Challenger"* (London: Sampson Low, Marston, Searle, and Rivington, 1876)
Summary of Results	John Murray, *A Summary of the Scientific Results Obtained at the Sounding, Dredging and Trawling Stations of H.M.S. "Challenger"* (Edinburgh: HMSO, 1895)
Thomson, Voyage	C. Wyville Thomson, *The Voyage of the "Challenger": The Atlantic,* 2 vols. (London: Macmillan and Co., 1877)

Tizard et al.,   T. H. Tizard, H. N. Moseley, J. Y. Buchanan, and John
Narrative   Murray, *Narrative of the Cruise of H.M.S. "Challenger"*
   with a General Account of the Scientific Results of the
   Expedition, 2 vols. in 3 (London: HMSO, 1885)

Wild,   John James Wild, *A Narrative of Experiences Afloat and*
At Anchor   *Ashore During the Voyage of H.M.S. "Challenger" from*
   1872 to 1876 (London: Marcus Ward, 1878)

At Sea with the Scientifics

Introduction

> [H]owever entertaining and well written these books [by naval officers and passengers] may be, and however accurately they may give sea life as it appears to their authors, it must still be plain to every one that a naval officer, who goes to sea as a gentleman, "with his gloves on" (as the phrase is), and who associates only with his fellow officers, and hardly speaks to a sailor except through a boatswain's mate, must take a very different view of the whole matter from that which would be taken by a common sailor. . . . *A voice from the forecastle* has hardly yet been heard.
>
> —Richard Henry Dana,
> *Two Years Before the Mast* (1840)

The exploits of navies have nearly always been told by the officers in command, rarely by the common sailor whose labors kept ship afloat and under way. Official reports of naval movements are, of course, the responsibility of the ship's captain, and subsidiary accounts have historically been the work of other wardroom members. In the case of voyages of discovery and exploration, the accounts of officers have sometimes been supplemented by the narratives of civilian men of science. But the voice of the able seaman is a voice seldom encountered in print.

These observations are valid for Britain's Royal Navy no less than for any other. Of course there have been a few British sailors, both real and mythical, who were famous enough that their names have survived time's fogging of the past. "Jack Tar" long ago became a generic handle for the typical British seaman.[1] But the perspective he represents, the unofficial, "below-decks" or (as Dana called it) "forecastle" point of view, the view not of the hero but of the common man, is rare in the annals of naval or maritime literature.

This situation is especially regrettable for those most exciting of peacetime expeditions, the voyages of exploration. What was it like to be a seaman sailing into unknown waters under Columbus, Cartier, or Cook?[2] Did the typical seaman understand—or care about—the larger purposes of the enterprise in which he, perhaps by accident, found

I

himself? Did he always share the attitudes of his superiors upon encountering foreign shores and peoples? Was his career ever advanced, as theirs so often were, by such adventures? And would our image of any one of these voyages, and of its leaders, be substantially transformed by an unexpurgated, firsthand assessment from "below decks"?

The possibilities of "below-decks history" have not gone entirely unnoticed. In *From the Lower Deck* and *Before the Mast*, Henry Baynham has given us excerpts from the diaries and letters of British sailors from the late eighteenth century down to the 1880s.[3] Baynham was able to draw on the (sometimes anonymously) published autobiographies of a few of his "Tars," but many were accessible only through manuscripts. He was not specifically concerned with scientific voyages, although his second work, covering the period from the close of the Napoleonic Wars to late in Victoria's reign—the Pax Britannica at its peak—does include the writings of several men caught up in Arctic exploration.[4] But its principal value lies in having captured British naval life in a period of great technological change: the replacement of sail by steam, of wood by iron, of cannon by gun turrets, and of cannonballs by shells.

Unfortunately, it seems improbable that many more fresh crew-member accounts of these voyages will come to light. Sailors were recruited, after all, for their willingness to perform physical labor and their ability to tolerate the special conditions of shipboard life, not for their literary abilities, which were in most instances nonexistent. The typical seaman had but a few years' schooling and often had to rely on superiors—the ship's writer or yeoman or a sympathetic division officer —to write the odd letter. He was not, by social origin or education, disposed to keep a diary, nor would his required labors or physical circumstances aboard ship have encouraged it. And with neither letters nor journals to record events in the freshness of their occurrence, the sailor could hardly be expected to produce a publishable volume, regardless of the significance of the venture in which he had taken part.

The *Challenger* Expedition

Until recently these obstacles seemed to be operating even in the case of the last of the great traditional voyages of exploration, the oceanographic circumnavigation of H.M.S. *Challenger*. The *Challenger* expedition, a joint venture of the Royal Society of London and the British Admiralty, was probably the most lavish instance of government-sponsored "big science" of the Victorian era.[5] Departing from British shores in December of 1872, *Challenger* sounded, dredged, and trawled the floors of the world's oceans at no fewer than 362 locations, before returning in June 1876. The results of the voyage, encompassing the physical, chemical, geological, zoological, and botanical dimensions of the oceans, were published in an unprecedented fifty quarto volumes during the ensuing twenty years. These *Challenger Reports* are still widely accepted as the bench marks for the science of oceanography.

Challenger was a spar-decked, steam-assisted screw corvette, 2,306 tons in displacement and 200 feet in length. Built in 1858, the ship had served previously on the coasts of North America, in the West Indies, and on the Australian station.[6] With three masts, plus a modest smoke-stack betraying the presence of a steam engine within, it was a hybrid of the eras of sail and steam. And with all but two of the original seven-

H.M.S. *Challenger* at Tahiti (Edinburgh University Library, Special Collections, MS Gen. 28). Reproduced by permission of Edinburgh University Library. Photo courtesy of the Natural History Museum, London.

teen cannon removed and the maindeck spaces converted to zoological, geological, chemical and photographic enterprises, Challenger was
as much a floating Victorian laboratory as a naval surveying vessel—or,
as Punch called the ship, "a peripatetic Polytechnic Marine Exhibition."[7]

Aboard Challenger at sailing time were 23 officers led by Captain
George Strong Nares, a crew of 240 "bluejackets," and a scientific staff
of 6—the "scientifics." The latter consisted of

> Charles Wyville Thomson (1830–1885), F.R.S.,
>   Regius Professor of Natural History, University of
>   Edinburgh—director and naturalist
> John James Wild (dates unknown; Swiss)
>   —artist and secretary to the director
> John Young Buchanan (1844–1925), M.S.,
>   Principal Laboratory Assistant, University of
>   Edinburgh,—physicist and chemist
> Henry Nottidge Moseley (1844-1891), B.A.,
>   Radcliffe Travelling Fellow of Oxford University—
>   naturalist
> John Murray (1841–1914)—naturalist
> Rudolf von Willemöes-Suhm (1847–1875; German),
>   Ph.D., Lecturer, University of Munich—naturalist

In addition to the official record of the voyage appearing in the Challenger Reports, the journals and letters of four of the scientific members
(Thomson, Buchanan, Moseley, and von Willemöes-Suhm) reached
print in one form or another. Moreover, personal accounts were
published by three of the officers: Sublieutenant Lord George G.
Campbell, Engineer William J. J. Spry, and Navigating Sublieutenant
Herbert Swire. Finally, the voyage was rendered pictorially in the work
of the official artist J. J. Wild, the watercolor sketches of cooper Benjamin Shephard, and the "wet collodion" prints of a series of photographers. (See Bibliography.)

This corpus made Challenger's one of the most extensively documented of all voyages of exploration. Still missing, however, has been
an account of the expedition from the most numerous constituency of
the ship, the crew. Such an account would be useful from several
aspects, not only as a corrective to the official and scientific versions of
the expedition's "gentlemen," but also insofar as it could provide
insights on the unusually arduous duty aboard the first oceanographic
survey vessels and on the attitudes of the middle and lower classes
toward scientific research, which by the late nineteenth century was
becoming specialized and arcane.

A Reporter for the Lower Decks

Enter Joseph Matkin, ship's steward's assistant and a member of the crew during the entire three-and-a-half-year voyage. Either on his own initiative or following the model of the officers and scientists, Matkin maintained a journal throughout the voyage. From this journal, eventually amounting to three volumes, Matkin composed the letters he sent home to his family.[8] He mentions his intention, once home, of copying all the letters into a single family volume,[9] but nothing further is known of this plan. The three journal volumes, mailed home, are now lost, but sixty-nine of the letters—surely the majority of those he wrote—have survived.

Of these sixty-nine letters, eighteen were written to his mother; most of these were intended for circulation among the immediate family members and nearby friends.[10] Thirteen letters were addressed to his brothers: Charles ("Charlie," 1852–1924), the eldest and heir to the family printing and stationery business, working in Bedford during Matkin's absence; William ("Will," 1855–1883), who was seventeen years old and learning a trade in the Cambridgeshire village of March when *Challenger* sailed (he later joined the army and was a fencing instructor for the Royal Brigade); and Fred (1857–?), aged fifteen and

Photo portrait of Joseph Matkin (Archives, Scripps Institution of Oceanography). Photo courtesy of Scripps.

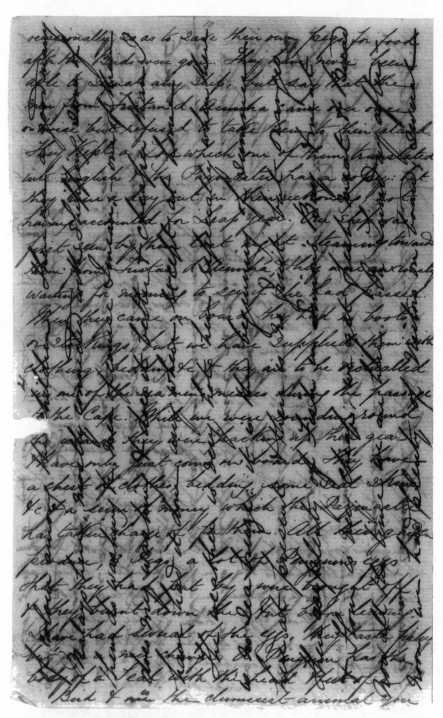

Matkin autograph letter. To save postage, which was based on the number of sheets to be mailed, Matkin often wrote both vertically and

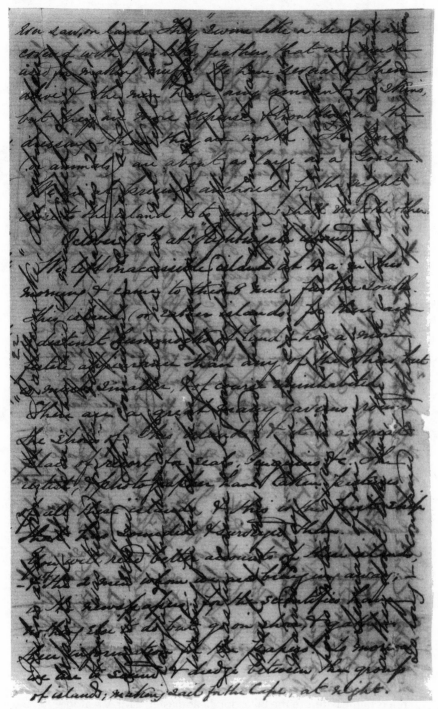

horizontally across each side of the page (Archives, Scripps Institution of Oceanography). Photo courtesy of Scripps.

working for a draper in Stamford during the voyage.[11] Two letters to the immediate family were too fragmentary to ascertain the addressee but may be assumed to have been written to his mother.

The remaining thirty-six letters were addressed to Matkin's cousin, John Thomas Swann of Barrowden, often greeted as "Tom" and referred to as "J.T.S." in letters to others. Internal evidence suggests that Swann was approximately Matkin's age and was preparing for the ministry. Swann himself copied the letters he received into a letterbook (846 pages in all), which remains in the possession of a Matkin descendant; the whereabouts of the original letters, however, is unknown. Matkin mentions writing to other individuals, including his father, who was not well at the time and died in 1874 while Matkin was in the Pacific; relatives living in Swadlincote, Leicestershire;[12] and family friends in the Oakham vicinity, especially a Mr. Daddo. None of these letters is known to have survived. The sixty-nine known letters, previously unpublished, constitute the principal substance, and the raison d'etre, of this book.

The extraordinary quality and volume of the Matkin letters owes much to the fact that Joseph Matkin came from a printing family. Born in 1853 in the village of Uppingham, Rutland—England's tiniest county, "a microcosm of rural England"[13]—Matkin was the second of four sons of Charles Matkin (1817–1874). The elder Matkin was a bookseller in Lincoln when, in 1851, he married Sarah Craxford (ca. 1824–1897) of Barrowden, Rutland, Joseph's mother. By 1855 the family had moved to the town of Oakham and established a printing and stationery shop. This enterprise became quite successful and was subsequently managed by three generations of Matkins.[14] In 1977 it was sold to new owners and now bears the name Matkins Printers Limited.

Young Joseph Matkin first attended school in Oakham,[15] then was sent to the Billesdon parish free school in Leicestershire. The latter was a one-room schoolhouse of venerable origin and reputation founded in 1650 by the generosity of William Sharpe of Rolleston "for the benefit of those who could not afford to send their children to one of the succession of private schools run in the village during the 17th century."[16] How long Matkin studied there and why he went so far from his hometown for schooling are two of the many mysteries of Matkin's life that remain unresolved. His parents evidently desired all the boys to have a strong academic foundation, for Matkin later wrote: "Few children in our station of life have had so much spent on their education & start in life."[17] The quality of the teaching at Billesdon school had apparently achieved some local renown; on more than one occasion, Matkin refers in the letters to the Billesdon headmaster, a Mr. Creaton.[18]

But what precisely had Matkin studied while in school? The letters demonstrate his competence in grammar and spelling and an acquaintance with poetry—a mastery of the English language quite respectable for his time and remarkable indeed in ours. The letters also exhibit a fascination with geography and foreign peoples and a concern for numerical precision. In contrast, Matkin appears to have been much less comfortable with natural history and the Latin nomenclature for things biological. These strengths and weaknesses may be just a matter of Matkin's personal preferences, but they do correspond to the emphases one would expect of his commercial class background. Moreover, they coincided with the orientation of the English school curriculum, which was just then on the verge of major changes. This context deserves a brief digression, so that we can appreciate more vividly Matkin's intellectual origins.

Mid-Victorian Science Education

Matkin's schooling occurred during a watershed period in the history of English education. Before the nineteenth century, natural science had played very little role in the primary or secondary schools of England. Mid-nineteenth-century Britain witnessed the growth of a movement for reform of the school curriculum, however, with a new emphasis on science and practical subjects, beyond the traditional curriculum of reading, grammar, arithmetic, religious knowledge, and, in the secondary schools, Latin and Greek. The beginnings of this movement can be seen in the 1830s, both in the countryside—at village schools like that at Hitcham, Suffolk, under the inspiration of the Reverend John Stevens Henslow (formerly Darwin's mentor in botany), and in the metropolis—where the newly created Committee of Council on Education adopted an orientation toward the sciences under the secretaryship of James Phillips Kay (later Sir James Kay-Shuttleworth). By the 1850s it had become clear that reform would depend on inexpensive scientific apparatus, science lesson books, and teachers trained in the sciences—and upon a dedicated cadre of Her Majesty's Inspectors of Schools, to see that government grants for curricular reform were properly utilized at the local level.[19]

The earliest efforts had advocated the introduction of subjects like mechanics and agricultural chemistry into the curriculum; simple experiments and tables of data were emphasized in one of the earliest guides for science teachers, Richard Dawes' *Suggestive Hints Towards Improved Secular Instruction Making It Bear Upon Practical Life*, published in 1847. Natural history and geology were less easily justified and more reluctantly adopted.[20] Thus, if the movement for science in the schools touched Oakham or Billesdon during Matkin's childhood (which is by no means clear), math, quantitative measurements, and physical properties might have received some emphasis, while zoology and botany likely would not have.

In any case, nationwide reform came only slowly, for these initial efforts generated more debate than action. What, after all, were the ultimate objectives of introducing scientific subjects into the school curriculum: early recruitment for a future scientific corps? or increased social mobility for the working class? or just moral elevation? And what were to be the emphases of the new curriculum: the practical "science of common things" advocated by Kay-Shuttleworth and by the Reverend Henry Moseley, inspector of schools (and father of *Challenger's* botanist)? or the abstract principles of pure science demanded by many in the scientific community?[21]

As a result of these disagreements, strong governmental support for school science throughout the land did not emerge until the 1870s. By the time a Royal Commission was finally formed to report on "Scientific Instruction and the Advancement of Science" (1867), Matkin was out of school. When the commission issued its second report five years later—on the eve of *Challenger's* sailing, the importance of natural science as "an essential component of elementary education"[22] still had to be stressed. Thus, it seems likely that Matkin's education had taken place just one decade too early for him to have acquired a layman's appreciation for the full range of *Challenger's* scientific undertakings. We will want to be all the more alert, therefore, to those scientific observations he does make.

Joseph Matkin Abroad

At about age twelve Matkin left school, and by 1867 he had entered the merchant marine. In December of that year he sailed for Australia aboard the *Sussex*, returning aboard the *Agamemnon*[23] the following summer. Soon thereafter he sailed again to Australia, this time aboard the *Essex*, but on this trip he remained in Melbourne for a year, working in furniture and upholstery shops. He returned to England by 1870.

On 12 August 1870, Matkin entered the Royal Navy—a step up from the "wavy navy," as the merchant marine was sometimes called. During the next sixteen months he served as ship's steward's boy aboard H.M.Ss. *Invincible* and *Audacious* successively. According to his service record, he was five feet seven and a half inches tall, with dark hair, blue eyes, and a dark complexion.[24]

Eleven letters from Matkin to his mother (not included here) survive from this pre-*Challenger* period.[25] They discuss routine domestic and naval matters for the most part, but there are a few items of interest: the construction of the docks at Sheerness, the *Great Eastern* steamship— "a monster, but very clumsy looking," and the sinking of his former ship, the *Essex*, near Melbourne, in memory of which he composed a poem.

What would attract a young, sensitive, educated lad from the Midlands, with no maritime background in his family, to embark on a career, or even just a cruise, in the Royal Navy? Could the romance of the high seas, the enticements of exotic ports, and the traditions of Nelson and the Empire compensate for the harsh conditions of life below decks?

> For months, sometimes years, at sea a man endured confinement, monotony, and isolation in an existence where he was subject to the dangers of exposure, disease, and rotten food. There was no privacy or female companionship. Overwork literally wore him out at an early age.[26]

By 1870, however, a young man might well consider a career at sea— not because the romance and tradition were more strongly felt, but because the conditions of naval service had been ameliorated appreciably.

Life in the Royal Navy of Matkin's day was markedly different from what it had been a generation earlier. Radical changes had taken place in the 1850s and 1860s, changes due only indirectly to advancing technology. In response to public and parliamentary opinion, recruiting difficulties, and—more specifically—problems experienced in the Crimean War and outbreaks of shipboard violence and disobedience in 1859, the Admiralty, gradually and tentatively, instituted a series of reforms which, in sum, touched every aspect of the bluejacket's existence, from signing on to pensioning off. The introduction of long-term enlistments ("Continuous Service") reduced the threat of involuntary impressment and raised the level of professionalism in the lower decks.[27] Discipline became more enlightened by the rationalization and regulation of punishments. The need for regular leave and liberty ashore and for rigorously observed paydays was recognized. Improvements were made in the naval uniform, the quality of food, and the availabilty of shipboard education. Hence, by 1870, the life and livelihood of the bluejacket had been transformed.[28] A career in the Royal Navy would have held far more attraction than it could have just a decade before. Matkin's prospects must have seemed quite bright.

Matkin was transferred to H.M.S. *Challenger* on 12 November 1872. Unlike many of his fellow crew members who deserted *Challenger* dur-

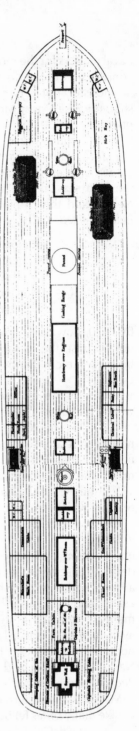

H.M.S. CHALLENGER,

as fitted for a voyage of Deep sea exploration . Dec.r 1872.

UPPER DECK.

MAIN DECK.

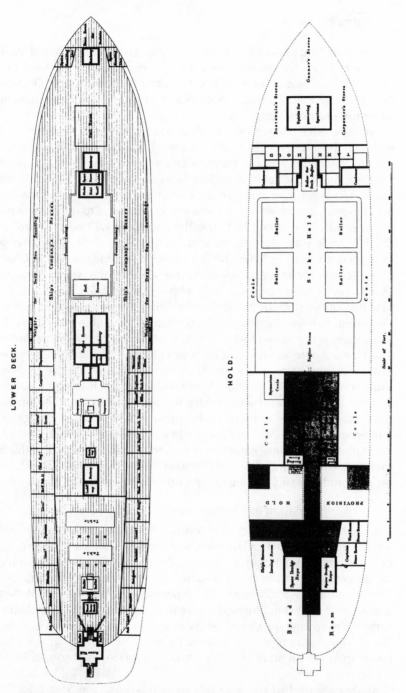

H.M.S. *Challenger* deck plan, from T. H. Tizard, H. N. Moseley,
J. Y. Buchanan, and John Murray, *Narrative of the Cruise of H.M.S.
Challenger with a General Account of the Scientific Results of the
Expedition*, 2 vols. in 3 (London: HMSO, 1885), vol. I, pt. I, at pp. 2, 18.
Photo courtesy of Scripps Institution of Oceanography.

ing the next three and a half years, Matkin, who often recorded their desertions in his letters, remained with the expedition until the ship's return to Sheerness in June 1876. One year into the voyage (2 December 1873), he was promoted to ship's steward's assistant, a permanent supernumerary position.[29]

Precisely what Matkin's duties were aboard ship cannot be known with certainty. References to his own work routine are rare. He does refer to having an "Issuing Room & Office" low down in the stern of the ship. And in a letter from Sydney he complained that "[o]ur Issuing Room is a dogs hole compared to what it was there [in his earlier assignments], it is down below the water line & I have to do all my writing by candle light, going up to breathe about every 2 hours."[30] Probably his work was identical to that on previous ships, and having described it in earlier letters he felt no need to repeat it in the *Challenger* correspondence. Apparently he assisted the ship's steward (Alfred Taylor, ship's steward 3c,[31] whose responsibility was the "victualling" of the ship) by managing the Ship's Steward's Issuing Room: procuring, stowing, issuing, and accounting for edible stores.[32] This assessment is reinforced by his frequent references to the edible stores purchased ashore and brought on board and his abiding interest in the availability and prices of foodstuffs at each port. It is possible that his storeroom contained additional, nonedible items, however. Matkin and the ships's steward were under the immediate command of the paymaster (formerly known as the purser), who was in charge of numerous items essential to the crew besides money and food, including clothing, bed linens, eating utensils, and tobacco.[33] With or without these nonedible stores, however, Matkin held a position of considerable responsibility and, among his peers one must assume, power.

Matkin as Correspondent

It was while standing at a desk in this Issuing Room, probably the largest space that he could call his own, that he seems to have written most of his letters.[34] What does a sailor on the world's first oceanographic circumnavigation write home about? Was he aware of the purposes and the historic significance of the expedition? As we will see, Matkin was surprisingly well informed about most aspects of the expedition. Early in the voyage, he described the oceanographic activities in some detail. This theme peaks some three months into the voyage with his transcription of an address to the crew by Wyville Thomson, chief of the scientific staff.

Matkin referred to the scientists most often as "the Scientifics."[35] Although not a prominent theme in the letters—Matkin was not a complainer—there are clear hints that relations between the "scientifics" and the below-decks crew were not always harmonious. Early in the cruise there were thefts of food from the officers' mess. Then there was dissatisfaction with the gentlemanly, nearly idyllic accounts of the voyage in Wyville Thomson's letters to the popular weekly *Good Words*. Moreover, dissemination of information to the men about the expedition's discoveries was evidently slow, with the result that the public would learn of *Challenger's* accomplishments before the ship's own crew. Were the scientifics trying to control the release of information, to ensure priority of publication? Or did they simply overestimate the chasm between the "two cultures" aboard ship[36] by underestimating the crew's interest in the oceanographic results? Finally, in the Pacific, morale reached its lowest ebb because the expedition had lasted so long and the crew had worked so much harder than that of ships with other purposes, yet without additional compensation.

On such a long and often tedious voyage, we could hardly expect harmony to have been maintained throughout. In fact, sociological analysis of long-distance voyaging suggests that such tensions are common, if not inevitable. Earlier in the nineteenth century, for example, the crew of H.M.S. *Beagle* became exasperated with Captain Fitzroy's repeated perfectionist's surveying of the South American coast; some crew members—less patient than Darwin—deserted the ship. Twentieth-century oceanographic cruises are typically of much shorter duration, and desertions are rare; but conflicts between scientific staff and ship's crew are nevertheless common. In fact, similar problems are being anticipated for scientific expeditions into space.[37]

Social theory has suggested two methods for reducing scientist-crew conflict: the ideal method of encouraging, in each group, greater understanding of and sympathy for the goals and values of the opposite group; and the more feasible (and in fact more common) method of maintaining a degree of isolation or "subcultural privacy"[38] between the groups. Recognition of this need for separation is especially important aboard modern vessels, where, at least in theory, a high value is placed on egalitarianism. In the Royal Navy, with its long-standing, hierarchical separation of officers and crew, strong isolating mechanisms were already in place to ensure that interactions would be rare except during official work time.

While these isolating mechanisms must have helped to counteract the build-up of tensions aboard *Challenger*, the very long periods at sea —much longer than for modern research—worked in the opposite direction. Then, as now, the sailor's principal goal was to finish the job and return to port, while the scientist's objective was to gather his data, no matter how long at sea the gathering might take. Therein lay the chief source of tension.[39] Fortunately, *Challenger*'s Captain Nares was not a moody tyrant like Fitzroy; nevertheless, his popularity could not compensate sufficiently for the arduousness of the voyage to prevent a great many desertions. Matkin's account brings out these tensions at last, for in all the other expedition sources there is hardly a mention of them. Modern-day oceanographers, meanwhile, may take some solace, and perhaps some instruction, from the presence of scientist-mariner antipathies even aboard the legendary founding voyage of *Challenger*.

Alongside the crew's view of the "scientifics," we should place the latter's perception of themselves. Wyville Thomson drolly cited the appellation

> "philosophers" as our naval friends call us,—not I fear from the proper feeling of respect, but rather with good-natured indulgence, because we are fond of talking vaguely about "evolution," and otherwise holding on to loose ropes, and because our education has been sadly neglected in the matter of cringles and toggles and grummets, and other implements by means of which England holds her place among the nations.[40]

As the voyage progressed, oceanographic operations became more routine, and Matkin referred to them less frequently. But then, even among the "scientifics" fascination for oceanography sometimes waned: botanist Moseley placed all his dredging information in the firal chapter of his book, "since the results obtained by deep-sea dredging, even in most widely distant localities, were very similar and somewhat monotonous."[41]

While Matkin's attention to oceanographic details declined with time, his accounts of places visited became increasingly elaborate. We cannot be surprised that his attitudes upon encountering foreign peoples were Anglocentric; what is perhaps surprising is that he manifested less chauvinism, less imperialism, than we might expect of a sailor in Queen Victoria's navy. He nearly always made an effort to present a balanced picture, to find both good and bad things to say, of a place and its people. Colonials he often judged more harshly than the local people, as in the Philippines, where he found "natives hard at work, and white folks doing nothing but smoke, perspire, and drink coffee."[42]

The statistics and the rather formal language frequently appearing in Matkin's descriptions of foreign ports make it clear that Matkin was drawing on the ship's log and on printed sources as well as on personal experiences and shipboard gossip. Books and newspapers were no doubt available in the ship's library,[43] which during a portion of the voyage was in the charge of Matkin's immediate superior, the ship's steward. In addition, a special collection of scientific and travel books was taken aboard explicitly for the expedition (see Appendix E), although these were probably reserved for the use of the scientific staff and may not have been readily available to Matkin. It is also possible that bulletins describing the ship's ports of call were posted for the crew's edification. Finally, Matkin himself on more than one occasion mentions visiting a library ashore.

Matkin the Man

One cannot avoid being struck by the apparent scholarly urge of this young sailor. At the same time, his efforts to convey a learned account of places visited, and the generally proper style of his prose, hint of a desire to make a good impression on his family. The Victorian spirit of "self-help" and "improvement" through education and perseverance, as articulated in the widely read works of Samuel Smiles,[44] may well lie behind these efforts. Testimony for this ethos is especially clear in Matkin's letter to his mother on the occasion of his brother Will's enlistment in the army: "His education ought to prove of great service to him in getting him promotion & he will have ample time & means of improving himself, for they have splendid reading rooms attached to those regiments."[45]

In any case, the amount of time he devoted to letter writing, under conditions that can hardly have been conducive to it, is surely testimony to a strong sense of family devotion. The intensity of his concern for the family is most evident in the correspondence following his father's death in 1874. The mails kept this news from Matkin for nearly four months.

A strong sense of religiosity, on the other hand, does not come out of the letters; indeed, Matkin's religious affiliation is not known with certainty. He writes of attending church services on a number of occasions, but more out of a sense of curiosity toward religious practices in foreign places than from a spirit of religious obligation. Judging from the fact that the woman he would later marry was an Anglican and that his parents were buried in the Church of England section of the Oakham cemetery,[46] we can assume that he had been raised an Anglican. His liberal views and expressed distaste for heavy ritual[47] and for missionary fervor suggest a Broad Church Anglicanism.

Similarly, Matkin showed respect but not reverence or awe for the royal family. This attitude may be a reflection of the flagging popularity

government employment and took his family back to Oakham, where the sixth and final son was born. From then until 1914, at a time of life when others are at the peak of their productivity, Matkin was listed in the Oakham city directory as merely "retired civil servant." How he supported a middle-class family of eight remains a bit of a mystery; income from both his own and his wife's families seems likely. In 1914, he moved back to London, but this time alone, as he was by then separated from his wife. He was a resident of Holborn at the time of his death in 1927, which resulted from a fall in the street. Family tradition states that he died of complications after being struck by a motorcycle.[53] Thus, like his more famous *Challenger* companion, "Scientific" Sir John Murray, Matkin was a victim of the "progress" of motor vehicle transportation.[54]

These later years seem sadly colorless by comparison with the heroic era of *Challenger*. The latter was the pinnacle of Joseph Matkin's life, and it is that pinnacle we ascend in the following chapters.

The letters that follow have been edited to eliminate repetitious passages; otherwise they are complete. Where two letters cover the same period and events, the more extensive one has been included, with any additional substantive material from the other interpolated in curly brackets ({}). Excision of substantial repetitious material is indicated by repeated asterisks (* * *). Uncertain or confusing punctuation has been corrected in accordance with modern practice, format of datelines has been standardized, and paragraphing (indicated by "¶") has occasionally been introduced for ease of reading, to compensate for Matkin's paper-saving habits. His spelling and usage remain unchanged. Deviations from Matkin's spelling and punctuation habits that appear to have been introduced in Swann's copying of the letterbook have been restored to conform with the style of his original letters.

1 Under Way in the Atlantic
NOVEMBER 1872–MARCH 1873

So never mind your Dredge, my boys,
Which you have lost below;
Our country now your power employs,
That man may wiser grow.
—Chorus to
"A Song for the *Challenger*'s Crew"

It was late November 1872—*Challenger* was still being made ready for sea at the government dockyards in Sheerness, at the mouth of the Thames—when Joseph Matkin penned the first of the letters to his cousin John Thomas Swann. We find him well informed about the makeup of the ship's crew and curious about the equipment being stowed aboard for the "scientifics."[1] And straightaway we learn of the drowning of a shipmate, an unsettling tragedy for the eve of the famous expedition.

H.M.S. "Challenger"
Sheerness
Nov. 22nd [18]72

Dear Tom,

We came into this ship on Monday morning 175 all told, including 19 Privates & one Sergeant of Marines, and the same evening one of the Marines [Tom Tubbs[2]] was drowned, walked right over the Ship's Gangway into the Dock Basin which is 27 feet deep, and was drowned in sight of several of his shipmates. It was about seven o'clock at night & pitch dark, and there was no light to direct any one walking on shore from the ship or anything round the edge of the basin to prevent them falling in. As soon as the alarm was raised all the ship's company went out on the Jetty with lights etc, and the Dock Yard Police went for the Drags & dragged for two hours, but were unable to find him so had to desist for the night.

In the morning a Diver was fetched and all the seamen from the Barracks, a crowd of Dockyard men assembling to see the Diver go down. As soon as he had his dress on with about 50 lbs of Iron slung round his waist, his mates began pumping air, and down he went to the bottom, and was there about three minutes walking over the whole space between the ship and the side of the basin without finding him, when just as he was coming up—he saw him standing bolt upright against the side of the basin, and after tieing a rope round his waist, he came up, and hauled him out of the water head first. He had only been immersed about ten hours, but his features were so altered, his messmates could scarcely recognize him.

He was buried yesterday afternoon with all the honours allowed by the service, the Flag ship's Brass Band playing the "Dead March" and the Blue Jackets and Marines of the "Challenger" following with arms reversed. The poor fellow's Mother and sisters came down from London to his funeral.

Another Marine has come today from Chatham Barracks in his place, and he turns out to be an old shipmate of mine, who was in the "Invincible" and "Audacious". I have about eight old shipmates on board.

We are to have 50 Boys when we get to Portsmouth, which will make the complement 230, all the Officers are on board except Lord George Campbell,[3] who, I expect is waiting until all the dirty work is done. We have two Steam Boats on board, and about 30 miles of deep sea line, and dredging line, the other six boats we shall take in when we go into the river next week. All the Scientific Chaps[4] are on board, and have been busy during the week stowing their gear away. There are some thousands of small air tight Bottles, and little boxes about the size of Valentine boxes packed in Iron Tanks for keeping specimens in, insects, butterflies, mosses, plants etc. There is a photographic room[5] on the main deck, also a dissecting room[6] for carving up Bears, Whales etc.

We had a grand funeral here on Sunday morning last, a Prussian Frigate came in to bury one of her seamen, the greater part of the officers and her crew followed him to the Grave, their band was composed of all seamen, playing a Funeral March, and the coffin was drawn on a Field piece by the seamen. They were dressed much the same as ours, but were not such good looking men, the Marines wore a Black and Silver uniform, and Helmets, which had a fine appearance. The Ship left again on Monday after saluting the Flag Ship with 18 guns.

We are to go out in the Medway on the 28th & expect to leave Portsmouth about the 2nd.

The Crest is a facsimile of the Ship's Figure Head[7] which is gilded.

Believe me,
Very Truly Yours,
Joe Matkin[8]

Two weeks later (7 December) *Challenger* departed Sheerness, arriving in Portsmouth after four stormy days in the Channel, a second dark portent to the superstitious. The tumult of that transit, unmentioned in most other accounts, is vividly recounted in Matkin's second letter to his cousin. The seagoing mettle of the "scientifics" was tested and, according to Matkin, found wanting.

H.M.S. "Challenger"
Portsmouth
Dec. 17th, 1872

Dear Tom,

We left Sheerness last Saturday week, and had awful weather round, having to put back three times, once into Deal on the Monday, into Folkestone on Tuesday, and into Dungeness on the Wednesday, arriving here on Thursday; the distance is 105 miles & we were 109 hours steaming full speed all the way. The wind was dead against us, and it commenced to increase as soon as we had passed Dover on the Saturday night, in the morning it had increased to a gale, and we found ourselves close into the coast of France. At 10 o'clock on Sunday the Captain gave orders to put back for the Downs, but the wind arose to a hurricane and we could make no head way against it. At Midnight the ship passed through a Cyclone, which caused the sea to come right over her and go down into the Engine room, through the hatchways, nearly putting out the fires; the Life boat cutter was smashed to atoms, and the Jib boom carried away with all the Head Sails, so that the ship drifted along under a single storm staysail. She rolled fearfully all night, I was pitched out of my Hammock two or three times, and think it was the most fearful night I ever passed in my life. Several ships were wrecked close to us, and it was considered the roughest night on the Southern coast for the last eight years.

We reached Deal on the Monday and waited until the storm abated, there were hundreds of vessels of all nations at anchor in the Downs. Directly we dropped ours the Scientific blokes made a rush for the railway station, and came on to Portsmouth by train

Scientific party from the Royal Society on board *Challenger* at Sheerness prior to the voyage (Natural History Museum, London, Photo No. 1). An engraving made from this photo appeared in *The Graphic* (London), 28 December 1872, with the caption "The Staff and Promoters of the 'Challenger' Deep-sea Expedition." Top row, left to right: James Young Buchanan, Philip Lutley Sclater, Henry Nottidge Moseley, Rudolf von Willemoes-Suhm, John Murray (publisher?); middle row: Professor William Crawford Williamson, John James Wild, Sir William Thomson (Lord Kelvin), Joseph Dalton Hooker, Professor Charles Wyville Thomson, Admiral George Richards, William Siemans, Professor George James Allman, Francis Galton, Professor George Stokes; bottom row: Dr. William Sharpey, Sir Henry Holland, Professor William B. Carpenter, J. Gwyn Jeffreys, Professor George Busk. Photo courtesy of the Natural History Museum, London.

arriving two days before us. They lay about sea sick during the gale and the sailors did make game of them. The damages are nearly all repaired except a new Life boat, which is to be sent out to the Cape Verdes in a months time.

The Chief Engineer made a sketch of the ship during the gale on Sunday night which is to be printed; this week's "London News" and "Graphic"[9] are also expected to contain plates of the "Challenger". Since we came in here we have been very busy drawing extra stores for the cold weather; the Lords of the Admiralty were on board on Saturday inspecting the ship, and ordered extras, such as Wines, pickles, jams, cranberries etc. We are to have warm clothing sent out to the Cape of Good Hope in a year's time, and it will be issued before going down amongst the ice, or rather lent, for it will be taken away as soon as we reach the latitude of Melbourne, and again issued at Petropaulosky[10] before going through Behring Straits. We are also to have extra pay in the cold weather.

The Marquis of Lorne was on board on Saturday, he is not as handsome as his picture. We also had Admiral Sir Leopold McClintock of Arctic notoriety,[11] and I can't tell you who has not been to look over the ship since we came here.

The Queen passed us to day in her Yacht on her way across to Osborne, and all the Marines were drawn up under arms.

Portsmouth is a very fine town larger I think than Hull.[12]

We expect to sail on Saturday next for Lisbon, and if we make haste shall most likely be there fore Christmas.[13]

Matkin may have passed through the busy terminus of Portsmouth on a previous voyage, although the brief mention of it above suggests the contrary. Had he taken time to record the view of it from the ship, his reactions might have been much like those of Sam Noble, "Able Seaman," who arrived in Portsmouth for the first time in 1875, at the age of sixteen.

> On the left, or Gosport side [Noble wrote], were ranged a long line of three-deckers, line-of-battle ships, frigates, and other trophies of the Spanish and French wars, in some places two and three deep, their high poops and fo'c'sle almost shutting out the view of the shore. Those in my time were mostly used as coaling hulks.
>
> The right, or Portsmouth side, was taken up with Government offices, building sheds, towering shears, cranes, and all the paraphernalia of a large naval dockyard; while

the piers and jetties were lined with transports and Royal
and Admiralty yachts. The wide expanse of water forming
the fairway was teeming with life and bustle—penny
steamers darting hither and thither; pinnaces, jolly-boats
and cutters, laden with the day's provisions, and pulled
along by brawny Jack Tars barefooted, in short-sleeved,
open-necked jumpers, showing off their hairy arms and
breasts brown with exposure under many suns, and with
their caps hanging at such an angle on their heads that it
was a wonder to me they didn't fall off, who went lum-
bering and tumbling back to their respective ships, like
plump jolly housewives returning from the market. . . . A
great scene, by jove! for a boy whose life had been spent
in a mill.[14]

Challenger did indeed "sail on Saturday," departing from Portsmouth
shortly after noon on 21 December 1872. Matkin's next letter, composed intermittently while en route to Portugal, ran to fourteen hand-
written pages. It contains abundant information relating to personal
finances and below-decks "housekeeping," no doubt of concern to his
mother. It is here that we learn that Matkin belongs to the Chief Petty
Officers' Mess, the senior mess below decks. He also presents a picture
of a less than optimal Christmas at sea, the first of four. Petty thievery in
the officers' mess is duly reported, as prologue to an accounting of the
man-of-war rations at sea.

> H.M. Ship "Challenger"
> Off Cape Finisterre
> Decr. 29, 1872

Dear Mother,
   We left Portsmouth on the Saturday morning & did not call at
Plymouth, so that the Post Card I had written with the future
addresses of the ship I was not able to send. We went out through
the Needles from Portsmouth & saluted the Queen at Osborne
House as we passed, we had a head wind down the Channel & we
have had one ever since—with very rough weather—until today—
we have a wind nearly fair—& the ship is a little steadier so that I
am able to write. We have had a week of awful knocking about &
half of us were sea sick for the first 2 or 3 days. The coal will only
just last into Lisbon, where we expect to arrive on the 1st January
or exactly a week longer than we should have been had we had
decent weather. I forgot to tell you in Miss W's[15] letter that we

were paid one pound the day before we came away & that it was all gone before night.

I am in the Chief Petty Officers Mess, there are 6 with me, & 2 Blue Jackets Boys to look after the Mess & keep it clean &c. We six each put 15/- into the mess to fit it out with knives & forks, cups & saucers, dishes &c which the service does not supply as well as 2 sacks of Potatoes & the materials for a X.mas dinner. So that after I had bought some collars & other little things my Advance looked very small indeed, & Father's 10/- that he lent me will have to wait until we get to New York when another advance will be paid us, & I hope to return it.

We had a miserable X.mas as far as the weather was concerned for the ship was pitching & rolling awfully & we had to hang on to our crockery ware like grim death, several of the messes lost all their crockery but we only lost a few cups—& a pot of Jam & a bottle of Pickles that broke & got mixed together. Our mess fared as well as any on X.mas day—for we had Ham for Breakfast & a good meat Pie & Plum Pudding for dinner; we made our pudding on X.mas eve; everyone did something towards it.

We had a short service in the morning, the Captain officiates for we are not allowed a Chaplain, only Ships carrying 295 men & upwards are allowed a chaplain & we have only 242 on board. In the evening the Captain gave every one of the Ships company one third of a pint of Sherry & very good wine it was. If the 3 or 4 ensuing X.mas's which we are to spend in the "Challenger" are no worse, we shan't hurt.

I thought of home about one oclock that day & how Charlie & Willie were spreading the Plum Pudding across their chests (as the sailors say), in a seamanlike manner.

The Officers had a grand dinner about 6 oclock & a Turkey very mysteriously disappeared just as it was ready to go on the table & it has never been heard of since. Again last night a Roast Goose, and 2 loaves of Bread were taken off the cooks table before the cook could turn round. Some fragments of a goose & some salt were found this morning up in the main top rigging where the goose had been taken & devoured. The Officers made a kick up about it but can't find out who takes them.

I should have liked to have helped pick a bit myself for I have never been so hungry as the last few days for we are now on regular man o war diet & it's bringing some of us down a good deal. I will tell you what the routine for meals is now: at 6 AM Breakfast of Cocoa & hard Biscuit, at 10.30 the Steward & I mix the Lime juice[16]

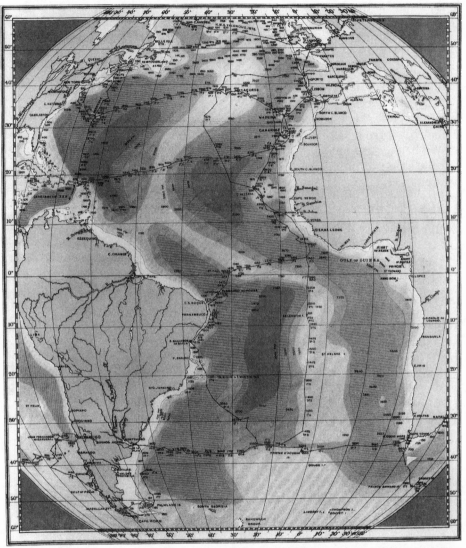

"Contour Map of the Atlantic," from C. Wyville Thomson, *The Voyage of the 'Challenger': The Atlantic*, 2 vols. (London: Macmillan and Co., 1877), vol. I, frontispiece. Photo courtesy of Scripps Institution of Oceanography.

½ gill in I gill of water for each man, at 11.30 dinner; one day it is salt pork & pea soup—the next salt Beef & Plum duff, the next salt Pork again & the 4th—Preserved Potatoes & Australian Beef in Tins, at 12 oclock we mix the Grog ½ gill of rum to I gill of water to each man & at 5 PM,[17] supper & tea together, Biscuit & Tea, with what is left from dinner.

If any one can get fat on that in 4 years they must eat more than their allowance.

We shall get Fresh Meat & Vegetables at Lisbon & most of the Ports we call at & we can buy soft bread if we have any money.

We are now off Cape Finisterre & just about the spot where the "Captain" went down,[18] the weather is quite warm already; to morrow we are to commence dredging or we might get into Lisbon by to morrow evening. On Friday last we passed a ship bottom upwards but it was too rough to lower a boat & she had been several days in the water for the keel was washed right away; again yesterday we passed the masts & remains of another wreck.

December 30th, off Vigo

Sighted the coast of Spain this morning but are still 150 miles from Lisbon. At 10 oclock this morning the first cast of the dredge was made & bottom obtained at 1500 fathoms but in hauling up the dredge & 100 fathoms of line was lost over board. The dredge was again cast but came up bottom upwards but on another attempt being made they succeeded in bringing up mud and several species of Fish from a bottom of 1125 fathoms, or nearly 1½ miles; on the mud being analysed numerous insects were found in it—the fish are preserved in bottles. The Dredge is of Iron & not unlike a pig trough with a net over it & weighs with the weights attached several hundred weight. The strain on the line is very great as it reaches the bottom & to ease it, several gutta percha ropes are spliced to it which will stretch when strained. The Dredging line is about the thickness of a man's two fingers. When dredging the Ship is hove to, & the Dredge is let down from the main yard & hauled up by a small steam engine & the line coiled away on the upper deck to dry; 1200 fathoms of line will take one hour in reaching the bottom & 3 hours in hauling up.

The first dredging station was off the entrance to Vigo Bay, coast of Spain. Interestingly, Wyville Thomson's account makes no mention of this first attempt in which the dredge and line were lost. He does acknowledge that the first leg of the voyage (England to the Canaries)

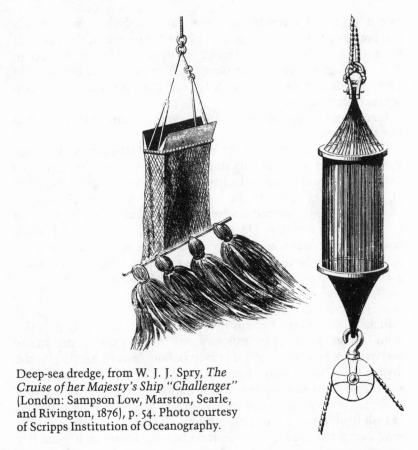

Deep-sea dredge, from W. J. J. Spry, *The Cruise of her Majesty's Ship "Challenger"* (London: Sampson Low, Marston, Searle, and Rivington, 1876), p. 54. Photo courtesy of Scripps Institution of Oceanography.

was regarded as a "shakedown cruise" by the scientists, because of the uncertainties of working aboard a ship with *Challenger*'s characteristics.[19] Of the other published accounts, only Lord Campbell's *Log-letters* (which tend toward skepticism and irreverence in scientific matters) mention this initial loss (pp. 1–2). Matkin continues:

We shall arrive at Lisbon on the 1st Jany '73 & it will take 2 days to Coal—we are to arrive at Gibraltar on the 8th the distance from Lisbon is 400 miles, most of the men will get a letter at Lisbon, & Gibraltar, but I suppose Madeira will be my first place for a letter unless you have written on spec. Did I tell you we had a Brass Band on board composed mainly of seamen & marines who volunteered. The Officers bought the Instruments & provided a Bandmaster to

The band of H.M.S. *Challenger* (Natural History
Museum, London, Photo No. 421). Photo courtesy
of the Natural History Museum, London.

teach them; there were 15 volunteers & 9 wanted to play the big
drum, they practice every day in the fore peak of the vessel & the
noise is something fearful & causes the Watch below to swear a
good deal. The Bandmaster expects to fetch tolerable music in
about 6 months.

We have a first rate library[20] on board & a good many of the magazines are sent out gratis; we have also a harmonium[21] but the ship
has been too unsteady to cast it adrift yet—I hope to have a time
now & then.

Dec. 31st, off Cape Mataplan
Or 130 miles from V̶i̶g̶o̶ Lisbon, under sail with scarcely any wind,
several steamers and Fruit schooners in sight—the ship is rolling a
good deal & I can scarcely write, we are to have a Lobster supper in
our mess it being the last night of the year; 12 months to night the
"Sussex"[22] was wrecked "off Melbourne."

January 1st, 1873, close to where the battle of Trafalgar was fought
This morning passed any amount of oranges floating on the water:
some fruit schooner wrecked about here lately. Cast the Dredge
again to day bottom at 1800 fathoms—nothing came up in the
Dredge—shewing the bottom to be rock. In hauling up a second
time the Lead & 300 fathoms more Line was lost over board.[23] Very
little distance made to day & we are still 100 miles from Lisbon but
shall get in to morrow if they have no more dredging. A good many
of the men complaining of the water which is condensed from the
sea at night & drank the next day & is scarcely cool. I felt ill myself
the other day but have improved by qualifying the water with a little Rum or Lime juice.

Jany. 2nd, off Lisbon
All day again we were dredging & obtained the deepest bottom ever
yet ascertained viz 2550 fathoms or nearly 3 miles, but in hauling
up, the dredge, & 1700 fthms of line was lost over board so that no
criterion could be formed of the nature of the bottom.[24]

January 3rd, Lisbon Roads
We entered the Tagus this morning at 10.0.clock & anchored just
abreast of the town at 12 o-clock. The scenery all down the river is
very grand & the river is a very large one wider than the Humber
but not as long, only 20 miles down to Lisbon. The land on each
side of the river is very high & the hills are covered with vine yards
& white mills (that turn round with a sail), for crushing the
Grapes. The houses are all white & look very pretty but none of
them have any chimneys. There are several old castles & forts on
the way down the river & from the fort which commands the
entrance—the Portuguese Admiral's boat came alongside & asked
several questions before we were allowed to proceed, such as "the
name of the Frigate, & the Captain, how many guns, how many
men, where are you from & how long out, any sickness on board?"
& he finished up with "The Portuguese Admiral sends his best
compliments to you, my Lord." Their seamen looked very dirty

although they were dressed like our men & their Flag Ship is not as good as this.

¶Lisbon is a large town & has a fine appearance being built on hills & all white houses. There are 300,000 inhabitants. I hope to go on shore on Sunday, then I shall be able to tell you more about it, there are some very fine churches & monasteries, & the Cathedral is a very fine one, at sunset the convent bells all ring for mass & it sounds very solemn on the river here. The King's Palace is built just above the town & has a very fine appearance, all white stone; to night it is lit up & looks very grand, so does the town but it has a very sleepy look not like an English town all bustle & business.

January 6th, 1873
The English mail does not go for a fortnight so I think I had better do as a great many are doing, put a 6d stamp on the letter & send it overland & write again by the mail in a fortnights time after I have been on shore—I am going to night.

Write "at once", to Gibraltar, Via South^ton (or elsewhere, & to Madeira 'Via Liverpool' before-the-2 when the next mail goes & mark it "By Africa S.S.Cos Steamer" also to "Teneriffe" Via Liverpool, by the first Febry mail, penny stamps will do, & ½ for newspapers. We shall leave Gibraltar about the 25th Jany for Madeira. Hoping Father is better & with best love to all,

From Your Affe^ate Son,
Joseph Matkin

P.S. I wish I had some of this foreign note paper   I must get some at New York.   I hope this will not be over weight. When ever you write, enclose one stamp for we can't always get them. JM
P.P.S. A Liverpool steamer is just going to start & will take our mail.[25]

The following day Matkin penned an abbreviated account of the Christmas and New Year's events to his cousin. The shipwreck remains receive additional attention.

H.M.S. "Challenger"
Lisbon Roads
January 7th, 1873

Dear Tom,
We left Portsmouth on Saturday the 21st of December, and only arrived in Lisbon on the 3rd of this month, true, we were three days

dredging in the Bay of Biscay, but it was the weather that delayed us so, gales of wind all the way, & always dead against us, we burnt nearly the whole of the coal coming this 700 miles.   *   *   *

On the 27th inst. we passed in the Bay of Biscay a vessel bottom upwards in the water, but it was too rough to lower a boat although we passed her quite close. Again on the 29th we passed the remains of another vessel, and a great quantity of oranges floating on the water; it was evidently one of the many fruit schooners running between here and England, and had foundered in the gale with all hands on board.

It was a very solemn sight to see that first vessel bottom upwards in the sea, with the waves breaking over her, she was quite entire too, and had not been many days in the water; every man in the ship came on deck to look at her, and it made our company quite dismal for a time.

I have not time to send you particulars of the deep sea dredging this time, but I have it all in my diary, and shall send it in my next letter. The dredge was cast in four different places, the lowest bottom was nearly a mile and a half, and the highest not quite three miles, the strain on the line is something wonderful as the dredge gets quite close to the bottom, and they have lost already in heaving it in, nearly two miles of line, and two dredges.

Only one dredge came up as it ought to, and contained mud, and some very rare species of fish, as well as a common crab, these have all been preserved, they were brought from a depth of 1500 fathoms or nearly 2 miles. The mud also contained numerous insects, and was successfully analyzed.

The weather here in the daytime is very hot, and we have all kinds of fruit very cheap, and I have eaten an enormous amount since we came in.

Our Captain and Professor Thompson[26] dine with the King of Portugal to night.

They have been coaling ship to day, and we sail for Gibralter on Thursday, leaving there again about the 25th of January for Madeira.

Our letters now have to be franked by the Commander, who is now busy with some hundreds and the mail closes in 5 minutes.

Believe Me,

Yours Very Sincerely,

Joseph Matkin[27]

A day later Matkin resumed writing to his mother. Here as elsewhere he commented at length on the mails: the costs, the delays, the uncertainties. More importantly, this was *Challenger's* first foreign port, and Matkin had much to report from his brief time ashore. Neither Lisbon nor the Portuguese impressed him very favorably at the outset; in fact a bit of English chauvinism is evident. By the time the ship departed, however, his feelings had moderated. He was especially enthusiastic about the local fruit, the port wine, the opera, and some of the public buildings.

> H.M. Ship "Challenger"
> Lisbon Roads
> January 8th, 1873

Dear Mother,

The last mail went all in a hurry, she was a chance steamer that called in from Brazil, on her way to Liverpool; & by taking our mail she will save all harbour dues when she arrives there. I was just going to buy a 6d stamp & send it (overland) when the Steamer came in so I tore open the envelope & addressed another, you see the commanding officer has to frank all letters or they would have to be paid for the other end, unless we put a 6d stamp on them.

The regular English mail goes on the 14th, so we shall have to leave these behind, as we sail for Gibraltar on the 10th or 11th.

¶I went on shore the other night with a messmate, we went on shore at 4 oclock & had a walk out in the country. The oranges are just ripe now in the orchards & are very cheap—2 Reis each or 5 for a penny. Apples are still plentiful here, so are all kinds of nuts & figs—if you could see all the fruit I have eaten since we came in, you would say I was a credit to you—and mackerel I don't know how many I have eaten, they are 1d each & they are brought off to the ship ready cooked, (Fried in Olive oil) 1½ each. The bread is sour & the butter is not up to much—cheese is good at 8d a lb. The women work harder than the men here, in the Fish market they unload the fish while the men smoke.

¶The town except in the principal streets is very dirty & ill-paved, it smells beastly of Garlic—so do the people; the town is well lighted with gas & there are omnibus's, & cabs, drawn chiefly by mules—as big as horses, drays & heavy carts are drawn by oxen. The Cathedral & several of the churches are beautiful buildings all built of white stone, & the Bells are very good ones, in the evening at 6 oclock they play opera tunes, & at 6 in the morning; we can hear them nicely from the ship. My chum & I were going in to the Cathedral but were stopped by a Priest, I suppose no service was going on—the Cathedral is something like St. Pauls in London—with the dome & towers but of course not as large, there is a ball &

cross on the top & being of white stone it looks cleaner than St. Pauls especially when the sun shines on it.

¶We walked up as far as the King's palace but could not get very near there were so many soldiers on guard—all the gates leading into the town are guarded by soldiers & they search every cart that comes in. The soldiers are dirty looking beggars except those round the Palace, & so are the sailors, they have 3,500 sailors—& no ships, they the sailors live in barracks on shore. They dress the same as our men, & so do the "Prussians", one of their ships is in here now & is going to Gibraltar & Madeira as well as us, she is a training ship for Boys, & is cruising round here for the winter. The channel fleet[28] always winter here & bring a good deal of custom to the place. There are 3,500 men in the channel fleet; they are at Gibraltar now—but will be here again shortly. One of their sailor boys belonging to the "Sultan" died in the English hospital on Monday & he was buried yesterday, half of our seamen attending his funeral; the channel fleet have a great many buried here in the cemetery—& they all have grave stones—erected by their mess mates.

¶The King & Queen of Portugal[29] are coming on board this afternoon to see the ship & the dredging apparatus &c, all hands are busy now cleaning the ship—& growling like anything at the trouble & honour. The Captain & Professor Thompson dined with the King at the palace the other night.

¶We finished up our ramble on shore by going to see Ruy Blas at the Opera & it was really grand, there were 250 in the orchestra, & the house was crowded, it is a splendid building—there are 7 tiers one above another, & the ladies were beautifully dressed & very handsome. The singing was grand, but in Italian, the same as in London, after the opera came the ballet a very airy affair—it was going on for one in the morning when we came out. The place is full of Theatres, & Henglers English circus is here, the same I saw in Hull; he opens on Sunday night the same as the Theatres, & Opera house; I suppose he believes in the maxim "When you are in Rome, do as Rome does"; he goes to Madrid from here.

¶Sunday afternoon & evening is the peoples' great holiday here, in the morning they go & confess & get absolution for the weeks sins, in the afternoon they go to the Bull fights out in the country & in the evenings to the Opera & Theatres. We had a peep at a masquerade ball but did not go in, every one wore a domino or half mask of black velvet & the costumes were magnificent; it is impossible to recognise any one who wears a domino; we were invited to put one on & go in but as we could not dance we went to the opera instead. There were more ladies than gentlemen I think & as you can't recognise any one you can ask any lady you like to dance. This masquerade Ball lasts for 10 days after X.mas & is something after the style of the Carnival at Rome. We enjoyed ourselves very much & the prices of everything was very moderate. The Portuguese are proud of Lisbon & say in one of their Poems that the man who has never seen Lisbon, has seen nothing. I should like them to see London or any large English town, the garlic eating fools.

¶We took in 150 tons of coal & are to fill up at Gibraltar, we shall be there about 10 days, the distance from here is 450 miles but we shall be dredging & may be a week from here on the way. I shall have another run on shore there & see what the Spaniards are like, a run on shore is very refreshing after being cooped up in the ship for a month. I never felt better than I do at present—I think it must be the Port wine; we drink it from tumblers at 2d a glass. I suppose it is pretty cold in England now—we shall have no more cold weather for 15 months, & then we shall have a freezer for 6 months.
8 PM

The King has been on board & the grand affair is all over, he was a fine looking man about 35 years old, and was dressed in a rich military uniform, several other officers came with him, & the English ambassador, their boats were very handsome, built like Chinese junks. The Kings galley pulled 26 oars, & the boats crew were about the pick of their navy & looked as if they had been white washed. All of our officers were in full uniform & the ships company were all in their best, & manned the yards for him. The Queen did not come, nor any ladies with him but the English ambassadors daughters were on board to [letter damaged].

The Captain & Professor Thompson dine at the ambassadors[30] tomorrow evening & I hear we are likely to sail when they return. I never knew that oranges sent to England are sent away green & they ripen on the voyage, but so it is for they are only just ripe here now.

Friday, January 10th

We should have gone to sea last night but it came on to blow—a regular dead muzzler[31] for Gibraltar & it is still increasing, however to morrow afternoon is the time fixed for sailing & if they don't go then we shall remain until Monday. The Captain has a large Lunch party at 1 oclock & we are to sail at 4 if possible. I hear that we shant leave any letters behind, if not this will have to go to Gibraltar with us & wait until the next mail leaves there for England. The Roads here are full of merchant ships that the gale has driven in for shelter, we had a heavy thunderstorm here yesterday.

I had a time on the harmonium to night for the first time—tried "Abide with me" & other favorite tunes—the school master[32] has promised to let me practise occasionally. I paid 1/6 for washing today & the way it was done was shameful, they wash them by smashing them down on to the wash tray, & break all the buttons on the shirts besides wearing them out.

Saturday, Jan. 11th
Still blowing hard & unable to go to sea.

Sunday, Jan. 12th
We left Lisbon at 4 PM for Gibraltar & left no mail behind, so I shall have to post this at the rock. It has been a beautiful summers day & the city looked splendid as we steamed down the river. I don't suppose I shall ever see Lisbon again for we are not to call on the way home—but it certainly is a beautiful place & has a splendid climate.

Jan. 13th & 14th, off the Spanish coast
The progress has been very slow as we have been sounding & dredging all day & only made sail at night with a very light wind, we shall not reach Gibraltar this week at this rate. When the dredge came up this morning full of mud & shell fish it was laughable to see the scientific gents with their sleeves rolled up overhauling the mud for Fish & insects[33] &c. The Captain's son Mastr Willie Nares, aged 9, was also very busy amongst the mud—I don't know whether I told you before that he was on board but he is going with us all the commission. The Photographer has taken several views of Lisbon & we are to be allowed to buy any we like 1/- each & have it charged to our accounts, I shall try & send you some of the different places we call at.

Jan. 17th
The most successful day for dredging, & fishing, was yesterday; a splendid net full of mud, crabs, fish &c, "that none of us had ever seen before" was brought up from a bottom of 2,000 fthms & a trawl net similar to that used in shrimping was towed over from the stern & after drifting along the sea bottom for some time was

hauled up & to the general surprise some tidy sized fish were brought up with eyes in the middle of their back & several other peculiarities about them. We made all plain sail for Gibraltar last night with a fine breeze & are now off the African coast at the entrance of the Straits, we shall get in early in the morning so I will close this letter ready for the first post when we arrive & commence another to morrow.

 With best love to all & hoping all are well. I remain,

Yours voraciously,
Joseph Matkin

P.S. Hoping for some fresh meat on Sunday next.[34]

Challenger finally reached Gibraltar on 18 January, remaining for eight days. On 20 January Matkin brought his cousin up to date, with further reactions to Lisbon and highlights of liberty at Gibraltar. The former city, it turned out, was much more to his liking than the latter.

I had no further opportunity of writing again from Lisbon, so must conclude my account of there as briefly as possible in this letter.

I had a run on shore the day after I wrote you, went with a shipmate in the afternoon and had a nice walk out into the country among the Orange Groves, we had a look at the King's Palace, which is a very fine one, built on the top of a hill overlooking Lisbon, from there we had a stroll round the city, and were very much surprised at the dirt and squalor in all parts except the principal streets and squares, in which are some really fine buildings, still there is no bustle and business like you see in an English town, the shops have no windows or fronts and are dingy looking holes, those only which can compete with the English are the confectioners and they beat us hollow in the varieties of sweets and pastry, and the style in which they display it.

The Cathedral is the most magnificent building in the city after the style of S. Paul's in London, with dome and towers, and being built of white stone looks of course cleaner.

There is also a very beautiful Convent there, built like Westminster Abbey on a small scale, and the peal of Bells, in that, and the Cathedral were the best I ever heard.[35]

At 6 in the morning and evening they play Opera tunes, which sounded beautiful from the ship. In the evening we went to the Opera, which was really grand, the singing was all in Italian. The Opera house is the finest (of the sort) I was ever in—seated 20,000, and there was a magnificent box for the King and Royal Family.

The Opera was "Ruy Blas" and the singing was truly splendid, there were 250 in the Orchestra, after the opera came a Ballet, and it was nearly 1 o'clock in the morning when we came out. The Portuguese ladies are very handsome; so are the Spanish, but about one in every five is marked with small pox.   *   *   *

We took in 150 tons of coal there, and were to have sailed on the 10th, but it came on to blow, and we were not able to start until the 12th (Sunday), we steamed down the Tagus just as the sun was setting over Lisbon, and very pretty the city looked as we slowly passed it; I should like to call there on my way home again, but think "Ascension" is the last place we stop at on our return.

We were five days on the way to Gibralter, distant from Lisbon 400 miles, but the best part of each day was spent in Dredging, Sounding etc, with Fishing, which, on the whole was very successful. No more line was lost, and a good deal has been done by the Scientific party, when the dredge is hauled up they stand round in their shirt sleeves, and commence overhauling the mud for fish etc, and as soon as they get any, down they all go to dissect and pickle them in glass jars. The sea bottom between Lisbon and here varied from 400 to 2,500 fathoms at which depth the dredge brought up mud & shell-fish, one, a regular common Sheerness crab.

The mud was analyzed and some baked in the oven in bread tins, numbered, and put away for the Museum when we return.

As is evident in the previous letter, Matkin clearly found some amusement in the activities of the "scientifics," but this private account was quite deferential compared to the published version of the irreverent Sublieutenant Lord George Campbell:

> The mud! ye gods, imagine a cart full of whitish mud,
> filled with minutest shells, poured all wet and sticky and
> slimy on to some clean planks, and then you may have
> some faint idea of how globigerina mud appears to us. In
> this the naturalists paddle and wade about, putting spade-
> fuls into successively finer and finer sieves, till nothing
> remains but the minute shells, &c.[36]

Matkin continues his respectful rendition.

A boat was lowered every day and a fishing party told off, which went a considerable distance from the ship, and succeeded in obtaining some very rare species, but the choicest kinds were by means of the Trawl net, which is the same as a common Herring net, & is paid out from the stern by a line of the requisite length to reach the bottom & towed along by the ship; it was only cast once and took about 6 hours hauling up from a depth of 1900 fathoms, and contained several fish varying from ½ lb to 4½ lbs, & of the most wonderful form and shape, heads larger than their bodies, and eyes in the middle of their backs, of course the Scientifics were in rare good humour and took them down to the analyzing room, where they christened them over some bottles of champagne; I forget the name with which they were honored but should know it by sight, as it was composed of about forty letters.[37]

The Scientifics keep all their information and specimens to themselves and I dare say you will know as much about the Expedition from the papers, by and bye, as we do who are in the ship.

We sighted the coast of Africa on Tuesday night (the 17th) at the entrance to the Straits, and during the night we passed right through into the Mediterranean sea, where we hove to until Saturday morning early, then steamed into the Bay of Gibralter.

The rock looks very grand and imposing from the sea, it is 1500 feet high, and has 1873 guns mounted, varying from 9 pounders to 18 and 25 ton guns (which command the Straits). The small guns protect it from the Spanish side, and there is one for each year, the one for 1873 is now being mounted—a 25 ton gun.

Deep-sea fish *Macrurus aequalis*, from Albert Günther, *Report on the Deep-Sea Fishes Collected by H.M.S. Challenger During the Years 1873–76* (*Challenger Reports, Zoology*, Section V, vol. 22 [London: HMSO, 1887], plate XXXII). Photo courtesy of Scripps Institution of Oceanography.

The Channel Fleet were at anchor in the Bay, and left the same afternoon for Madeira & Teneriffe, where we shall see them again. It comprised the Ironclads "Minotaur", "Agincourt", "Sultan", "Hercules", "Bellerophon" and "Lively"; the other one, the "Northumberland", has gone to Malta to be docked for repairs, having been accidentally rammed by the "Hercules" and seriously damaged. We came alongside the jetty on the South Mole as soon as the Fleet had gone ready for coaling, and today, Monday, all hands are busy filling up with it.

On Sunday morning we all went on shore to the English church; the congregation was nearly all soldiers, and Officers, with their families. In the afternoon I went on shore and had a long ramble— first to the top of the rock; the guns are mounted nearly to the top, some small ones are on it, and the rock is regularly honeycombed all round for them; the Royal Engineers have done a good deal of this, but the heaviest part was done by the Spaniards, before it was taken from them in 1704.

The view from the summit of the rock was magnificent, opposite lay the coast of Barbary, an awfully rugged and mountainous country; some of which are twice as high as Gibralter.

The only towns in sight are Tangiers and Sutor [Ceuta], and on the Spanish coast across the Bay is Algerias [Algeciras], a very pretty place, and higher up the Mediterranean is Malaga. The Artist made a sketch of the former town as we came in, and the Photographer took the south side of the Rock, he also took views of Lisbon; we are to be allowed to purchase any we like at 1/ each.

From the top of the rock I had a walk into Spain, past the line of British pickets across the 500 yards of neutral territory, past the Spanish pickets and into the adjoining towns. The Spanish soldiers are worse than the Portuguese, they wear blue coats, green trousers, and cocked hats, and look about as fierce as maggots; you enter the town by a gate, at which stand about twenty Beggars, the most wretched looking objects, you ever saw, most of them deformed, and the stench from them is something fearful, they stop every one for alms and the Reis come in handy there being twenty to an English penny. There are any amount of Moors here from the Barbary States, dirty looking beggars most of them, and dressed like Arabs. I like the Portuguese the best after all and Lisbon is 10 to 1 to Gibralter.

There are two towns, English town, and Spanish town both under the English governor. There was plenty of gambling going on yesterday morning among the Spaniards.

We received a Mail as soon as we came in, and I had one letter from Mr. Blackwood,[38] congratulating me on being selected to accompany the "Challenger" on her wonderful expedition.

This letter goes to-morrow & you may expect another from Madeira or Teneriffe.[39]

Madeira, reached on 2 February after a week under way, was the first of *Challenger's* numerous island stops. From here Matkin wrote his mother at length.

Last Sunday afternoon, at this time, we were leaving Gibraltar, & this Sunday afternoon we are just about to drop anchor at Madeira, distant from there 700 miles, so that we have averaged 100 miles a day & have been dredging fishing & sounding every day for 7 or 8 hours. The deepest bottom yet obtained was at 6 oclock this morning just as we sighted the Desert Islands, 2,700 fathoms was the depth, or rather more than 3 miles. We could not get in to Tangier it was blowing too hard.

¶On Tuesday the Trawl net was hove overboard & allowed to drag along the bottom for some distance before being hauled up; it took 1 hour going down & 3 hours hauling up, the depth being 2000 fathoms, & it contained some very rare species of Fish. The scientifics have given them a name (only 29 letters[40]) & stowed them away in the Museum. On Wednesday a small Turtle was caught & eaten by the Officers. On Thursday 2,100 fathoms of sounding line was lost overboard in hauling up; the sounding line is not half the thickness of the dredging line. While the sounding &c is going on the ship is hove to, & a Boat is lowered for Fishing purposes, one of the scientifics going in her to preserve the fish.

¶This morning we picked up a large Bale of cotton which had been a long time in the water, it is now being dried in the sun, ready for sale. We started a Canteen at Gibraltar, the Captain advancing the money; they sell Beer, Cheese, Butter, Sardines, Jams, & will sell a great deal more bye & bye; it is a great convenience in the ship & the profits all go to the Ships company.

¶The Desert Islands belong, with Porto Santo, to Madeira, & the whole group belong to Portugal. They are situated to the North East of Africa & are distant from Lisbon 540 miles. They have an area of 330 square miles & the population is 120,000. The greater portion of the islands is immense mountains 6,200 ft high or nearly 5 times as high as Gibraltar. The productions of the Islands are grown at the foot of these mountains, & consist of Sugar, Coffee, arrowroot, oranges, grapes, pomegranates, plantains &c, the vine is the chief product & the principal export is the Wine the Island gives name to, Madeira. Funchal is the principal city & here all the British Merchants reside & there are also a great many invalids here from England for the winter season. The climate is warmer than Gibraltar but the Trade winds blow here nearly all the year round & make it more temperate. We sighted the Desert Islands, (which are 20 miles from Madeira) at 6 oclock this morning & were 8 hours coming up to them steaming 6 miles an hour all the time. I have seen these islands before in the "Sussex" but we kept clear of them; most of the Indian & Australian ships sight them.

February 3d

The Captain would not anchor last night but kept on dredging & trawling all night & the result was the Trawl net, (worth 100£) & 2,100 faths of line was lost in hauling up. Of course we have more nets & lines; but the Ships company set the accident down as the result of dredging & fishing on Sunday.

¶We anchored off "Funchal" the chief city of the island at 10 AM & received the lost mail. I got a Grantham, & a Mercury,[41] & have read them through as well as some of the London Papers; the death of Napoleon[42] was the chief news, there was also a letter about our ship written from Lisbon in the "Standard" of Friday the 10th Jany. Another mail is expected in shortly but I am afraid we shall be on the way to Teneriffe, however it will be sent on. Teneriffe is only 2 days sail from here but we are to sound all the way & may be a week. I hear we are to go to sea tonight so that the stay here is very short, we are to be longer at Teneriffe, surveying the other Canary Islands. I shall write again from there & after that you need not look for a letter for 2 or 3 months for the distance to St. Thomas is 2000 miles & we shall be dredging all the way. You can write to St. Thomas until the middle of March & anything you see in the Telegraph about future addresses you may rely on, send plenty of Newspapers & enclose a stamp now & then likewise, don't forget to address, J. Matkin, Ship's Stew^ds Assist. or S.S.A.[43] will do, & if you are writing to Aunt Lizzie you can tell her so. I expect "Bermuda" will be the address after St. Thomas but we have not been told yet. St. Thomas belongs to the Danes but all the mails to the West Indies call in there (about once a fortnight), so that you can write more than one letter, the Canary Islands with Teneriffe belong to the Spaniards. We don't go to sea until the 5th, I hear.

¶As soon as we dropped anchor there were scores of Boats round, some to take passengers, & others loaded with Oranges, Nuts,

Bananas, Sweet Potatoes, Apples, Figs, & a lot more fruit, some boats had Fish & Turtle, all alive, for sale & in one boat there were several naked boys who commenced diving for money, they would dive right to the bottom, 30 ft, after a penny. We have taken in 3 days Fresh Meat & soft Bread for the Ships company which is a treat after 2 months Biscuit.

¶"Funchal" is built at the slope of the mountains & is a beautiful place to look at, something like Gibraltar, but more like Lisbon —the houses are, & there are Churches & Convents here the same as in Lisbon. A great many of our men went on shore yesterday afternoon & came off staggering, the Madeira didn't agree with them they said. They say it is a beautiful country on shore, the sugar canes look very nice, some of them took Horses & Donkeys & had long rides into the country & up the mountains, one fellow did nothing all the time, but hired 2 men to carry him about in one of their mountain carriages, a sort of Hammock on 2 poles, & you can lay back in it & smoke as this man did. He says he tired 4 men out carrying him about & it cost him 4/6. They all came off at 8 oclock & had to fall in for inspection, some of them did look comical & slept on deck all night. One man kept trying to say "Dismiss", without any s's. I did not get on shore myself but shall try & go this afternoon.

Feby. 4th
All hands are busy now filling up with coal, which is 37/- per ton. When we received the mail here there was a Parcel containing Songs, for the "Challenger's" Crew & addressed to the best Singer on board. I send you one to read & keep. I don't know where they came from. On the back of the song I have written a copy of a "Notice" just put up, as to the address for Bermuda. My Log Book will last me just about half the commission, I shall send it home when filled & get another.

The English mail is expected in today, but these letters we shall have to leave behind. Send the letters on to Charlie, Willie, & Fred, & tell Fred to let Uncle Joe read them, & you may let Walter Thornton[44] read any you like, you did not tell me how he was going on this winter. I suppose you have no Frost, or Skating, this season.   4 PM   I am just going on shore so must post this before I go as the Homeward Mail is in sight now, of the island, & will be here by 6 oclock.

The Mayor & Mayoress of Edinburgh[45] are on board now with Professor Thompson—they are staying here for the winter. I hope to have a nice walk when I get on shore for I don't get any exercise in the ship. I have not got such a fine Issuing Room & Office as I had in the "Audacious" for this place in this ship is below the water & all my writing has to be done by candlelight, & I have to stand up all day as there is no room for a seat. The Stewd does his writing up in the Paymaster's Office.

I hope this will find Father a good deal better, I wish he could winter here like the Mayor of Edinburgh. Give me all particulars how he is getting on now & with best love to you all, & the Boys when you write.

From Your Affect^e Son,
Joseph

Enclosed with the letter was a broadside containing the following lyrics, still of unknown origin and published here for the first time.

A Song
for the
Challenger's Crew
Dedicated to the Best Singer on board
H.M. Ship *Challenger*

Old David Locker stole our Dredge,
To study well its form,
Whilst we were fishing on the Ridge,
Where all his Imps are born.

CHORUS

So never mind your Dredge, my boys,
Which you have lost below;
Our country now your power employs,
That man may wiser grow.

That smutty, sulphurous, envious king,
Is but a myth indeed;
Then out upon that School and bring,
Fresh facts for men to read.

So never mind, &c.

When we have done our work abroad,
And ocean beds, are land;
Our Rope, and Dredge will then afford.
A fact, unique and grand.

So never mind, &c.

Some "Dear old Fogy," then will say,
When delving deep for stone,
A Fossil fish we've found to day,
Whose length will ne'er be known.

So never mind, &c.

Our mission is to teach the world,
What man ne'er knew before;
All truth, by science is unfurl'd,
Which nature has in store.

So never mind, &c.

Our deeds will win immortal fame,
When home our ship returns;
And time, immortal truth will gain,
When all her deeds it learns.

So never mind, &c.

England now expects each man,
His duty to perform,
To carry out our Captain's plan,
The future to adorn.

So never mind, &c.

Three cheers my boys, three jolly cheers,
Our captain to inspire;
His glorious staff, knows no fears,
There souls are now on fire.

So never mind, &c.

So never mind Old Neptune now,
We'll dredge his shifting bed;
That science from our deeds may grow,
When we have join'd the dead.

So never mind, &c.

January 14th, 1873[46]

A week later *Challenger* had arrived at Santa Cruz, capital city of Tenerife, the largest of the Canary Islands. Matkin wrote his cousin that Madeira was to be preferred to Santa Cruz.

We left Gibralter on Sunday the 26th of January, intending to call at Tangier, Morocco, but it was blowing too hard, and we had to get up steam to keep clear of the African Coast. On the following *Sunday* we were just ready to drop anchor at Madeira after a very pleasant passage with sounding and dredging every day, the deepest bottom was 2,600 fathoms, or rather more than 3 miles, obtained off the islands of the Desert (belonging to Madeiras).   *   *   *

On the Monday early we anchored off Funchal, the chief city of Madeira, and made preparations for coaling. The Madeira islands are in Lat. 32° N. & Long. 17° W. they all belong to Portugal, and are distant from Lisbon 540 miles, they consist wholly of mountains 6,000 feet high, and we could see them 45 miles away. The productions of the islands are grown at the foot of these mountains, and consist of dates, plantains, pomegranates, bananas, figs, alligator pears, sugar canes, coffee, grapes, arrowroot etc.

The climate is the best in the world, and there were a great many English visitors staying the winter there, of whom we had several on board, including the Mayor of Edinburgh and his family. The islands have an area of 336 sq. miles, and a population of 120,000. Funchal is a very clean & pretty place, built something after the style of Lisbon. I had a fine run on shore at Madeira, walked nearly to the top of the island, and drank some of the wine as well. We expect to call there again in July, when of course it will be much hotter.

We coaled on the Monday, and sailed on Tuesday at 3 o'clock, just as we were leaving H.M. Troop Ship Orontes came in from England, and brought us some packages. As soon as we got clear of the island, we picked up a splendid breeze, the fires were put out, and every stitch of Canvas set, driving the ship 12 miles an hour.

The Canary islands are distant from Madeira 250 miles, and we sighted the Peak of Teneriffe, at 3 PM yesterday, exactly 24 hours from Madeira.

We remained out sounding all night and came into Santa Cruz early this morning. Being so misty we are unable to see the Peak to day, but I think some of the Scientifics are going to try and reach the summit before we leave, it is 12,300 feet high, or nearly 2½ miles, and on the top, is at present covered with snow, although here at the bottom it is quite hot & sultry. Teneriffe is not such a pretty island as Madeira, and Santa Cruz is not to be compared with Funchal for cleanliness.

Its population is 300,000, and it has an area of 878 sq. miles. [Las] Palmas one of the other Canaries is nearly as large. They all belong to the Spaniards, and one of their Gun boats is at anchor close to.

The productions are the same as at Madeira, with the exception of Cochineal, but I think the Bananas are finer. Santa Cruz is very strongly defended, and there are a great many Spanish Soldiers here. It is famous as being the only place Nelson was unable to take, but he failed in consequence of the great surf which runs here capsizing the Boats before the men could land. He lost his right eye, and a great many men here.[47] An Ironclad of the present day could knock the place into ashes in about 2 minutes.

The Spaniards fired a salute for us at 12 o'clock to-day, but as this ship is not fitted out as a man of war she did not return it.[48]

Sunday, Feby. 9th

The mail for England closes in five minutes, so must cut this short. I am going on shore this afternoon so must describe the place in my next letter.

We can see the famous Peak quite plainly to-day, a party of three Officers and Scientifics with four seamen went away yesterday to scale the mountains. They have taken five days provisions etc. Tomorrow we are going to cruize round the other islands, and are to come back on Thursday, for the surveying party, when we shall start for St. Thomas.

Last night one of the men caught a Shark, about 6 feet in length. He was a long time dying, for after his tail was cut off, and his entrails (including 14 young ones) taken out of him, he was as lively as ever. The young ones were all alive and the Scientifics took charge of them. The old gentleman[49] is now being eaten for dinner in the mess of the men who caught him.[50]

On the 14th of February *Challenger* set sail across the Atlantic, destination St. Thomas. With no opportunities for posting mail for the next thirty days, Matkin strung out his letters over the period. The first, to his cousin, begins on 27 February with a summary of the final days at Tenerife.

Dear Tom,

As I have time to spare this afternoon I will get the "S. Thomas's" letter under weigh. We left Teneriffe before the mail came in from England so have had no letters yet, but shall expect to receive some at S. Thomas, and Bermuda. The last I posted to you just before I went on shore at Teneriffe. I went on the following

Shark on board, from T. H. Tizard, H. N. Moseley,
J. Y. Buchanan, and John Murray, *Narrative of the Cruise
of H.M.S. Challenger with a General Account of the
Scientific Results of the Expedition*, 2 vols. in 3 (London:
HMSO, 1885), vol. 1, pt. 2, p. 560. Photo courtesy of
Scripps Institution of Oceanography.

Sunday afternoon, and had a splendid walk over the mountains
from Santa Cruz to La Laguna, the Spanish Capital, 5 miles in the
interior. At Santa Cruz the heat was tremendous, but on climbing
the mountains the air got gradually cooler, and when I left La
Laguna, after sunset it was quite chilly. I could not make it out at
first why the men and women coming down from the mountains
all wore a large blanket each for a cloak, and on arriving at Santa
Cruz took it off, but found out it was through the change in the climate. The men and women all went barefooted, and the latter

appeared to do the best part of the work, some of them were carrying enormous baskets of fruit from the country into Santa Cruz. La Laguna was the largest place of the two, but the Cathedral was the only building worth looking at in it, in which there was a very fine group of statuary. The sides of the hills were covered with a pretty shrub called the Prickly Pear on which the Cochineal insect feeds. That is the principal export, the fruit of the Prickly Pear is not up to much, but Alligator Pears are very fine eating; Santa Cruz or the "City of the Holy Cross" is where the Spanish Governor resides, but that is not such a nice place as Madeira. The next day after going on shore we got under weigh early for a four days cruize round the other Canary islands, with occasional soundings, dredging etc.

After we got clear of the island we had a fine view of the Peak of Teneriffe, and could see it from all the other islands. They are all of volcanic origin, and most of them rise for 8 & 9000 ft above the sea level; some of them rise perpendicularly, and have a very wonderful appearance.

We steamed round Palmas, Fuerita Ventura, and two or three smaller ones and were away four days. The soundings varied considerably, in one place it was only 75 fathoms, a little farther on it was 1900, then 2,500, and so on, showing the bottom to be as varied in depth as the islands in height. The dredge brought up a sort of petrified cinders, which had every appearance of being thrown up from volcanoes ages ago. Off the island of S. Christopher we fell in with a Spanish fishing boat, and bought all the fish they had for 3 dollars. There were 3 large baskets full, and they were of the most wonderful color I ever saw, being gold and silver, green, violet, blue, and every shade you could mention, I never thought the sea contained such beautiful creatures. We were away four days, after which we anchored off Santa Cruz, and picked up the exploring party, who had been away five days attempting to scale the Peak, but they only reached 9,000 feet in height, out of the 12,360, when the cold was so intense that the Spanish Guide refused to go any higher, so down they came again. For 1200 ft the Peak was covered in snow, which lies on the summit all the year round.

The day after we returned we took in all the coal there was in Santa Cruz, 25 tons, and got ready for sea; the same evening one of the Boys fell from the main yard into the sea, but was immediately picked up by one of the seamen who jumped overboard after him, he was stunned at first, but soon recovered, the distance was 37 ft from the main yard.

The next day Saturday, Feb. 15th, we got under weigh for S. Thomas, distant 3,100 miles, and to-day we are 1300 miles from Teneriffe, and have 1800 more to go. The Peak could be seen the best about 70 miles away, rising above the clouds like a pillar of snow.

We have had the Trade winds all the way, and might have been there by this time, if it were not for the sounding and dredging etc every day, which occupies 8 & 9 hours each day besides the labour of furling, and making sail after it is all over.

The scientific operations, I believe, have been very favorable, the lowest bottom has been 1900 fathoms, and the greatest depth 3,650, or rather more than 4 miles. Several thousand faths. of line, and a few dredges have been lost, but of course that counts as nothing.

In some places the dredge brought up black coral and beautiful sponges, in others, shell fish, mud, gravel &c. Various experiments have also been made by the Scientifics, such as telegraphing down to the bottom of the sea, and finding out the strength of the electric fluid at such great depths.

This is a reference to the "resistance deep-sea thermometer" invented by Sir William Siemens for "telegraphing" submarine water temperatures to the ship, first used by *Challenger* on 15 February 1873.[51] Matkin was apparently misinformed about the purpose of this "experiment," as is clear from a similar passage in the succeeding letter to his mother. Apparently electricity was still a quite mysterious subject to the layman in the 1870s, despite the efforts of physicists like Michael Faraday to popularize their successful research. Matkin continues:

By means of a certain instrument water is also brought up from the bottom; at the greatest depth it was quite fresh and drinkable, the temperature was 34°, or 2 above freezing point.[52] Boats have been lowered occasionally with an instrument for testing the strength of the current;[53] yesterday it was ascertained to set at the rate of 4-½ miles per hour in a N.E. direction, or in the direction of Barrowden; of course this is at some distance below the water that the current is felt. We entered the Tropics on Sunday last, and it is hotter now than in the middle of summer in England. The time slips away though much quicker than with you, and the health of the ship's company is excellent at present, though our living is far from flattering, and I should enjoy some fresh Butter, as from one week's end to another it is Biscuits, Salt Horse, Pork, and Australian Meat, which is very far from gay living. The Officers had their meat safe broken open last night, and everything it contained was eaten before 4 o'clock this morning, so they had a short allowance for breakfast, I expect it was taken and ate up aloft, (like the Turkey at X.mas). I will write some few more (as Artemus Ward[54] says) in a few days.

March 12th

Within 250 miles of S. Thomas, W. Indies, and expecting to get in on the 15th. The dredge has just come up full of mud etc from 3000 faths. of water, and all hands are making sail again for the night. We had a most interesting lecture given last week by Professor Thompson to the Ship's company, on the object of the expedi-

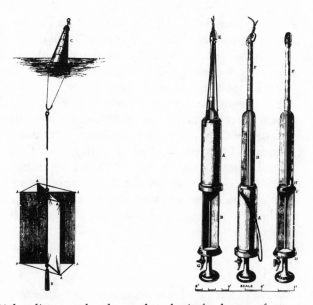

Right: slip water-bottle, used to obtain fresh water from the sea bottom, from T. H. Tizard, H. N. Moseley, J. Y. Buchanan, and John Murray, *Narrative of the Cruise of H.M.S. Challenger with a General Account of the Scientific Results of the Expedition,* 2 vols. in 3 (London: HMSO, 1885), vol. I, pt. I, p. III. Photo courtesy of Scripps Institution of Oceanography. *Left:* current drag, used to measure the strength of the current, from T. H. Tizard, H. N. Moseley, J. Y. Buchanan, and John Murray, *Narrative of the Cruise of H.M.S. Challenger with a General Account of the Scientific Results of the Expedition,* 2 vols. in 3 (London: HMSO, 1885), vol. I, pt. I, p. 80. Photo courtesy of Scripps Institution of Oceanography.

tion, which I have written as well as I can remember, and send for your perusal. The Professor has promised to lecture again on the same subject ere long.[55]   *   *   *

We passed a Spanish Schooner, called the "Virgin Mary" while we were in Church on Sunday, she was bound to Cuba, and had one Lady on board; great excitement on board the "Challenger" to have a look at her, not having seen one of the fair sex for a month.

We have had an enormous Shark following us for the last 1100 miles, and the men say he won't leave until some one dies, and is thrown overboard. They have tried everything to catch him, and the Officers have fired at him with Rifles, but to no effect, he only goes under the ship's bottom for a time.[56]   *   *   *

March 16th

We have been delayed by the fog, but are just about to drop anchor in S. Thomas' Harbour, and the mail leaves in about 2 hours, so I must wind up, and describe the place in my next letter. We shall soon have some English news now, and letters too, also something to eat I hope in the shape of Bananas, & Alligator Pears.

We caught three sharks yesterday about six feet in length, but the monster one is not yet caught, and is most likely under the bottom of the ship now. The young ones were eaten for breakfast this morning, I tasted a small quantity, and thought it very good indeed, though rather dry. We were sounding yesterday off the island of "Sombrero" & the dredge brought up a young Sea Serpent[57], which was collared by the Scientifics. The bottom within the last 120 miles, ran up from 3,075 faths. to 400, a difference of 3 miles.

Yours as ever,
J. Matkin[58]

The last of Matkin's letters during this first Atlantic transit was begun to his mother on March 3 and completed upon arrival at St. Thomas two weeks later. His oceanographic descriptions are more detailed than in the previous letter. Moreover, appended to the letter was a carefully copied transcript of Wyville Thomson's first and only lecture to the crew on the purposes of the voyage, delivered during the first week in March. Curiously, there is not even the barest mention of this event in any of the other *Challenger* accounts.

We are within a thousand miles of St Thomas to day so I will get a letter ready for you by the time we arrive. Of course I shall get a few letters there, & at Bermuda; everyone is looking forward for the same reason, although a great many received letters at Madeira. We are sure several mails have miscarried & we expect to pick them up at St Thomas. The "Teneriffe" letters were sent on there by the Admiralty through some mismanagement on their part, & the latest English news we have at present is the 17th Jany.

We are 17 days from Teneriffe today & expect to be there in about 12 more days for we furl sails every day for 8 or 9 hours, & dredge, & take soundings &c which of course makes the journey much longer, & tedious.

We left Teneriffe on the 15th Feb with the Trade wind right behind us & we have kept it ever since & have had splendid weather all the way; the ship has averaged 100 miles per day & could have done twice as much if it was not for the dredging &c. The scientifics are getting good sailors now (wear white trousers, & blue serge jackets) & their labours have been satisfactory I believe. The bottom of the ocean it appears is as varied as the land for there are valleys & mountains, hills & plains all across the Atlantic. One day we got soundings at 3,800 faths or about 4-¼ miles & the next it is only 1900, or about 2-¼ miles in depth. The average depth is about 2,500 fathoms, nearly 3 miles & in some places the dredge brings up mud, & clay, ooze, &c; in others it brings up gravel, &c. The greatest bottom from which anything was brought up was 3,175 faths when the dredge brought up black coral, & some magnificent sponges, but nothing living, that I could see with my *naked eyesight*. By means of a scientific instrument attached to the end of the line, Water was brought up from the bottom & found to be quite fresh & palatable, the temperature too was very low 34° or 2 above freezing point. They also telegraph down to the bottom (I don't know the object, but I think to test the strength of the current of electricity at such great depths). On fine calm days while the dredging is going on a boat is lowered with an instrument for ascertaining the direction & force of the current at certain depths below the surface: on the last occasion the current was found to set North East (or in the direction of England) at the rate of 4-½ miles per hour. This was some distance down & would have very little effect on a ship. We are now in the tropics & wear white trousers, & white caps all day, the heat is about equal to Melbourne at X.mas time, & I need not say that I have stowed my sleeping Rug away until we leave the Cape for the Antarctic. For the last few days we have been dogged by a regular old Tiger shark about 20 ft long & the scientifics have tried all they know to catch him but he's a regular old soldier & up to every move on the board. He swims up to the hook & smells of the Pork on it then turns round, & swims away, sometimes we lose him for half a day but as soon as anything is thrown overboard John comes out from under the ships bottom & grabs it like a lawyer. He's been with us a week now & some of the men say it's a sure sign that somebody will be thrown overboard before we reach the West Indies & if they do catch him they're going to cut him into 4 lb pieces with their Knives (while he's alive). * * * The "Challenger's" Brass band played selections on Saturday night last for the first time, they have got on very well indeed considering that they have had to practise in their (Watch below) all the time. Two Marches & "God save the Queen" was the programme & when they had finished the officers gave them plenty of Wine & the Ships company clapped them as if it was the first time of hearing—instead of being treated with the same old tune every day for the last 3 months. I need not tell you that the "Diet" since we left Teneriffe has been very far from gay, regular old Navy, Salt Meat & Biscuit everyday—with an allowance of Lime juice to prevent scurvy: however there is no sickness of any sort in the ship & we are fast coming down to our precise fighting weight. The Officers live now nearly as well as in Harbour, their Meat Safe was broken open the other night & cleaned right out before morning although they keep a marine sentry on it night & day.

March 11th
We expect to be at St Thomas on the 15th & the mail leaves there for England on the 16th. We had a very interesting Lecture the other night on the main deck, given by Professor Thompson, all about the expedition, & different animals brought up in the dredge. The animals (magnified) were beautifully drawn by the Artist & the Lecture altogether was a great success, the Professor promised to continue the subject at some future time. I have written the Lecture out as well as I can remember & I send it you by this same Post; after you have read it I want you to send it on to J. T. Swann for him to read, then he can send it you back again. I have written to Mr. Daddo[59] by this mail & forwarded him a copy, too. * * * I have taken up a good many clothes since we left England, tropical clothes I mean, but I shall not require any more for a long time. I shall not be able to save a deal for the first year but after I get my increase of Pay I shall commence putting in the Saving's Bank & I intend to save 50£ or more by the time we return.

The Saving's Bank is a Government affair & the Interest is 2-½ per cent per annum, several of the men have commenced an account already. One of our men was spinning a yarn the other day about doing 4 years in China, & returning home he wrote to his wife & told her to get plenty of Provisions in for he should be home on Friday with his 4 years pay. He drove up to the door in a cab, & when his wife asked him for his pay (before he's been 2 minutes in the house), he laid down 1/9d, all he had left of his 4 years pay.

March 12th

The dredge had just come up full of mud &c from 3000 faths of water & we are under plain sail again, we shall anchor at St Thomas on the 15th & are to remain about a week before sailing for Bermuda, distant 990 miles. We give orders as to what Provisions & repairs we want done there & while they are getting ready we go on to New York, distant 1000 miles & are to remain there a week, then back to Bermuda for repairs in the famous floating dock there. From Bermuda we go right across the Atlantic again to Madeira, dredging & sounding all the way. The future addresses were posted up to day on the main deck. Bermuda letters to arrive by the 8th April, New York, letters to arrive there by the 25th April, Bermuda again to arrive by the 8th May, after which Madeira. Letters for New York can be posted in England up to the 12th April I think. Write often & send plenty of newspapers not forgetting a stamp occasionally, I have not one left for this letter unless I get one at St Thomas. Tell Charlie & Will to write sometimes & if I can find time I will write to them; but of course you always send them my letters to read.

Sunday, March 16th

Just going to drop anchor in St Thomas's Harbour, we were delayed by the fog off the island of "Sombrero" yesterday & shall only just manage to catch the mail which leaves for England to night. We caught 3 sharks yesterday about 6 ft in length. The old shark of all we have not caught, he is under the ships bottom now & will follow us to Bermuda most likely. The young sharks were eaten for Breakfast this morning & I had a small portion, it went down high for we were all half starved, one of the sharks kept looking about for a snack after he had been cut in a half more than an hour, & some of the men said this morning in Church that when the Harmonium struck up the shark moved in his inside. We shall soon have the letters on board now & the English Newspapers also. The island is about the size of Madeira & the land is rather high. I shall tell you all about it in my next letter, we have not anchored yet. With best love to Father, Miss W[ildman], & the Boys, & likewise to yourself.

From Your affec^te Son,
Joseph Matkin

P.S. Remember me to everybody in the county.

Substance of Professor Wyville Thompson's Lecture, to the ship's company of H.M. Ship "Challenger," on the Geography of the sea, & the object of the "Challenger" Expedition. With remarks on the progress hitherto made:

I have been asked by the Captain to try & explain to you, as well as I am able, what is the object of our expedition & what we are doing from day to day. I need not remark that it gives me pleasure so to do, for we are to be common shipmates for the next few years, & doubtless, each one has some interest in the work, & the results, if successful, will be creditable to us all. In the first place I must tell you that the bottom of the sea occupies an area of 3/4 three quarters of the globe, & this immense portion has hitherto been as a sealed book to the human race. We have a comparatively accurate notion of the Land; we know the geology & the Natural History of most of the Countries of the earth; even Africa, & Australia; are becoming annually more known to us; & the indomitable energy of man is slowly but surely bringing each country into what I may call the regular routine, & causing it to contribute somewhat to the comfort & happiness of the rest; inasmuch as their productions (whether natural or artificial) whether as necessaries, or more generally as luxuries, are spread in this manner over the world, & in this way conduces to the general happiness of mankind. One reason why our ancestors did nothing toward lifting the veil from the sea bottom was because it was thought that no object could be gained by so doing; & the difficulties in the way were deemed insurmountable. For it was thought, and with reason, that nothing living could exist at a greater depth than about 400 faths. Now you all know that where an empty bucket is put over the Ship's side & allowed to sink down a little distance what difficulty there is in hauling it up, & what a resistance is offered by the weight of water on the top of it. That resistance increases the lower we go; so that if a man was placed at the bottom where we were to day sounding, 2,800 fathoms, he would have a weight pressing on him equal to the weight of the *"Challenger."* To explain this, I must tell you that a man's body presents a surface of 375 square inches, & that at 2,800 faths the weight of a column of water would be over 3 tons to the square inch. If the island of Teneriffe was sunk in the place we were to day the Peak would scarcely reach the surface of the water. These are among the principal reasons for the former statement (which we, as well as others, have proved to be false) "that no animal life was to be found lower than about 400 fathoms." It is now less than 20 years ago that scientific men began to talk of a scheme of Ocean telegraphs, whereby the continents of America and Europe might be placed in almost instantaneous communication. How this scheme has succeeded we all know, but in the outset it was resolved by all practical men that some knowledge of the nature of the sea bottom was imperative, to enable the cables to be laid with any degree of accuracy & safety. Soundings, therefore, were at once taken across that portion of the Atlantic where the cables were to be laid. The Americans, as well as other nations, commenced a series of Atlantic soundings, which have been continued more or less ever since. The first dredging in deep water with anything like success was obtained by the Americans; a clever Lieutenant[60] in their Navy, invented an instrument similar to the one we now use in sounding, whereby a very small portion of the bottom could be brought up. Their dredging, however, was in comparatively shallow water, and it was reserved for an English expedition, in 1863,[61] to bring deep-sea dredging to something like the system we are now working on. For that purpose H.M. Gun Boat *"Porcupine"* was fitted out for a series of soundings, Dredgings, &c., in the Bay of Biscay & the home waters. A colleague of my own (Dr. Carpenter[62]) was among the scientific party on this occasion, & the account published by him on his return[63] awoke the first desire for further information on the subject. The greatest depth from which they succeeded in bringing up anything from the bottom, was 2,300 fathoms, in the Bay of Biscay. Other countries have carried on the research in a desultory sort of manner, but it was at length decided by scientific men that no country but England, & none but British seamen, could solve the problem in anything like a satisfactory manner. For this purpose the Chancellor of the Exchequer was consulted & the result was that the *"Challenger"* & her crew, ~~was~~ were selected for the purpose; & I am happy to say that we have thus far succeeded tolerably well. We have obtained the deepest dredging yet known in the world, viz. 3,175 faths. & from that depth we obtained specimens of animal

life, thus proving (contrary to all our opinions) that living creatures do exist in these great depths, although they are very low in
the scale of animal life. I am of the opinion that the ocean is no
where much deeper than "4000 fathoms"[64] & that the valley, as I
may call it where we obtained our deepest soundings, the other
day, is of no great extent; however we shall know more about that
bye & bye. The mud that the dredge has more commonly brought
up from the bottom, is of a grey colour, &, to our general surprise,
on being placed under the microscope revealed nothing but the
shells of what had been living creatures. This mud we found to
contain carbonate of lime; & the cast-off shells of these animals is
causing a gradual formation of rock at the bottom of the sea, similar to the chalk cliffs on the South of England. As regards animal
life in the ocean, we find at from 200 to 400 fathoms a class of creatures of the same form and character, & the animals down to that
depth are tolerably well known to Naturalists; at 400 fathoms they
are very scarce, & at 500 cease altogether, thus causing us to
believe, until very recently, that below that depth no living thing
existed. This, within the last 10 years, has proved to be a fallacy;
for after 600 fathoms down to 800 the sea is teeming with animals
of a sort hitherto unknown to man: animals nearly transparent,
but which have eyes, & lungs, & hearts the same as we have, &
which reveal, under the microscope, all these organs to perfection.
We see them going through the whole of their animal functions
from their cradle to their grave; we see them seizing their prey; we
see them digesting it; we see also their eggs being formed; & we see
them (under the microscope), when fully developed & the creature
springs into existence. I hope some time to shew any one who
cares, these things for themselves. Among the creatures that we
have brought up in the dredge, I purpose shewing & describing the
most curious. You see them here drawn by Mr. Wyld, the Artist, &
you can form some idea of their beauty & delicate formation. This
creature half Plant & half animal possesses the power of emitting a
strong green light from its body;[65] we obtained it off the Canary
islands in 1700 fathoms of water, & when hauled up, gave a sufficient light to enable Captain Maclear to experiment on it.[66] This
makes the 5th fifth creature of this species on record. The peculiar
phosphorescent light which emanates from these & similar creatures, is the only light existing below 400 fathoms of water, & is
wholly independent of the light of the sun. This other vase-shaped
animal, half glass & half sponge is not unknown in England, but
has never before been obtained elsewhere, than off the Phillipine

'A good portly man, i' faith, and a corpulent; of a cheerful look, a pleasing eye, and a most noble carriage; and, as I think, his age some fifty, or, by'r lady, inclining to three-score; and now I remember me, his name is . . . .' (*Henry IV, 1st Part, Act I, Scene 4*).

Sketch of Professor Wyville Thomson, director of the expedition, from Herbert Swire, *The Voyage of the Challenger*, 2 vols. (London: Golden Cockerel Press, 1938), p. 79. Photo courtesy of Scripps Institution of Oceanography.

Islands.[67] From their great beauty & rarity they were, on their first appearance, sold in England for £50 apiece, as chimney ornaments. We obtained this off the coast of Portugal. This gelatinous creature, which looks like a ship's anchor, we obtained off the Canary Islands; & these peculiar excrescences in the shape of an anchor which protrude from the creature's body serve for exactly that purpose, in enabling it to cling with wonderful tenacity to any substance, "especially a man's fingers." What caused these excrescences to assume this particular shape has puzzled many. Whether the animal first saw a ship's anchor, & took a fancy to its form & shape, or whether we first saw the animal & borrowed its nautical appendage for our anchor, is more than I am able to decide. To-day the dredge brought us up an immense Prawn, or Lobster,[68] from 2000 fathoms of water, some of you perhaps saw it, & noticed that it lacked eyes. Now a common Lobster has a pair of very bright eyes fixed to the end of a sort of twig, & you may have seen it occasionally sling its eyes over its shoulder & look behind. Not so our crab, where his eyes should have been, there was nothing; & the reason, I take it, why there was nothing, is, that as there is no light where the gentleman lives; no eyes he requires, as they would only be an encumbrance. It is now getting late, & I feel that I must draw my discourse to a close, observing that I shall be most happy to resume the subject at some future time, if you care to hear me.

(Loud and continued Applause.)

JM[69]

Sounding, dredging, and trawling operations, which often went on for hours at a time, were arduous and tedious work for the crew. The captain, and Thomson, undoubtedly recognized the value, for morale, of bringing the crew members, even if only superficially, into the circle of knowledge to which *Challenger* was contributing. As we have seen, Matkin was occasionally miffed that the British reading public was better informed than the crew about the *Challenger*'s activities.

For most of the men, of course, arrival in the New World was an even more welcome event.

2 North Atlantic Transits
APRIL 1873–OCTOBER 1873

> While my Shipmates are dredging the Sea,
> I'll indite a letter to Thee.
> —Joseph Matkin, 1873

Challenger arrived in the West Indies on 16 March, remaining at St. Thomas for eight days before proceeding north to Bermuda on 24 March. There is a gap in Matkin's letters until 7 April, when he wrote to his cousin from Bermuda. An earlier letter to his mother is missing.

> H.M.S. "Challenger"
> Bermuda Dockyard
> April 7th, 1873

Dear Tom,

I received two letters from you at S. Thomas, the day after I wrote you from there, one being addressed to "Madeira", and the other "Teneriffe" so that they always manage to reach us finally, and none the less welcome for the delay. I am expecting another or two by the same mail which brings this to you. I read in the papers about the "Northfleet" catastrophe.[1] We lay at Dungeness on our bad passage down Channel, and I should think there were 300 vessels at anchor in the Downs. We get most of the newspapers sent out, and a great many magazines. "Good Words"[2] will soon receive accounts of our doings from Professor Thompson. Did you see the picture in the "Graphic" of the King of Portugal in our ship?

I have given instructions for a letter sent home to be forwarded you to read, containing an account of the island and the colored *pussons*,[3] & about our picking up a large Iron Merchant Ship belonging to Greenock,[4] that had been dismasted in a gale soon after leaving New York for Liverpool. We left S. Thomas on the evening of March 24th, under all plain sail, and the dredging & sounding at once commenced, and has continued up to our arrival here. On the 25th we obtained the deepest soundings we have had hitherto, and I believe the deepest on record, 3,875 fathoms, or about 4-½ miles. The dredge was hove overboard, and the strain on the line was so great when it reached the bottom, that when they

63

commenced hauling it in, it carried away an iron block that was screwed in to the Deck, and had all the strain to bear. The block as it flew up struck a sailor boy, named [William] Stokes, on the head, & dashed him to the deck with such a terrible force, that his thigh was broken, and spine dreadfully injured. He was carried to the Sick Bay and attended to by the Surgeons, but he was insensible the whole time, and only lived two hours. At 5 PM the next day the Bell tolled for his funeral, all the Ship's company and the Officers and Scientific gents, attending on the Main deck. The captain read the Service, and at the appointed place in it, the body was lowered into the sea, by the lad's messmates, three 36 lb. shots were attached to it, to sink it, the depth was over four miles. The boy came from Deal where his Father is a Channel Pilot. All his clothes, and effects were sold, and the money, with his wages, a few Photos, letters, and his Bible will be sent to his friends by this same mail.

¶We had soundings right across and the average depth was 2,800 fathoms, nothing of importance was brought up in the Dredge. On the evening of the 1st of April they piped "Hands to Bathe" and about 80 of us went overboard and had a fine swim. A Boat was lowered to keep away Sharks &c, the water was over 3 miles in depth. A few Sharks have been caught & eaten since we left S. Thomas and the other morning one of the men Harpooned a Dolphin, which is the most beautiful fish in the sea, and best eating.[5] When dying they keep changing color, and after they are dead are of an ordinary color. They are about 3 ft. long and live chiefly on Flying Fish, which are about the size of a Herring, and skim the water like a swallow; but have not very good times for they are pursued in the water by Dolphin and other fish, and in the air by Gulls and Sea birds. On a dark night if a lighted candle is placed in the Ship's port they will often fly in. At 2 PM on the 3rd of April the Bermuda islands hove in sight, the land being wonderfully low for islands in the middle of the ocean, and looks very bleak and bare from the distance, but when we arrived in close it had a better appearance.

We dredged round the islands all day on the 4th and in the Evening the Pilot (a darkie) brought us in.

The navigation on entering is very intricate and difficult, as there are so many shoals & rocks; the pilot had to stand in the foretop, where he could see the rocks, and give orders for the steering &c. The Bermuda islands all belong to Great Britain, and are about 300 in number, but the whole lot are no larger than S. Thomas. Bermuda itself is the largest, and Hamilton the chief town is built on it, the whole population is about 12,000, and the greater portion are colored people. The islands were discovered in 1527 by Bermudez a Spaniard, who named them after himself. In 1609, Admiral Sir George Somers, was wrecked on the islands and founded the Colony; and have belonged to the British ever since. There is a fine fortified Dockyard, on one of the islands, and an immense iron floating dock, for the convenience of re-fitting the Fleet stationed in North American waters, without sending them home. This immense dock was built at Woolwich, and towed across the Atlantic a few years ago, by four of our largest Ironclads. This place would be of great importance as a rendezvous for the Fleet &c, in case of a war with America; it is 5 days sail from New York, and 800 miles from the nearest land, Cape Hatteras in North America.

The Governor of the island is Major General Lefroy, and there are plenty of soldiers here for its protection. The climate is not nearly so warm as at S. Thomas, but we are able to bathe every evening. The greater portion of our North American Fleet is here, including the "Royal Alfred" flag ship of Admiral Fanshawe, for whom we brought a heavy mail from S. Thomas.

Their head station is Halifax, but they generally winter here. The Governor with a party of Ladies & Officers are at present on board, lunching.

On the night of April 4th we had another awfully sudden death on board. I left the upper deck at 10 PM to go down into my Hammock, leaving the Ship's Schoolmaster, Mr. Ebbells and the writer[6] (who sleeps next to me) walking the upper deck together. Soon after I turned in, they also came down, after which I went to sleep, I had not been asleep more than half an hour before the Writer woke me, and said that the Schoolmaster was dead. He was heard to be breathing hard, and one of the men tried to rouse him, but not succeeding went for the Doctors, before they came he was dead. On Saturday morning they held an inquest, the Verdict being "died from Appoplexy".[7] In the afternoon, and less than 7 hours after he died, he was buried in Bermuda Cemetry. Nearly all the ship's company attended his funeral; I should have gone but could not be spared, the Band of the "Royal Alfred" played the "Dead March in Saul", to the grave; our band not having practised it yet. A very singular & affecting scene occurred at the Funeral, the Chaplain of the Dockyard buried him, & did not know just at the time whose corpse it was, but happening at the conclusion to glance at the Coffin plate, read the name backwards, as he stood at the head of it. "Adam Ebbels, born April 4th 1837, died April 4th

1873," and for some time was so much affected that he was unable to continue the service.

After he had finished, on addressing the men at the grave, said he had sailed round the world with the deceased in H.M.S. "Liffey", and had been intimate with him for 3 or 4 years, during which time he had always found him an honorable man, and was never so surprised before as he was on reading that Coffin plate. The deceased was a very strong and healthy man, & clever & intelligent. He was not married and only had one sister living at Woodbury near Xeter,[8] and what was very singular, died on his 37th birthday.

The Captain has his only son on board, a little lad of 10 years, he intended taking him as far as Sydney with us where he was going to school. The Schoolmaster was educating him in the meantime, but now he is dead, I hear the Captain will send his son back to England from New York. This makes the fourth sudden death on board since we commissioned, and hope we shall have no more. The 1st Lieutenant[9] played the Harmonium yesterday in the Schoolmaster's place, he is a splendid player, and has a fine little instrument of American invention, I think it is called a Melodian and much sweeter than a Harmonium.

H.M. Ship "Tamar", a troop ship sailed for England yesterday morning with time expired troops, their Band playing "Auld Lang Syne" as they steamed past us.

I went on shore yester afternoon, & had a fine walk on 5 or 6 different islands, which are connected either by a Bridge or Ferry Boat, and have some splendid walks on them. The flowers here are beautiful, and there a[re] plenty of English singing birds, as well as native ones of most brilliant plumage. The walks are shaded by trees on each side, and there are several Churches, schools &c on the islands. Hamilton, the chief town is 8 miles from here, I walked nearly there yesterday, and intend going all the way before we leave, about the 20th inst.

I had a look in at Somerset church while the service was going on, there was a good organ, and the chanting & singing excellent, the congregation was nearly all black, and some of the ladies, "oh! they just was dressed", would beat Barrowden all to smash. The best bit of fun was the morning after we arrived, at about 6 AM about 50 black washerwomen came on board, and there was some talking and jabbering, they have plenty of cheek too, went into the Officers Cabins while they were turned in, and with a grin on 'em that you'd have thought it impossible to get their mouths shut again, asked "Any Vashing Sare" (answer) "go to the Devil"—"Yes sare, after you sare." They are not as spruce and good looking as the Ladies of S. Thomas.

We returned on board about 7 PM and nearly half the men were intoxicated, and all but Capsized the boat coming off, fighting all the way. When they get on board, they drop down on the deck & sleep sober before morning. We came into the Dock basin this morning, and are now close to the Iron floating dock.

The dates of departure have been again altered, you can write from England, up to April 28th and address New York, after which Bermuda again, by the mail leaving England on May 6th as we shall not sail from here the second time before the end of that month. Don't forget to put S.S.A. on the address or it may be charged as an Officer's letter.

April 8th 9 AM
No sign of the mail yet, our Band is playing away now first rate on the quarter deck.
2 PM
The mail has not yet arrived from England, but ours closes at once.[10]

The next letter follows the ship from Bermuda to Halifax, where she arrived, after a stormy passage, on 9 May. The dockyard cemetery was Bermuda's most vivid memory for Matkin. He found the English ambience of Halifax more comfortable despite wintry weather.

> H.M.S. "Challenger"
> Out at Sea
> May 7th, [18]73

Dear Tom,
"While my Shipmates are dredging the Sea, I'll indite a letter to Thee" (one of their poetical specimens and quite extempore). Your last letter reached me on April 9th and being the only one by that mail was all the more welcome; on Good Friday the other Mail came in "Via Halifax" and brought me some epistles, all yours have reached me quite safely.

On Good Friday we had Church in the morning, & in the afternoon I went in the Ships Steam Launch over to Somerset, and Hamilton, the Capital of the Bermudas. The distance is 18 miles, and we passed about 80 islands on the way, most of them covered with Forest & Shrub. The shops were all closed so I went for a walk into the country, and saw two cricket matches being played, which made it seem more like Old England. One was between the soldiers, and the other the Negroes, I liked the niggers best, they made such a sweet row, and stopped the balls with their hands occasionally, but they are very weak about the shins. I had some tea with half a dozen eggs &c before coming back, some of our men who were having tea, ate a dozen each, but I did not feel equal to that. The Landlady's daughter who waited upon us was a trifle blacker than a crow, the Sailors called her Miss Julia Snowball.

¶We arrived on board again about 8 PM and found the Halifax mail had just arrived from England, it brought us Halifax papers containing accounts of the loss of the Atlantic steamer[11] but think the one I read was greatly exaggerated, so am anxiously waiting for news from England about it. We shall pass the scene of the wreck in another 2 days. Easter Sunday was very wet at Bermuda & I never left the ship. On the 14th inst a seaman belonging to the "Royal Albert" ironclad, fell from the main rigging to the bottom of the great iron floating Dock, a distance of 70 feet, and was picked up dead. On the 17th we took in an 100 tons of Coal, & caught an Octopus, or Devil Fish in the bathing place. This fish has 8 arms of feelers capable of clinging to anything, a man or a rock, & is the most horrible thing to look at in the world. His eyes and mouth are in the middle of his body, and he looks something between a Dragon, Star Fish, and the Devil. A large Octopus is capable of holding a man under the water until he is suffocated, they are not often met with, but there is one living in the Aquarium at Brighton;[12] our specimen died.

On Sunday 20th all hands went to the Dockyard church, and in the afternoon I had a walk to the Dockyard cemetry, & spent nearly two hours reading the inscriptions on the Stones &c. You would be surprised to see what a quantity of seamen & Naval Officers have been buried there, during the last 80 years. There are 80 or 90 different Ship's gravestones; each ship has a large stone with the names of all the officers, seamen, and marines, that they have buried engraved thereon. Drowning, falling from Aloft, and Yellow Fever appear to have caused the most deaths; some of the stones were beautiful, and had the Ship's crests carved on them. There are 3 Admirals buried there, and a large monument to the 350 Officers & men belonging to H.M. Ships Acorn & Supply; which foundered with all hands between Halifax and Bermuda about 40 years ago. But the prettiest stone of all is one dated 1809 to the memory of 2 midshipmen, brothers belonging to the same ship, one 17 years old, the other only 13, who died within 3 days of each other from Yellow Fever.

We are to bring a Stone back with us from Halifax to Bermuda, in memory of our Schoolmaster, and the sailor boy who was killed off S. John. The subscription was made for it the other day and over £15 was obtained. On Monday April 21st we left Bermuda for Halifax, but remained off the island dredging & sounding on what is called the Great Bank of Bermuda, extending out to sea for more than 50 miles. A few days after the Captain decided going to New York, but when we were within 80 miles of Sandy Hook, he again altered his mind, and are now within 220 miles of Halifax. Expect he was afraid of losing some of our men at New York.[13] One of our Surgeons & the Chemist went in the mail boat from Bermuda to New York, and are there still, from where they will bring our letters, & go on to Halifax by train.

On the 25th we passed a French Barque and a few days later an American. On the 26th & 27th we had a heavy gale of wind again, the ship pitching & tossing frightfully & scudding before the wind under close reefed top sails, the weather was bitterly cold, but has improved within the last few days. During the gale one or two accidents happened, on Sunday 27th a seaman fell from the fore-yard arm down on to the deck & was very severely injured. For a long time they thought he could not recover, but is improving now, and will go to the hospital at Halifax. On the day following another accident occurred by the Wheel taking charge—a tremendous sea struck the rudder suddenly & caused the wheel to vibrate with great velocity; 2 men & 2 boys were steering & the latter were dashed insensibly to the deck, the men were not hurt as they let go. The Boys are still in the Sick list but getting better fast. During the gale there were hundreds of Mother Carey's chickens [petrels] about, but they disappeared with the wind, we also had several Land birds come on board the ship in an exhausted condition, and died soon after. They were a species of lark, and had been blown out to sea from the coast of America, a distance of 280 miles.

On the 28th we lost overboard while sounding 2,100 fathoms of line. The average depth up to yesterday was about 2,500 fathoms, but we are getting shallow again now. We spent 2 days taking temperatures in the Gulf stream & trying to sound; the water on the surface was 10 degrees warmer than the atmosphere but there was such a tremendous under current flowing in the direction of England, that no soundings could be obtained, although the water had not the appearance of being so deep as off the West Indies; 2,500 faths. found no bottom. One or two experiments were made as to the effect of the pressure of a column of water on any substance at 2,800 faths; 2 large iron Buoys were sent down and came up twisted & smashed out of all shape, after that an empty ginger beer bottle was sent down carefully corked, it came up full of water with the cork forced inside.

On May 2nd, the dredge brought up a live star fish and mud, on the 3rd, more star fish & 2 small sea serpents.[14] On Sunday we had church in the morning & kept under sail all day; we sang Hymn 222, A & M, which we have every Sunday. Today the sounding & dredging is all over, but don't know with what result. At present 7 PM the Band is playing splendidly, & the ship steaming away for Halifax, and the shaking caused by the engines is such that you would be unable to write.

May 9th, Halifax docky'd

All day yesterday we were fishing and some scores of fine Cod were caught, who we had for dinner, tea & supper. We sighted the coast of Nova Scotia early this morning & steamed up the harbour, mooring alongside the Dockyard at twelve o'clock. The passage down the river was splendid, regular English scenery, Lighthouses, Forts & Batteries, and Halifax itself is quite like one of our towns at home. It is the principal Naval station in British North America & the dockyard is very extensive. The Harbour is considered one of the finest in the world, & the fish market more plentifully supplied than any other known, cod, salmon, halibut & mackerel are very abundant, & lobsters are only 1d each. The weather at present is very cold, more so than in England in May; although it is in the same latitude as Paris, the winter is more severe than it is in the north of Scotland. We saw nought of the wreck of the "Atlantic"[15] coming in, but nothing is spoken of on shore here but that at present, and bodies are being brought every day for identification or burial. The walls on shore are placarded with descriptions of the drowned people, & very large rewards are offered for some of the bodies.

May 13th

I have not been on shore yet but am going in a day or two, Sunday & Monday were wet, cold & miserable, to day has been fine but is still [not?] very hot. Several of the Officers have gone to New York by train; distant 400 miles. On Monday one watch went on 48 hours leave, & are gradually coming back again, they like the place much & say the people treat them well, but the military and police

are very smart on them when they get 3 part[s] 7 eighths intoxicated. One or two have been fined 7 & 8 dollars for drunkeness &
assaulting the police, and you would be surprised to see what a
quantity there are with broken noses and black eyes. One man said
it took 12 bobbies all night to take him to the lock-up, he had a terrible cut across the forehead, & was fined 7 dollars, which is paid
by the Paymaster & charged against his wages. He says he rather
thinks he shall give 'em "what for" if he can get on shore again
before we sail. Another of our scientific seamen brought a young
Nova Scotian youth on board to see the "Mermaid" we have
caught, all alive O!

¶To day we have been taking in six months provisions &c, we
have had any amount of visitors off to see the ship already[16] and
the Halifax papers are crazy about her. Yesterday the Diver went
down & examined the ship's bottom, which is done every quarter
if we are in harbour. We have 2 Divers, a carpenter & a seaman, & 2
sets of diving gear. The Flag Ship "Royal Alfred"[17] came in yesterday from Bermuda, & is moored just ahead of us, she was only 3
days and a half on the journey, while it took us 19 days, she brought
the Admiral, his wife, family, servants, carriages &c. from Bermuda; an Admiral on a foreign station[18] lives on shore and has his
wife and family with him, this one has a house here, and another
at Bermuda where they pass the winter. An Admiral's pay is £6 per
day, with an allowance of another £6 for table money, servants &c.
The "Royal Alfred" carries 750 men, and is larger than the "Audacious." We sent two seamen home from Bermuda, who were invalided, also one to Gaol, having three others in their place from the
"Royal Alfred"; three more have deserted since we came here,
which we shall have to obtain others in place of.

May 16th
We have taken in six month's provisions, & 200 tons of coal, and
are to sail on Monday afternoon for Bermuda. I have been on shore
this afternoon & had a good walk out into the country, the place is
very strongly defended, and there are plenty of troops here for its
protection. The town is about the size of Leicester but not such a
business place. The streets are all laid with tramways but are very
dirty and ill-paved, & 9 out of every 10 houses are built of wood, the
chimneys only being of brick, and you may not be surprised to hear
some day of Halifax being burnt like Chicago. Living is pretty
cheap, but the climate is much colder than England, yesterday we
had snow, and the other night the Northern Lights were very brilliant for about three hours.

¶While on shore I saw the mail come steaming in from England,

and she dropped anchor an hour ago, so we shall receive the letters to-morrow, we shall only be 8 days on the way back to Bermuda. I told Willie to send you his letter from S. Thomas, not having time to write you from there, so that you will have a description of each place we visit. From the papers I see Roger[19] is in full swing again, one of our seamen says he will give all his pay to hear the conclusion of the trial when we reach home in 76.

May 19
We sail at 4, the mail closes at 3. The ship is full of visitors who have come all the way from New York to see it, one Yankee professor[20] wishes to go to Bermuda with us but not sure if he will. The dates for postage are from Bermuda to Fayal in the Azores, & only 1000 miles from England, to arrive by the 28th of June, after the S. Michael to Madeira, then to S. Vincent, Cape Verdes, arriving there July 22nd, Bahia in Brazil until the 27th of August, from thence to the Cape of Good Hope, by mails leaving England every 10 days, the last to write by being Nov. 5th.

   We are 3 weeks behind the time appointed for leaving Bermuda by the Admiralty, & are to make up for it before reaching the Cape. The Admiralty have the whole voyage marked out for us, with dates & departure from each place, the time for arriving in England being May or June [18]76.[21]

Challenger remained in Halifax only ten days before heading south for Bermuda on 19 May. While at sea Matkin replied to a letter from his brother Charles, and also wrote to his cousin again. Charles was living and working in Bedford.

> H.M. Ship "Challenger"
> Out at Sea
> May 26th, 1873

Dear Charlie,
   It was scarcely worth while to write from Halifax, so I thought I would have a letter ready by the time we got to Bermuda & send it by the first mail, I can write to Mother by the second mail—as we shall be there 2 or 3 weeks. Your letter dated April 10th was sent on from New York & only arrived at Halifax 3 days before Mother's—dated 3d May. I also had a long letter from Geo Barlow, written Good Friday—& addressed New York—which arrived the day after yours. George was going on all serene & had had another increase of Salary; Seaton was home ill. I wrote a long letter to Seaton from

Halifax & told him that Fred would like to get in a business like theirs at Hull & asked him to look out for him should a chance occur; & I know he will for he rather liked Fred when he saw him at Stamford with me. I rather expected a letter from Will at Halifax —but none came so I am looking out for one at Bermuda, I rather like his choice in young Ladies, I should like to see him on a fine Sunday afternoon in his (go to meeting Crabs) taking her out for a walk. You spoke of a chance of *your* being married before I get home again. I don't think it myself, even if you can find any one to have you. If you do marry early, you must marry for money; for myself I mean to wait at least 10 more years, & look about for a good one, White, Black or Brown, as long as they're the right sort.

¶There were plenty of good looking girls in Halifax, more like English girls than at any place we have yet been, but the town is not to be compared to Melbourne for size, business, or fine buildings. Things were pretty cheap in Halifax, but not as cheap as Melbourne. We were regularly swarmed with visitors at Halifax for we lay alongside the wharf there & they had nothing to do but walk on board. The ship's departure from Halifax was livelier even, than leaving Portsmouth, the Wharfs were all crowded with people cheering, & the "Royal Alfred" was crowded with visitors to see us off, & as we steamed past, their Brass Band played "Auld Lang Syne" & their 750 seamen all stood in the rigging & gave 3 cheers as we went passed. Five of our seamen deserted at Halifax, one was discharged, & one went to Hospital, so we had 6 men from the "Royal Alfred" before we left.

We sailed on Monday May 19th at 4 PM & the weather was bitterly cold; 3 days after—we were in the current of the Gulf Stream & the weather was so warm that all the iron in the ship was dripping with damp, & the change was considered very unhealthy, a great many are even now on the Sick List with Rheumatics, & low Fever. It has been gradually getting warmer every day, & to morrow all hands are to wear white trousers, white caps &c. We can't bear any bed clothes on at all now & only 7 days ago I had a Blanket & my Rug in use. We shall be 5 months now before we get any more cool weather, & the greater part of the time we shall be in the Tropics.

¶We have been under steam ever since we left Halifax, & that makes the lower deck (where we live & sleep), almost unbearable, to day we are 280 miles from Bermuda, but are not going in before Friday as a good deal of Dredging has yet to be done. The distance from Halifax to Bermuda is only 730 miles, but the course we steer

would make it about 1100 miles. On the 20th the depth was 80, &
200 fms close to the coast of Nova Scotia, the bottom was rocky &
the Dredge brought up star fish, anemones &c. On the 21st depth
1250 fathoms, bottom of mud, more star fish, insects &c. On the
22nd bottom at 2,200 fathoms, but in hauling up, the dredge & 1800
fathoms of Line was lost over board. The Line gave way—so that
day's work went for nought. On the 23d bottom of rock at 2,800
faths (over 3 miles), nothing brought up. On the 24th Gale of Wind
blowing, & dead against us, ship steaming against it but only made
40 miles & rolled tremendously, no soundings could be taken.

¶Being the Queen's birthday the Captain issued to "all hands"
one third of a pint of the Sherry,—supplied to the ship as "extra
surveying Stores". Being Father's birthday I drank his health
instead of the Queen's, which is an offense amounting to Mutiny
(if known), & is punishable by Death or such other punishment as
is hereafter mentioned, according to the Articles of War. Our Band
also played selections while the Wine was being Drunk. Yesterday,
Sunday, Gale was still blowing & no Service was performed on
board, they had smoking instead, which is the regular routine in
the Navy—if there is no Church the men smoke until dinner &
read the Service if they feel disposed.

¶When we turned out this morning the wind had dropped & the
sea had gone down, the Sun was very hot even then (5 AM) & the
Dredge was overboard already. The depth was 2,200 faths, bottom
of Sand; at 4 PM they commenced Steaming again, going only half
speed & as I write now in our Issuing room—right in the stern of
the ship I can hear the screw propeller grinding & boring the ship
along at the rate of about 5 miles per hour. (9 PM must turn in.)

May 28th, 9 PM
Yesterday, Soundings were taken at 2,200 fathoms, & the Trawl
net was sent down, it took 2 hours going down, & 8 coming up &
contained nothing but a few insects—which Professor Thompson
said were worth their weight in Diamonds. To day we took soundings at 1876 fathoms,[22] the weather is very hot indeed. The Revolv-
ing Light on Bermuda island has just hove in sight about 20 miles
distant, but they are keeping away from the land again, & are going
in to morrow. We have burnt 150 tons of coal since we left Halifax.

May 30th
All day yesterday we were dredging on the East of the island,
depth varying from 700 to 1500 faths, the dredge brought up star

fish, insects &c. In the evening we anchored in 30 fathoms of water —about 2 miles from the Land. This morning at 4 o'clock the anchor was got up & we went round to the West of the island, soundings were taken at 1800 fathoms. The dredge was sent down in the morning & came up empty. This afternoon the Trawl net was sent down, & it has just carried away & 1700 fathoms of line with it. So this has been one of the most unlucky days we have had.[23]

May 31st

A black Pilot came out early this morning to take us in, for the entrance to Bermuda is the most intricate & dangerous of any in the world; at present there is a large merchant ship on shore to the West of the island—which is not likely to be got off, & in the Dock-yard there is a large Steamer called the "Petersburgh" of Leith being repaired, she also ran on a shoal. About 80 or 90 years ago a great many vessels were lost on these islands, fragments of which can be seen to this day on several of the uninhabited islands. If you stand up on the rising ground on shore & look out to sea in any direction—you can see rocks & shoals for several miles out—lying about 10 ft under the water. The water all round the shoals is green while the deep water is light blue & you may see hundreds of these green patches when the sun is shining.

¶At 4 PM we entered the Dockyard basin & moored in our old place—to the jetty. At present there are here belonging to the North American & West Indian Squadron, H.M. Ships "Sirius", "Fly", "Vixen", "Viper", "Britomart", & "Terror", as well as a great Line of Battle Ship called the "Indefatigable"[24] which is used by the crews of ships while their own is being repaired in the Float-ing Dock. We expect to turn over to her while our ship goes in dock.

¶As this has been rather a stiff working day, Wine has been issued to "all hands" again to night. No other ship in the Navy is supplied with this Wine, neither with the Pickles & Preserved Mutton & Vegetables &c, because they have not as many changes of climate or such long sea trips as the Challenger. These provi-sions & a daily issue of Limejuice will prevent Scurvy, the worst disease a ships company is liable to. Captain Cook's expedition round the world about 100 years ago lost nearly half of their officers & men from scurvy, & Cook you know was murdered at Hawaii island in the South Pacific, the second in command Captain Clerke died at sea & was buried at Petropaulofski in Kamschatka;[25]

H.M.S. *Challenger* near the floating dry-
dock at Bermuda, 1873 (Natural History
Museum, London, Photo No. 138). Photo
courtesy of the Natural History Museum,
London.

we are going to Hawaii on the way to Japan—after which to Petro-
pauloski, & the Aleutian islands—15 in number, all volcanoes. The
Wine is issued at the Captain's discretion generally after a hard
days dredging or sounding. If the dredge comes up empty or carries
away, the men don't get any wine—although the work is equally as
hard. The weather is a great deal hotter than it was when we were
here before & the flies are more numerous, & the mosquitoes more
troublesome. I have been & had a good bathe this evening, &
intend having one every morning & evening while we are here—
which will be 14 days at least.

Monday, June 2nd
    Yesterday we all went to the Dock yard Church, which is a very
pretty one, Venetian windows all open to the sea, a pretty good par-

son & a nice organ & singing. I went again in the evening by myself, there was a storm while we were there, & after the sun went down the sky changed about to all the colours of the rainbow, such a sunset as you never see in England. There were several coloured people at Church dressed very smartly, the negresses try to imitate the white Ladies, but the chignon rather tries them as their wool is seldom more than 2 inches long, still some of them manage it with the aid of a net & some padding. You should have seen them in their white muslins & tiny parasols running for the church to get out of the rain, & holding their skirts up considerably.

¶To day they are getting the great floating Dock ready for us & I think we are going in to morrow. The mail is expected in to day via St. Thomas, & I am expecting a letter from Willie; she will sail again on Wednesday & I expect will take this. At present there are about 50 black washerwomen in the ship beating up for washing, they come to you & say: "how am you my lub. I wash for you before, you let me have your washing sare." Yes says the sailors & you can have me & all. I have just heard that the mail is in, but we shan't get the letters on board yet. A notice has just been posted up to tell us that the mail will close in 10 minutes time, so I must cut this shorter than I intended & send it off, another mail goes on Friday, via New York & I must write to Mother by it or to Willie if there is a letter from him to night when the mail arrives. Write whenever you have time. Mother will give you future addresses & dates. Send this home as soon as you have read it & they can send it on to Willie at March.

Best love to all,
From your Affec^{te} Brother,
Joe[26]

Concurrent with the above letter was one written to his cousin containing much of the same material. Matkin reported that the gravestone for the late schoolmaster had been taken on board at Halifax, along with "an American scientific gentleman"[27] who would go as far as Bermuda. Dredging, which we learn occupied about forty men, and the sailor's reward for success come in for additional attention.

May 27th, 1873,
out at sea

. . . We have come an entirely different course to what we steered when going, keeping further out into the Atlantic, & dredging &c,

every day; but the depth is not so great as the sea is deeper in towards the American Coast—the way we steered going to Halifax. The depths up today are 80—200—1250—2,200—1800—2,800— 2,200—& today 2,675 fathoms.

. . . [Sherry] is supplied to the Ship as extra surveying stores, no other ship in the Navy is allowed it. It is issued at the discretion of the Captain to the men & boys who are engaged in dredging, sounding &c, $^1/_3$ of a pint for men, & $^1/_6$ for boys, about 40 are so employed, and they generally have it about twice a week, and they say that if nothing comes up in the dredge, no wine is given, though they work just as hard. The Steward and I have to serve it out—and being greatly interested in the dredging operations, generally drink ourselves to its success. "No more to night".

A few days later Matkin concluded to his cousin.

June 5th
The mail closes at 4 PM, so I must finish at once. The S. Thomas mail brought us no letters, so shall have no more until we reach Fayal at the Azores. All the curiosities &c, that have thus far been obtained, are going home in the same mail which brings this. We are not going in Dock now before we get to the Cape of Good Hope, or Sydney. The coal is coming in this week, and 6 months Biscuit next week. Some of the old is quite mouldy & contains Maggots & weevils &c, but of course it has to be eaten, & I expect we shall get worse, before we get better.

The rest of the pleasure party, consisting of Lord Campbell, 2 or 3 Scientifics, & other Officers returned from New York in the mail boat yesterday, some of them have been away nearly 2 months. We expect to sail hence to the Azores, about June 13th and shall most likely be at Fayal by July 8th, Madeira the 18th and S. Vincent, the 30th &c. But the Admiralty have the letters altered at the Post Office when the dates are incorrect, so that if you addressed a letter to Fayal, Azores, and it was not likely to arrive there in time to catch us, it would be crossed out, and put Madeira, so that you may always depend on its arriving some time or other.

We had a letter in the last mail, addressed to a seaman, for H.M. Ship "Challenger" on her voyage round the world, Meditteranean, or elsewhere, and it was at Halifax as early as if it had been directed there. On the 3rd, H.M.S. "Fly" sailed for England to pay off, having been out here 4 years. Our Brass band played her out of the harbour, they have improved wonderfully, and by the time we get back will be good musicians. The same day a merchant seaman belonging to the "Petersburgh" in dock here, was crushed between the ship & a boat, by which his thigh was broken in 2 places. One of our surgeons attended him, and had him conveyed to the Naval Hospital. The same night one of our seamen came on board three parts, seven eighths,[28] and fell from the hatchway on to the iron cable, causing a dreadful cut across his left eye.

¶The Gravestone we brought from Halifax, went to Cemetry yesterday for erection, & I intend going to see it on Sunday afternoon. It is a large Marble cross of pretty design, costing £16, it bears the following inscription—"This stone is erected by the Officers & Crew of H.M.S. "Challenger", to the memory of Adam Ebbels, Naval Schoolmaster, who died at Bermuda, April 4th: also to Wm. H. Stokes, 1st Class Boy, who was killed March 25th 73, off the West India Islands. In the midst of life we are in death."

¶The Governor of Bermuda & family were on board yesterday; and in the evening they gave a Ball to our Officers.

¶As these letters have to go through New York Post Office, we have to put 2 stamps on them. By the time you get this we shall be in mid Atlantic, & much closer to England; Fayal & S. Michael, Azores, is where most of the Oranges come from that are sold in England, but I think we shall be too early for them. The islands belong to Portugal.

4 PM

Mail just going to close—another man went to Hospital to day with Cholera, brought on through eating too many Cucumbers, Tomatoes, bananas &c. 2 other seamen are breaking their leave, and there is £3 reward out for them, we think they have left the island in the steamer "Petersburgh" which sailed this morning for New York, if not, we shall soon have them on board again, as they can't walk off the island.

> Believe me to be,
> The great Circumnavigator,
> Joseph Matkin[29]

Two weeks later Matkin began a lengthy letter to his mother which would continue until the ship's July 16 arrival at Madeira. The disease of travelers had caught up with him at Bermuda, but its passing, and that of Bermuda itself (a port of call Matkin had little love for), left him in an improved state of mind and body.

H.M. Ship "Challenger"
Mid-Atlantic
June 19th, [18]73

Dear Mother,

We are halfway across to the "Azores" so I will commence another letter. The last mail for England left Bermuda the same day we sailed, but I was not able to write—in consequence of being doubled up with an attack of Dysentry. Nearly every one in the ship, including the officers, had a touch of it, some in a mild form —others very severely; I was very bad indeed with it for 5 days but got rid of it soon after we cleared the islands. I ate nothing the whole time & was on the Sick List, diet—beef tea & Arrowroot, the Beef Tea is made from "Liebigs Extract of Beef"[30] & is called by the sailors "Animal Fluid," Pick me up &c. It is a very weakening sickness & has made me look quite thin, but I feel so much fresher & livelier since I recovered that on the whole I think I would rather have had it than missed it.

Of course you received Charlie's letter to read & was not over-anxious at receiving no further news from Bermuda; I wrote to JTS by the 2nd mail & intended you to have the last, but the sickness came on the day before the mail left. Bermuda is a very unhealthy place in summer, & June is the worst month, the Dysentry came with the South wind & the white people on shore were attacked as well as us, the Negroes it don't affect as they are pretty well acclimitised to it. The surest way of recovery is to leave the island & push out to sea for a breeze—for unless the wind changes to the North or East, it often proves fatal. We did not go in dock at Bermuda but the dockyard Artificers repaired the Condensers &c so that we are not coming home from Fayal—as was supposed for repairs.[31] I received neither letter or newspaper from anyone at Bermuda but we shall receive a mail at Fayal or St. Michaels. We took in 6 months Biscuit at Bermuda, which will last us to the Cape of Good Hope, & we took on 200 tons of coal which will last us to Madeira. We left 2 men behind in Bermuda hospital suffering from Dysentry, so that through sickness, & desertion at Halifax, we are 5 men short; 236 is the total number borne, instead of 241 the proper complement,—which will be filled up again at the Cape.

¶We left our moorings in Bermuda dockyard at 3 PM June 12th but anchored for the night off the island & early on the 13th we sailed away with a fine fair wind & soon had the islands out of sight. They are islands I never wish to see again; 2 days after we had been to sea all the mosquitoes & flies disappeared, the sea air does not agree with them, so they go to sleep somewhere & wait for land. Some of the sailors were covered with mosquitoe bites, but I was scarcely touched by them, having been sucked pretty well dry by them in Australia. I think there is no mosquitoe nourishment left in me for my messmates who sleep on each side of me were almost driven mad by them every night & could not understand why they didn't give me a turn.

¶We have had such fine trade winds that we are half way across to day & have sounded &c nearly every day, but the wind has fallen light again so that we shall most likely be another 14 days, & shall have to steam the other 1100 miles. Yesterday the depth was 2,700 fathoms (over 3 miles) but nothing was obtained. To day it was 2,875 fathoms & the dredge brought up some fine specimens of zoophytes &c, also what was more wonderful, a large piece of Amber.[32] This has been such a successful day that the Captain issued this evening to all hands, one third of a pint of Madeira, the wine supplied expressly for this ship. We are to take in at Madeira enough of it to last us all the commission.

June 25th, 340 miles from Fayal

We have had a good wind again & expect to be in on Saturday 28th, the day I told you the letters were to be there, so we have made a very good passage considering that we are hove to all day dredging &c & only sail at night. The depth on the 20th was 3000 fathoms, on the 21st 2,700—22nd Sunday, 2,750—23rd, 2,700—24th 2,170 & to day 2200 fms.[33] The Trawl net has been over all day & was 6 hours coming up, it contained several fine specimens including a blubber fish[34] 5 ft in length which is at present on deck & emitting a very powerful Phosphorescent light (the sun has gone down of course) 8 PM. But the specimen which is most valued is a large species of Prawn[35] of a brilliant scarlet colour, with a spike protruding from its head, and other appendages in the shape of wings. So this has been a very successful day & the Scientifics are in great glee. We passed 2 vessels this morning but they were a long way off.

June 27th, evening

The great blubber fish that was left in a large tub of water on deck last night was not to be seen as a fish this morning, but as atoms of one, floating about in the water. Several of these species of fish possess this wonderful power of falling to pieces in the water, or violently exploding themselves, & each separate atom has life & motion, & I believe eventually developes into a perfect fish again. To day we sounded at 1,675 fathoms & hove the Trawl but it came up bottom upwards & the apparatus that was overboard to test the strength of the current, carried away with 2000 fathoms of fine line, so this has been an unlucky day.

June 30th, off the island of Flores, Azores

On the 28th we sounded at 1250 fms, & had no dredging. Yesterday, Sunday, we went to Church in the morning & kept under sail all day. The wind being nearly ahead & taking us 30 miles out of our course, so that at 4 this morning the island of Flores hove in sight, which is some distance to the north of Fayal. We have kept away from this island all day & never went nearer than 8 miles, at present the island is nearly hull down although the land was very high, as high as Madeira. We sounded this morning at 1000 fathoms, & are now under steam, half speed.

¶The Azores are a group of 9 islands all belonging to Portugal, called also the Western islands, there names are Flores, Pico, Fayal, Corvo, Gratcioso, Terceira, St. Miguel, St. Jorge & St. Maria. St. Michael is the largest & is where the greater part of the oranges are grown that come to England, 250,000 boxes being the average annual export. They are all fertile & all inhabited, the total population is 252,000, & the total area 1150 sq miles. The islands are nearly all over 5000 ft high & the Peak of Pico, the highest point of the lot is over 8000 ft high. They are 800 miles West of Portugal, & only 1100 from Southampton so that we might be home in 5 days if we wished. Ships homeward bound from Australia generally sight these islands & consider themselves nearly home when they see them. Perhaps 3 years hence we may sight them again & whistle for a fair wind up channel.

¶We shall pass out of the Gulf stream in the morning early, for it runs to the north of these islands into the Bay of Biscay. This is only a branch of the real gulf stream which breaks off near Bermuda & flows straight across here, the way we have come. The main currents of the famous gulf stream runs up past Halifax & as far as the Banks of Newfoundland where it is met by another current of icy water from the North Pole & turned out of its course right across the Atlantic past England, Ireland, North Scotland, Norway, & can be traced as far north as Spitzbergen, where its waters get cold again & return to the equator as an under current over 2000 fathoms down. We found such a current in the tropics at 3000 fathoms deep with a temperature of 34°, or only 2° above freezing—while the water in the surface was between 70 & 80°. As this cold under current nears the equator it rises to the surface & the tremendous heat of the sun soon raises its temperature to 70 or 80°, after by which it loses its equilibrium & starts on this wonderful journey to the north to regain it. First it runs into the Gulf of Mexico, out through the Florida channel, past Bermuda, where this branch gets separated by the Bahama reef, & comes across to the Azores. This stream of tropical water is like a river in the ocean & loses scarcely any heat on its journey, although it raises (in winter) the temperature of England, Ireland & Scotland several degrees. Without the Gulf Stream England would be as cold as Newfoundland which is in the same latitude & Lisbon would be as cold as New York which is in the same latitude again. This branch of the stream runs in towards Lisbon, but is not very extensive & gets mixed up with the other waters in the rough bay of Biscay. It is this Gulf stream that has kept us so long in the North Atlantic but we have finished with it now, & in another 6 weeks I shall be south of the Equator. I have been reading a good deal about the Gulf stream lately,[36] so I have given you & the rest of the family a dose of it, perhaps you knew all about it before. I hear that Professor Thompson will shortly give another Lecture,[37] when of course there will be further information about the Gulf Stream, & I shall chalk as much down as I can remember & send it home for you to criticise. It is a very fine night & the ship seems to be steaming through fire, all the blubber fish, cuttle fish, nautilus &c shine like stars in the water at night.

July 1st

Passed out of the Gulf Stream & sounded at 1000 fms, the sea was swarming with all kinds of blubber fish, nautilus, jelly fish &c some in the shape of immense snakes 8 & 9 ft long, but not snakes of course. Several fish were caught & eaten for breakfast. At 11 AM the islands of Fayal & St Mie Jorge hove into sight, & at 12 Pico came out of the mist, but at present the Peak is not to be seen. We are pretty close in now & in a few minutes the pipe will go "all hands bring ship to anchor", & out will come the Portuguese Boats with Fish, Fruit &c. I don't think we shall get any letters until we arrive at St. Miguel [St. Michael]; our latest news direct from England is May 8th, so we are nearly 2 months in arrear.

9 PM

We dropped anchor at 4 about ¼ of a mile from the chief town Horta, a very pretty place built all along the beach & looking like Brighton seen from the sea. The Houses are all white without chimneys, & away far up in the mountains a few convents can be seen. At 8 o'clock the bells were ringing for evening mass & sounded very solemn indeed. They don't ring a peal like we do but have one deep toned bell which is rung very slowly like a passing bell. These islands are all cultivated & at present the corn is just ready for cutting, it is grown all along the slopes of the mountains & looks more like England than anything else.

¶As we soon as we anchored the Quarantine boat came alongside to see if we had any sickness on board, after considerable delay they gave us pratique, which means permission to land &c, but they also informed us that the small pox was very bad on shore & also at the other islands & Madeira, so our stay at these islands will be as short as possible. The disease was brought here by a British merchant ship from Boston, America, but the boatmen told us that none but children were attacked as yet. However the Captain allowed no one on shore but himself & a few of the scientific gents & the boatmen were not allowed in board, but we might go down to their boats alongside & purchase what we liked. They had fish, eggs, cheese, bread, green figs, apricots, plums &c. Apricots & green figs are a penny a dozen, I have bought 6 dozen of each, & half a dozen eggs, which is the total amount I have expended here. There is no safe anchorage at any of these islands; in case of a gale ships must push out to sea. Fayal is better than the other islands, the harbour of Horta being protected on 3 sides by jutting peninsulas of land. Soon after we anchored the British vice consul came off & brought 3 or 4 letters for the captain, they came over land to Lisbon & hence by the Portuguese mail boat. Our letters are either at St. Michael or Madeira.

July 2nd, midway between Corvo & Fayal

The Peak of Pico came out very plain to day, looming right above the clouds, it looks almost as high as Teneriffe, but has no snow on it. The islands of St. Jorge & Graciosa are to windward making 4 now in sight & they look like black clouds resting on the water.

¶We took in a live bullock this morning & he will be killed tonight, so we shall get 2 days fresh beef. The Portuguese brought him alongside in a barge & he was hoisted on board by a whip from the main yard. I don't know which made the most row, the bullock or the Portuguese boatmen. Nearly all the messes bought a cwt of new potatoes, price 7/– so we shall do very well to Madeira.

¶At 2 PM we got up anchor & steamed away from Fayal, having been there just 24 hours, it is a beautiful island, like coming to an oases in the desert, after crossing the Atlantic & very different to the Bermudas, for climate & fertility. We sounded twice between Corvo & Fayal, depth 875 & 1000 fathoms. The dredge was also hove & some good specimens obtained. The distance to St. Michael is 150 miles but as we shall sound & dredge all the way, it will be Friday night or Saturday before we get in. At present we are under all plain sail with a light wind, & the island of Fayal is getting very faint on the horizon.

July 3rd

The peak of Pico was still in sight this morning but the island could not be seen, we were over 50 miles from it. They sounded at 900 fms this morning & hove the dredge, a few sponges & aneroids[38] were obtained. We are at present under sail again. The Captain issued wine again tonight, so I have had Figs, Apricots, Biscuits & Madeira wine for supper & am getting visibly stouter.

Challenger finally put in at St. Michael (San Miguel) for long-needed liberty. As Matkin repeated to his cousin, "Reaching an island like this after 18 days at sea, is like coming to an oasis in the desert." Liberty had been *too* long awaited, it seems, for much of the crew took to drinking and brawling with the inhabitants. When tempers had calmed down, Matkin went ashore and had a much more positive experience.

On the ensuing passage to Sta. Maria, Matkin reported briefly on his visit to the natural history workroom, finally heeding the invitation issued by Professor Thomson in his lecture to the crew.

July 4th

Sounded at 875 fms & proceeded onward. The island [St. Michael] hove in sight at 11 AM and is as large as Madeira, though not such high land. At 7 PM we anchored off the chief town, which was about as large as Horta and very similar in appearance. The Quarantine boat at once came out and gave us Pratique informing us that the island was free from sickness of any sort. Our letters are at Madeira we are told. Oranges is the chief cultivation here, & at present they are green, other sorts of Fruit are not so cheap as at Fayal. Eggs are 24 a shilling, so I have invested a bob.

July 5th

Nearly all hands on shore, they will be off again at 6 PM. I am to go to morrow or Monday. The men have just returned on board, and it was worth 6d to see them come up the ship's side and fall in for inspection. Very few could walk straight and several were rolling; being Saturday night a great many had brought off the materials for a Sunday's dinner, some had half a sheep on their backs, some had pigs heads, some goats flesh, large cheeses & lots of other things; very few were properly dressed, a great many had lost their shoes, & hats. All who could not walk along the deck without staggering were set down as being drunk, several lay down there and then went to sleep; some excused themselves on the ground that the ship was unsteady, these were all reported and had a days leave stopped.

Sunday, July 6th

I was unable to go again to day, for the Steward wanted to go, & I am going to morrow. I am very glad I was not able to go for there has been such fighting and disturbances on shore between the Portuguese soldiers, & our men—that very likely will cause serious trouble. We had Church on board this morning & after dinner leave was given to all hands that could be spared, & only a very few remained in the ship. I told you before that in these Catholic countries the Sunday only lasts until 12 o'clock, after which the Wine shops &c open as usual. I suppose our men considered it was Monday like the rest, and drank the strong wine as if it had been beer. The wine here is very strong, and the effects very injurious if partaken too freely. The men called for pints, and soon got to fighting. Several of them were fighting in the principal square, & the soldiers had to be called out. Of course this brought on a fight between the sailors and the soldiers. The soldiers used their bayonets, and our men used stakes, wheelbarrows, & anything they could get, & several civilians joined the soldiers. At last the whole regiment was called out & eventually our men were all embarked & have been off about 2 hours. The return was worse than last night, & there has been any amount of fighting on deck since they came off, some are raving mad, & require 2 or 3 to hold them, but they will be all right to morrow. One or two have been pricked with bayonets, but they have dressed their own wounds, so that the Doctor shall not see what it is. A good many of the Portuguese soldiers were hurt, & several civilians who interfered. No more leave is to be given except to the very few who have not already been, so I shall have a quiet day to morrow.

¶The Captain, Commander, & most of the officers & scientifics went away on Saturday into the interior to see some boiling springs.[39] They will be back before to morrow. The 1st Lieutenant is in command and he read the Service at Church this morning. We had the hind quarters of a Goat for dinner this morning, but I did not care about it, something between mutton & dog's flesh.

July 8th

I went on shore yesterday with 2 messmates & spent a very pleasant day, we hired a donkey each for 1/6d & went a long ride into the country. Each donkey had a Portuguese boy behind it armed with a long stick with a spike in it. We did not let them use the stick, for each donkey had a sore place where his spike is always thrust in, & if we touched that place with our finger the donkey would start off galloping, while the boys would be a long way astern shouting out "in Portuguese" to the Donkeys to stop. We kept singing out "woa woa" but they only understood their own countrymen and having no bit in their mouths we could not stop them until the Boys were near enough to make them hear. I should have liked you to see us.

¶The country was beautiful but the oranges are all green & will not be ripe until September, we saw whole forests of orange trees, they are as thick as blackberries in England. The corn was nearly all in, they pull it up by the roots & instead of threshing it—oxen head it out in the open field. We saw the Cemetery, such a nice one, monuments all after one pattern and the ground laid out with such taste. The town of Ponta Delgado is quite a large one & has some fine squares & public gardens. The streets are something like Lisbon. There are enough Churches here to accomodate all the islands, the bells were ringing for evening matins at 6 o'clock, but only a few sisters of mercy, & ugly old women seemed to be going. The soldiers amused me most. They stand about 4 ft 10 in. high, & are armed with old muskets. The whole kit and arms would fetch about 1/8d in England. One of our men said that on Sunday night about 45 of them in single file were driving him down to the Boat & every time he turned round & pushed the end man they all fell over. I think they are hired out by the day by anyone that wants them—for at the Portuguese Hotel where we had dinner, a Sergeant waited on us. Every one on shore was very civil to us & I believe are quiet if let alone; we heard that one of the Portuguese civilians was killed on Sunday night in the squabble—but we have heard nothing about it on board. We were on board by 7 PM & I need not say that I kept clear of the wine for I only had Half a pint all day.

¶The Captain & party also returned last night from their 3 days excursion to the hot springs. To day the Portuguese Governor, the British Consul & a party of Ladies & Gentlemen came on board, & Lunched with the Captain, after which they look'd round the ship. We have had any amount of visitors off to the ship here, especially last Sunday afternoon—we were crowded. The women here are far from handsome & don't dress in the latest style. I hear we are to sail for Madeira tomorrow night—where I hope we shall find our letters all right. The distance is 490 miles (south east) & we shall be nearly a week on the way, as we shall have sounding & dredging every day.

July 15th, dredging 30 miles from Madeira

At 5 PM on the afternoon of the 9th we left San Miguel under sail with a light wind. On the 10th we were midway between San Miguel & San Maria, dredging. The depth was 1000 fathoms & some fine specimens of coral were obtained.[40] The same night being about 30 miles from San Maria we sighted some wonderful small rocks that were shot out of the water by a marine volcano only about 20 years ago. The water all round them is very deep

The naturalists' workroom aboard *Challenger*, from T. H. Tizard, H. N. Moseley, J. Y. Buchanan, and John Murray, *Narrative of the Cruise of H.M.S. Challenger with a General Account of the Scientific Results of the Expedition*, 2 vols. in 3 (London: HMSO, 1885), vol. 1, pt. 1, p. 509. Photo courtesy of Scripps Institution of Oceanography.

indeed. On the 11th we dredged up a few starfish from a depth of 2,025 fms. On the 12th sounded at 2,625 fms & took temperatures at various depths. On the 13th—Sunday, we had Church in the morning & after dinner sounded at 2,650 fms, & made sail again. Yesterday the depth was 2,200 fms, & at present the dredge is overboard. The island can be seen now through a glass & the Captain will get in to night to catch the mail from the Cape of Good Hope, which sails for England in the morning. We sighted 4 or 5 vessels this morning outward bound to the Cape, India, Chinas, & Australia; some of them we signalled.

¶I had the privilege of examining some of the curiosities in the Analyzing room the other night, & was very much surprised & interested with what I saw. The mud that comes up from the bottom of the sea is softer than velvet & passes through the fingers like so much cream or butter. The wonderful Prawn was in spirits of wine in a glass jar & was almost as large as a small Lobster. He had a pair of wings folded over his back like a pigeon's. I also saw several things through a large microscope, even more wonderful.

7:30 PM
The depth to day was 1500 fathoms & the dredge brought up a few star fish &c. I don't think the Captain is going in to Funchal before morning, when the mail goes. I hope to send this & to receive a lot of letters & newspapers from different people. We are at present under sail.

6 AM, July 16th
Just anchoring off Funchal, Madeira. English mail going at once —not able to answer any letters until we get to St Vincente—distant 900 miles. Our letters are not on board yet.

> With very best love to all,
> From Your Affectionate Son,
> Joseph Matkin

Remember me to all friends.[41]

By this time Matkin had also completed a letter of slightly greater length to his cousin. Here as elsewhere, although the substance of concurrent letters was generally much the same, those to his cousin were a bit less formal in language and more likely to contain items of a gossipy nature, as in his description of his drunken mates returning from liberty —"It was the most amusing scene I ever saw. * * * Some had lost their hats, some their shoes, and one the seat of his inexpressibles, which he was trying very hard to keep secret from the Officers"—or in

the following passage of 1 July, about Sir Roger Tichborne's impersonator:

I hope we shall get some fresh Meat & Bread here, for I am a long
way behind in body yet, & as thin as a rat. We are now running into
the anchorage off the capital town "Horta" by name, and the Boat
is already to go on shore to see for letters, and hope we may get
some for we have had no English news later than May 8th. We
heard a strange rumour just before we left Bermuda, about "Sir
Roger" that he was proved to be Arthur Orton, and had murdered
the real Sir Roger in South America 20 years ago, and was now
indicted for it. Considering that there is no submarine Telegraph to
Bermuda from any other place & no mail having arrived there from
England, or any other place, I, for one, did not believe it; and
expect to hear something fresh about "The Claimant" by this
mail, and think his career of deception (which must have cost
him many a sleepless night during the last few years) is nearly
played out.[42]

Ten days later, at Funchal Roads, Madeira, Matkin wrote his mother
that smallpox was "frightfully bad" and might cut their stay from a
week to a day. The long-awaited mail had arrived, he reported with
relief. Not surprisingly, his letter is devoted to reactions to events at
home, and though it tells us nothing further about the voyage, it does
finally give us some sense of circumstances at the other end that must
often have been on Matkin's mind.

IO AM
Have just received 8 letters & Mercury, & am satisfied. Read yours
first, & was very sorry to hear that Father had not lost his pain yet.
I hope the summer may improve him. . . . Hope you & Father will
go to Uppingham & get about more this summer, also Miss W[ild-
man]. My love to Aunt & Uncle Sharpe if they come. Fred is of
course at home now, I was pleased with his letter, his writing is
good. I will write him long letters when he gets out tell him. Had 2
letters from Will enclosing 2 photos, he appears to get on well &
looks on very good terms with himself, tell him I will answer from
St Vincent. Next letter was Walter [Thornton]'s, full of news &
very interesting, I shall answer his from St Vincent or Bahia. Next
2 letters from JTS one sent to Fayal; lots more news about Bar'dn
[Barrowden], the county & the country in general. The next letter
was from Mr. Daddo, which I was very pleased with & must

answer from St Vincent where we shall remain 10 days. He said Mrs. D. had been very ill & was at present out recruiting her health, & was coming to Oakham during the summer. The last & longest letter was from Seaton at Stamford, which I have not had time to read yet, indeed I have only glanced over them *all* at present & shall read them all over again this evening.

2 PM

Seaton's letter I have just read, & am very sorry about his eyes. He says the Doctor tells him that he must not be a clerk any more, & he is trying for a traveller's[43] place of some sort, but he says he finds it more difficult to obtain than he thought. He says he would not recommend Fred to go to Hull as Cashier, & he thinks the Drapery would suit him best. I think the Drapery the best trade for him & always told him so, he'd be a regular little Dandy bye & bye. However let him please himself as much as possible & try & get away before Xmas & winter comes on. Charlie, I have not heard from this time. Should like to see his pleasure boat very much, has she any money at all? Seaton's letter is dated June 28th & has only been here a few days. The rest have been here some time. I posted a letter to J. Swann with your other letter this morning, & before I received his two letters. So I think as it will be perhaps a month before he gets an answer from St Vincent, you might drop him a Post Card to say I received them all right. We sail in the morning & perhaps may call at Teneriffe, as it is in the track. The mail has her steam up & will start in half an hour. So, Goodbye & best love to Father, Miss W. the Boys & yourself.

> From Yours &c.,
> J. Matkin

P.S. My compliments up aloft.[44]

P.P.S. Received one Mercury only, also 6 stamps from you, 3 from Will & 6 from Mr. Daddo, which I am sure he need not have sent. I shall be able to see the Home news in our reading room papers— but of course there is no Rutland news. There has been a quarrel already about Tichborne since the mail arrived. The Cornish paper never came so I suppose I shall never see that.

Remember me to Mr. & Mrs. Sleath & family & to all enquirers.[45]

Because of the shortened stay at Madeira, *Challenger* was passing the Canaries by 19 July en route to the Cape Verde Islands. On 21 July Matkin again wrote his cousin.

Dear Tom,

Through the Commander's impatience in closing the mail bag before the Steamer was ready to sail, I was unable to answer the Fayal & Madeira letters. Within 10 minutes after they were sent away, I received two from you and six others, and was then informed that the mail bag was again opened for two hours, so had time to write a short note home. Yours was the only one sent me to Fayal, but it was no fault of yours I did not receive it there, as the British Consul sent it with others on to San Miguel & Madeira, by H.M.S. "Adriane." I suppose the blackberries will be looking up by the time this reaches you, and I would rather have a walk along Barrowden pasture again than at all the Madeiras, S. Michael's or Teneriffes, we have been to, or shall visit. I was very pleased to read of Joe Pepper's[46] success in the Mercury, he is certainly the best & most creditable pupil that Mr. Creaton has turned out.

I read in the papers all about the Shah's visit,[47] & should have liked to have been in the "Audacious" at the time. If ever the Shah visits England again, I should think they won't make such a fuss about him, or waste so much money, for I have never read or heard that he has done us any good, and don't expect he ever will. Still I suppose like the Siamese Twins and the two headed Nightingale, he was a novelty for about 10 minutes. As regards Roger,[48] I should think he will soon be in fashion again, & expect by the time we reach Bahia it will be all up with him. There is generally a pitched battle among the sailors about him after we receive a mail.

We only stayed 24 hours at Madeira for the smallpox was very bad there, so of course no leave was given, so am glad I went on shore when there in January. We took in a whole lot of splendid wine for the ship's company, also Fresh Meat, Vegetables, Bread, Fruit &c. Peaches, Bananas, and Figs were very plentiful there. The Captain gave a party just before we sailed, and took them in the ship over to the Desert Islands, 25 miles off, but the ladies were very sea-sick and glad to return. Our brass band was playing all the afternoon, and the Ladies and Officers were dancing on the upper deck. At 7 PM the same evening the Captain issued Wine to all hands, after which we set sail for S. Vincent, Cape Verdes, distant 1040 miles.

We have had the N.E. Trade winds behind us and at present are only 570 miles from S. Vincent, and expect to arrive there on the 27th; at present we are off the Canary Islands, and have been dredging and sounding all the way. On Saturday we were dredging close to the island of Palma, & some fine black coral was obtained, at a depth of 1100 faths.

We shall get letters at S. Vincent, & are to have a fresh Sub Lieutenant, & Schoolmaster there, from England by mail boat, since the other poor fellow died I have had a lot of his work to do.

"Work" passed down by the ship's steward, no doubt. The latter had taken over the seamen's library at Bermuda the previous April upon the death of the schoolmaster, for which he was paid an additional allowance.[49] This arrangement was destined to continue, as we see below.

Sunday, July 27th, dredging off S. Vincent

On the 23rd we sounded at 2,300 fathoms, the 24th at 2,425, 25th 2,075, 26th 1,975, and today the depth is only 500 fathoms.

Several specimens have been obtained by the Trawl net, and Dredge, and at present it is coming up. We sighted the islands of San Antonio and S. Vincent at 4 AM this morning, and at once stood in between them, & commenced dredging, in consequence of which we had no Service today, excepting the usual prayers at 9 o'clock. The land is tremendously high, & summits of the islands enveloped in mist. About six of them are at present in sight, there are 14 altogether, but only about 8 are inhabited, for they are rocky, barren, and unfruitful, scarcely worth occupying, the whole population is only 40,000, of whom 3 parts are Negroes, and half castes. They all belong to Portugal, and take their name from Cape Verde on the African Continent, distant 340 miles.

The Portuguese send all their convicts, and political prisoners there, I believe. The island of Fogo is 9,159 feet high, nearly 3 times as high as Ben Nevis. S. Vincent is a great place of resort for mail boats & steamers for coal, and we can see several vessels in the harbour now. The dredge has just come up empty (judgment for dredging on Sunday) and we are standing in to the Harbour which is a very fine one, nearly land-locked.

9 PM, at anchor of [off?] Mindelli, the capital of Cape Verdes

As soon as the Quarantine boat had made the usual enquiry, the letters came on board, a few indeed, with only one for me from yourself.

The British Consul has just been off to the ship, and informed the Captain that a Sub. Lieutenant [Harston] & Schoolmaster [Briant] arrived 8 days ago, by the same mail that brought your letter, but that the latter left the Hotel the day after he arrived to go for a walk, & has not since been heard of. The Consul is of the opinion that he has been murdered, so search parties have been sent out, and a reward of £20, offered for information, but nothing has yet been heard.

August 1st
We have finished coaling having taken in 250 tons at £5 per ton, & I hear we are to sail for Bahia on Sunday Evening the 3rd inst. On Monday H.M. Troop Ship "Smoom"[50] came in for coal sailing again on Tuesday for Cape Coast Castle, distant 700 miles. She took in 500 tons of coal, and was 10 days from Portsmouth, and has on board 300 Marines & lot of stores for the Ashantee War.[51]

No information concerning the fate of the missing schoolmaster has come to light, but his clothes & effects were brought on board yesterday & overhauled. A more disgraceful affair I never heard tell of than the loose manner he was sent off to join the ship, and the unchristian treatment he received on his arrival here from the British Consul. From the date of his appointment to the ship, he kept a daily journal, which closes on the day of his disappearance, I read it all through last night, and this is what I was able to educe from it. His name was Briant, & he came from near Bristol, he was 33 years of age, and had been a schoolmaster in the Navy for 10 years. Was a supernumary on board the Royal Adelaide, when he was ordered to proceed to Southampton for passage to S. Vincent by mail boat for this ship. A second cabin passage was given him from the Admiralty, and he sailed from England early in July calling at Lisbon for some hours when he went on shore, and of course spent some money.

¶He arrives on this barren island on the 19th with only £2 in his pocket, and after describing the kind of people on shore here, in his Journal, he goes on to say—"Having such a small sum of private money with me, and not being furnished by the Admiralty with any funds to defray my expenses on disembarking, I consider my best plan is to consult the British Consul at once about it. The Consul kindly informed me, he says, that the "Challenger" might arrive in 10 days, or she might be a month, he really could not say, but he told me I might manage to subsist at an Hotel until her arrival, at the rate of about 8/ per day. I then enquired—"Can you furnish me the means for subsistence until the Ship's arrival for I have only about £2 with me, and you shall then be reimbursed everything". His answer was—"I have no instructions that would justify me in so-doing, and to me, you are no more than any other British subject".

¶After this pleasant information (he says) I walked about the place to collect myself and consider what was best to be done, and I have calculated that I shall be able to manage at least a week if I only sleep at the Hotel. I must pass away the time during the day by walking about the island, and must do the best I can in the town for food". This is the last entry in his Journal, for on the same afternoon (8 days before we arrived) he left the Hotel to go for a walk, and was never again seen alive. He was wearing his watch at the time, and had his money about him, and it is considered by people who know the place well that he has been murdered for his money and buried, which is no uncommon circumstance on this island, for I have been told that 6 British subjects have disappeared in the same manner during the time the present Consul has resided here. The poor Schoolmaster had 7 or 800 pounds in Shares at home, so that his embarassment was only temporary, & I suppose he considered he had funds sufficient for his journey expecting to find the ship here on his arrival. I consider the Consul ought to be superseded at the least; he is getting £400 pounds a year for looking after the interests of British subjects, and that is the way he treated one of his own countrymen. Sub Lieut. Harston came out in the same mail boat, and was nearly in the same predicament on landing here, but the Consul invited him to his house, and then kept him until our arrival—merely because he was a Commissioned Officer. People say a Freemason can find friends in any part of the world, but poor Briant could not though he was one of the fraternity, and had been a considerable time.

August 4th
The mail boat sails in an hours time, and this will reach you about the 15th. If you write to Bahia by mail leaving Liverpool, Aug. 27th, I shall receive it there, where we shall be by September 16th, and stay a few days, the distance is 3,100 miles.

¶I was on shore yesterday, and climbed to the top of one of the highest mountains, where a flag staff has been erected, upon which scores of names have been engraved, and of course added mine to the number in letters an inch deep, also putting Oakham, Rutland, so that if any one else belonging to the large county ever ascends that mountain, his eyesight will be gladdened. The island is totally destitute of any sort of vegetation, every thing is scorched up by the sun. The view from the top was grand, nothing but mountains and sea to be seen, the climbing was difficult and dangerous, and very few would care to attempt it. The people are mostly negroes of a poor description, & are nearly all employed in coaling the barges, & getting out coals from ships, the women working at it the same as the men, & smoke their pipes equally as well. They are very ugly, and not all bashful, thinking nothing of undressing on the beach and bathing without any sort of gown whatever. The children run entirely naked, and can swim like fishes.

We sail in the morning for S. Jago, one of the other islands— thence to Bahia. The mail boat that brings this has come from there, where Yellow Fever was very bad indeed, so at present she is in Quarantine.

I have another letter to finish so must conclude, wishing you to observe that I'm all serene, and still unmarried.[52]

The other letter was probably to his brother Will. Only the final pages of it remain, and it adds little to the description of St. Vincent, except to reinforce his distaste for the natives and his disdain for the behavior of the British consul.[53]

St. Jago, reached on 6 August, Matkin found to be a vast improvement, as he told his cousin in a letter begun ten days later off the West African coast. This letter would follow events southward to the equator and across the Atlantic once again, via St. Paul's Rocks and Fernando de Noronha to Bahia, Brazil.

> H.M.S. "Challenger"
> Sierra Leone
> August 16, [18]73

Dear Tom,

It will be at least a month before this letter leaves the ship, but must write it by instalments as the information comes to hand. We left S. Vincent on the 5th for another of the Cape Verde islands, called S. Jago, 200 miles further south. No more information concerning the fate of the missing schoolmaster turned up, and every one on board considers that he was murdered, and buried long before our arrival. Between S. Vincent & S. Jago we sighted 2 or 3 islands belonging to the same group, one of them Fuego is a volcano 9,200 feet high. We were 2 days on the passage and found S. Jago quite a paradise compared with S. Vincent; we remained there 2 days, and quite enjoyed it. The island is the largest of the whole group, and the chief town is the seat of government, it was quite an extensive and well built place, and we found more Portuguese than at S. Vincent, still the majority of the population were negroes. The island abounds with cocoa nut & date palms which grow quite to the waters edge. Bananas and plantain were the finest we have yet seen, and pine apples, mangoes &c, were very plentiful.

There is any amount of Fish in the Bay, a party of our seamen caught enough to supply all hands for breakfast and dinner. Of course I went on shore & spent a very pleasant afternoon among the Palm groves. I bought a cocoa nut off the tree for 1d, but it was not so good as when kept some time. When the green shell is on them they are twice as large as those you see in England, they grow in enormous clusters at the top of the tree, just underneath the branches.

There were Monkeys & Parrots on the island, one of the boys bought a Monkey for 9/6, & let him run up the rigging, when he was made to chase it for three hours until he caught it, and then had to throw it overboard, so it was not a good speculation for him. Monkeys &c, are not allowed in the ship until we are homeward bound. The Governor of the Cape Verdes came on board before we sailed, and looked over the ship. We took in 2 Bullocks, 300 Cocoa Nuts, and 5000 Limes for the ship's company, and sailed on the 9th for S. Paul's rocks. The Captain is going to run the African Coast down as far as Cape Palmas, then sail along the Equator to S. Paul's Rocks. This is almost doubling the distance for we shall be further away at Cape Palmas than we were at S. Jago. Today we are off Sierra Leone (or the white man's grave[54]).

For the last 2 or 3 nights the sea has presented a wonderful phosphorescent appearance in consequence of certain minute animals flourishing most abundantly in this particular portion of the Atlantic. The ship appears to be literally ploughing her way through fire, and the surface of the water is lighter than I have ever seen it on the brightest moon light night.[55]

Aug. 29th
Made fast to a rock in mid Atlantic. The depth between S. Jago and these rocks of S. Paul is on the average about 2-1/2 miles, and some magnificent specimens have been obtained in the Trawl net. We sighted these famous rocks on the 27th and have been here 2 days surveying and sounding round them.

When we first sighted them they looked like a Railway train rising out of the water, and at 6 PM we steamed close up to them and made the ship fast to one of the rocks by a hawser. There are seven distinct masses of rocks, and rocks such as I have never seen before, for all the world like rough pumice stone, and the edges as sharp as a knife, the whole surface of them would not cover a square acre in extent. They are 850 miles from the African, and 650 from the American continent, and are only 90 miles from the Equator. The sea all round them is 2 miles in depth, they only rise 60 feet out of the water, and as it breaks all over them, landing is very difficult.

There is not a particle of vegetation on them, and with the exception of thousands of sea birds, no animal life whatever, neither is there a drop of Fresh Water. They are out of the track of any ships, and as nothing is to be obtained, no vessel ever comes, the last known to call here was a man of war in 1845. The birds are called Boobies and are about the size of a Goose, but have very long sharp beaks & claws, they make no nest, but lay their eggs on the bare rock, and feed the young ones on Flying fish. The rocks are swarmed with them and are certainly their owners. When the boat first landed they were so tame that the men could knock them over

H.M.S. *Challenger* secured by hawser to St Paul's Rocks, 1873; from a drawing by J. J. Wild. Reproduced by permission of the Master and Fellows of Christ's College, Cambridge. Photo courtesy of the Natural History Museum, London.

with their sticks, but are much wilder now. The sea round the rocks abounds with fish and some scores of fine ones called Cavalho [Cavalla] have been caught, enough to last all hands 2 days.

Yesterday was a grand holiday on board, and all who wished could land, & go fishing. Of course I went & stretched my legs, & also had the skin burnt off my neck by the great heat of the sun, but spent a pleasant afternoon in fishing and teazing the old birds, and should think am about the only Oakhamite who ever landed there. To day the Boats are away sounding, and we sail to-night for the island of Fernando di Norhana [Noronha], 400 miles distant. Pass on to the next island—

September 3rd, at anchor off the island of Fernando di Norhana

We crossed the Line at 1 PM on Aug. 30th, no shaving &c, was allowed,[56] but the Captain issued Wine to all hands, in honour of the Event. We have now finished operations in the North Atlantic, having crossed it in 3 different directions, and visited most of the islands south of England. I see by a book I am now reading that the Canary islands were the first to be discovered by the Spaniards in 1390, next came Porto Santa & Madeira in 1410, after which the Azores & Cape Verdes in 1420, all by the Portuguese. The Canary islands was the final point of departure by Columbus on his voyage of discovery, and at that time Teneriffe was throwing out Fire & Ashes from its Peak.

On his second voyage he sailed from the Cape Verdes, & the last land seen was the Peak of Fuego, also an active volcano at that time. The town of S. Jago where we were the other day was burnt down by the English under Francis Drake in 1553. This island [i.e., Fernando di Noronha] hove in sight on the 1st of September, and we anchored off the town the same night. We sounded at 2000 faths. just off the island. It belongs to Brazil and is used as a self supporting Convict settlement, it is about 10 miles long & 2-½ broad, & has 2 or 3 high peaks on it. The highest about 1500 feet, hangs over on one side, and looks as if it would fall with a slight push. The island is very fertile; but we have been unable to get a single thing here, & all the Officers can manage to buy is a few Turkeys, Melons, & Bananas. We came on purpose to survey the island & Harbour, and yesterday morning all the Boats were got out and provisioned ready. Just as they were on the start the Governor of the island, an obstinate old fool, took it into his head that we had come to see what sort of an island it was, whether it would be a desirable possession at any time for England; so he refused to allow the boats to land or take a single thing from the island. The Captain went on shore and tried to explain what we had come for, but the old fellow was too stupid to understand, so the Captain returned on board.[57]

¶The Boats are all in again, and in half an hour we sail for Bahia, distant 700 miles. It appears that every one on the island is a convict, male & female, except the soldiers and officials, and they have to support themselves by cultivation &c. The nearest land is 500 miles distant, so there is no chance of escaping. The Convicts live in houses of their own during the day, but all sleep under a Guard at the Fort at night, and any one later than 8 PM is flogged. They wear what clothes they like, and are allowed to traffic among themselves but not with strangers without the Governor's permission. Some of them are allowed to go fishing in tiny canoes made out of hollow trees & propelled with a small paddle but they are not very safe, and afford no means of escaping, unless it is on board a ship. For that reason they dont care for a ship to come here at all unless it is a Brazilian vessel with supplies &c. They speak the Portuguese language but a few know English; one of them asked one of our Boats crew if he had any old European newspapers, as he had seen none for 10 years. Most of them are here for life, the murderers & worst criminals are strictly confined.

Sunday, Sep. 7th, 500 miles from Bahia
We have had bad winds & weather, and at present are just this side Pernambuco, under steam. We sighted the American Continent last night off Cape S. Roque, and saw the sun set behind the land. Several vessels about. We expect to be at Bahia by the 11th or 12th, there to receive our letters and get in some fresh Provisions.

Sunday afternoon, Sep. 14th, just steaming in to Bahia
We shall be at anchor again in an hours time, and this evening hope to be reading one of your letters, and to morrow have a look at the newspapers &c. This has been a long passage, 60 days from Madeira. We have been steaming down the coast of South America all week, & in sight of it nearly all the way, the land is not so high as the coast of Devonshire. We sighted Pernambuco Light house the other night, and have been sounding and Trawling all the week, depth sometimes 2000 faths, at others only 14, some splendid specimens have been obtained in the Trawl. We are burning the last ton of coal now, so shall have to fill up here for the Cape,

where we expect to be about the 5th of November, and to remain until the 6th of December. I hope there is no sickness here as they had Yellow Fever when we were at Madeira; the weather is very hot although it is only the beginning of summer yet. Bahia is in Lat. 13° South of the Equator. This morning many miles from land we had some splendid Butterflies come on board, & at present the ship is swarmed with them, & suppose we shall soon have the Flies and Mosquitoes round us.

Sep. 15, anchored in the Bay of Bahia
We had not enough coal to bring us in, so had to sail in, and was too late to get any letters last night, but this morning the mail from England came in 30 hours before her time, and I received seven letters, two being from you. I like to hear of Joe Pepper's success. You can tell him that the Naval surgeons are almost all Irish & Scotch men of very doubtful abilities,[58] promotion is very slow, and the pay small until they are old men, but don't suppose he has any idea of joining the Navy, as it is only fellows who can't get on anywhere else that apply for an Assistant Surgeon's commission in the Navy. Regarding "Roger" I am more in a fog then ever, and shall read no more of his trial until its culmination. The "Daily News" sent is the latest I have seen, as the mail was only 18-½ days on the journey, 5,200 miles, & calling at S. Vincent & Pernambuco for coal, she stayed here 8 hours, and sailed again for Rio, Monte Video and Valparaiso. From the British Consul at S. Vincent she brought news that the body of the missing schoolmaster had been found in a deep gully at the foot of a mountain, but owing to its having lain there for a month, they were unable to say whether the man was murdered or accidentally killed.[59]

Bahia is a splendid city seen from the sea, looks very like Lisbon, and is quite as large, but shall tell you all about it after I have been on shore. The Bay is splendid and the Harbour full of shipping, the oranges & cigars are beautiful & very cheap. The city is the 2nd in Brazil, & was once the Capital. Rio [de] Janeiro which now takes that honor is 850 miles further south, & where the Emperor lives; the country of Brazil is almost as large as Europe, but the population including negroes & native Indians is only 12 millions.

Sep. 21st
The mail leaves to night so I must close up at once. I have been on shore & had a good walk about the town & into the country. I was

more surprised at the luxuriant vegetation than with anything else here. The Creepers & Tropical plants grow as thick as a jungle, and the flowers & trees are beautiful. The Banana tree is the most prolific of all in the world, and I believe it bears 15 times more abundantly than Wheat. The Cocoa Nut Palm & Bread Fruit Trees are most handsome, the latter also bears most plentifully & the Fruit is very pretty, & about as large as a Child's head, the rind being irregular like Maize when ripe. Guavas, Custard Apples, and Alligator pears also grow here to perfection, & Pine apples to an enormous size. The Birds & Butterflies are magnificent, I think Humming Birds most attractive, they are like flying gems, and their plumage shines like gold, perching on to flowers as bees, & some of them are no larger if it were not for their tails. I shall bring some stuffed ones home with me from Rio Janeiro, it is much larger than Bahia, having 300,000 inhabitants, the latter has about 150,000, of whom about $^4/_5$ are negroes, who are of the same species as those at the Cape Verdes, & originally came from the same part of Africa. They are very ugly, the women coarse & fat, the native Indian women are much more handsome.

The greater part of these negroes are slaves, and are let out for hire by their masters for about 6d per day. We had about 60 of them getting in our coal, while the ship's Company went on leave, and they got in 200 tons in a day. They live chiefly on Manioc flour, the same root from which Tapioca is made. Bahia is a well built town, the houses are of stone, & 5 or 6 storeys high, but the streets are very narrow, & stink fearfully, the negroes empty all their refuse &c into the streets from the top story, and if ever you come here be sure to walk in the middle of the street. No wonder that Yellow Fever makes such havoc in this country, when the streets are so narrow & badly drained. The negroes occupy the greater part of the town, the European Merchants &c, all live at one end of the place facing the sea. Very few of the people can speak English, the money is all paper, excepting copper pennies, the loss is considerable in changing the money for English. The principal streets are all laid down for tramways, the cars being drawn by Mules. Instead of Cabs & Omnibuses they use Sedan Chairs, carried by negroes; some of our seamen spent a good deal in riding about in these chairs, one or two are absent, and most likely deserted, as there is a tremendous lot of shipping here, a great many from America. The town is not very wide, but nearly 6 miles in length, & I never was in such a place for churches, which of course are all Roman Catholics, and such gaudy rubbishy affairs inside, all candles, pictures and gilding. There is a Cathedral which is a little better than the Churches, inside & out as well, but even that is nothing com-[pared?] to one of our own at home. I looked in at Evening Mass, and it was crowded with negroes chiefly women, & in full evening dress of course; the singing was very fair; the shops all close at six, & after that the town is very quiet, there are very few amusements here only one Theatre & Opera House I believe. The soldiers are almost all negroes about 4-½ feet high, and are 3 times worse to look at than the Portuguese. The Brazilians also have a bit of a navy, there is a Turret ship of theirs here now, which was built on the Thames, & the seamen are nearly all negroes, and are thrashed a good deal at times. Just ahead of us is a large American Man of War called the "Lancaster" but their men are not allowed on shore, as half of them ran away at Rio; they are paid very irregularly, and bolt at every good opportunity. As far as we have gone in this ship the deserters have been the very worst among the ship's company.

The American Admiral was on board yesterday looking over the ship; & we have had a great many visitors off to the ship. Our Brass Band plays "Hail Columbia" after "God Save the Queen" every evening at 8 o'clock.

We sail on the 25th of Sep. for the Cape, and arrive about the 1st of Nov. remaining until the 6th of Dec. I enclose the latest instructions as to future dates & addresses. You will get this about the 12th of October, quite getting on for winter in England, the summer will be just commencing there & finishing at Australia, but there is a little cold weather between the two. Must now conclude, as the mail leaves in an hour, and shall write you by the next from where we are.[60]

To avoid quarantine for yellow fever, *Challenger* left Bahia two days ahead of schedule, on 23 September. Matkin was not remorseful for the early departure: despite the exotic natural surroundings Bahia had little allure.

By mid-October the ship reached Tristan da Cunha Island, where Matkin began letters to his mother and cousin. Only a portion of the former letter survives; it is largely repetitive of the latter. Both conclude in early November, upon *Challenger*'s arrival at Cape Town, South Africa. Between Bahia and Tristan da Cunha Matkin found little excitement to report except the activities and seamen's lore of the albatross. At Tristan, however, *Challenger* rescued the two Stoltonhoff brothers, marooned on Inaccessible Island for nearly two years. Matkin took pleasure in reporting from *their* journal, but he complained to his mother again that too little information was passed down to the crew: "You will read better accounts of these islands—& of the 2 men whom we are bringing away—in the newspapers, for the scientifics have nothing else to do but go on shore & gather their information for the papers."[61]

> H.M.S. "Challenger"
> Tristan d'Acunha
> Oct. 15th [18]73

Dear Tom,

We are lying at anchor again, but at a far different place to where I wrote last. The temperature at Bahia was 82° in the shade, & at this wonderful snow covered island—to day—it stands at 45°, rather a sudden change in 3 weeks—a'nt it.

Still the change has driven the Yellow Fever out of the Ship, which was the object of our Captain, in running so far South before standing to the East. However I will commence my letter from the date of closing the last—Sep. 23rd.

I was not over struck with Bahia and did not care to make a second visit on shore there; nearly every night while we remained we used to see a display of Fireworks &c, in the Cathedral Square, & the Cathedral itself was generally illuminated, all in honor of the Saints & to please the niggers. Our Naturalists procured some large serpents, & native animals, before we left—to practice on, but they are all dead now, and preserved in spirits of Wine. There was a large Iguana, or immense Lizard, which the negroes are very fond of in a stew, also a Racoon, an Ant Eater, & a large Sloth, which was the sleepiest looking customer you ever saw, more like a bundle of old hay, he has a very thick skin which protects him from the bees when robbing their honey; is more active in climbing, than at anything else.

¶We lost two of our worst characters at Bahia by desertion, & the day before we sailed sent a man on shore to Hospital, suffering from Yellow Fever. This hastened our departure for we should have had to go in Quarantine, had we stayed any longer; on the 23 Sept. at 5 PM we steamed away from Bahia, & steered due south for 1100 miles so as to get into cool weather as soon as possible. One more man—a Marine—was attacked, but by being kept on the upper deck in his hammock soon recovered. Captain Nares had also a severe attack of some sort of Fever, but is all serene again now.

¶We had a pair of Goats on board, that we were to land on the island of Trinidad[62] for breeding purposes, but the old gentlemen died before we arrived so they ate the old lady, & did not call at the island, but are to do so on our return trip in 1876. We kept down the South American coast for some distance, below Rio Janeiro, then stood across for this island, sounding, dredging &c about every 3rd day. The average depth thus far—from Bahia—has been about 2,100 faths, but the dredging has been very unsuccessful, and the bottom must have been very rocky, for we lost a Dredge, a Trawl Net, and 2,100 faths of Dredging line, which costs 8d per fath, & the Trawl a lot of Money, the Coal also for each day's dredging amounts to about £10, beside[s] the time & labour lost.

We have had some very weath rough weather too this trip, been under double reefed topsail for 2 days. Only 2 vessels have been passed since leaving Bahia, for this part of the ocean is not often crossed, the only living companions we have had are Albatrosses, Cape Hens, Cape Pigeons, Petrels, & a host of other sea birds. The Albatross is the monarch of sea birds, frequently measuring 17 feet from wing to wing, & in flying does not flap his wings but keeps them spread out. These birds will follow a ship for miles in cold latitudes, & have very powerful wings & beaks.

Some years ago H.M.S. "Sutlege"[63] on her way to Vancouver's island, lost a man overboard in a gale of wind; a boat was lowered but just before it reached him—an Albatross swooped down & drove its beak into his skull, killing him on the spot.

Perhaps you have read Coleridge's quaint poem "The Rhyme of the Ancient Mariner"—describing the calamities which occurred to a ship among snow & ice—through one of the seamen shooting an Albatross with his bow. (Quotation) "Is it he, quoth one, is this the man, By Him who died on the Cross—With his cruel bow, he laid him low, the harmless Albatross"—(this is all at present in reference to Albatrosses).

At 4 PM yesterday morning the Peak of Tristan d'Acunha, hove in sight being a mass of snow & ice, rising 8,300 feet above the level of the sea. We were 80 miles away then; in the early morning objects at sea can be seen much plainer than in broad daylight. We hove to, & sounded at 2,150 faths. then dredged, but obtained nothing. At 5 PM we steamed in towards the land, but only anchored this morning at about ½ a mile from the settlement.

¶The island was discovered in, I believe, the year 1530, by a Portuguese Navigator named Tristan da Canta; but owing to its isolated position in the midst of an unsheltered and stormy ocean, no one ever thought of settling on it until the year 1812—a few English, Scotch, & American families settled there, & have themselves and their children remained ever since. There are two other islands forming the group, all three in sight of each other, and rising very abruptly out of the water, this one almost perpendicularly to the height of 6000 ft. A goat could not ascend it; so of course there is no sea beach at all, but a terrific surf breaks all along the shores making it very difficult for a boat to land except in the finest weather. It is about 3-½ miles in length, & 2 in breadth, but nearly all mountain, except a small tongue of land which runs about ¼ mile to the West, here the settlers have built their cottages & graze their cattle. Corn does not thrive but potatoes, and many sorts of vegetable are plentiful; sheep, cattle, & poultry are the most profitable articles on the island, & the principal inducement for a ship to call. Fish, & sea bird eggs are very abundant; in the winter Seals, and Sea elephants come to breed, from these they obtain their oil. Flour, tea, sugar, coffee, &c, they procure from any ships that may call, which are chiefly whalers, & English Men of War.

¶The inhabitants are very primitive & sociable in their habits, living almost like one family. The profits derived from bartering &c with ships are placed in a common fund, & equally divided among the children when they are sent out in the world to shift for themselves. The island will not support many, and as the population gets too numerous, some of the children are sent away to the Cape in passing vessels. Only 3 men of war have been here during the last 8 years, the last one in 1867, was the "Galatea" under the Duke of Edinburgh, she took some of the children to the Cape, & her Captain baptized all the children, & married what eligible couples there were, for no clergyman resides there.

¶As soon as we anchored the whole male population came off in their boats, bringing with them potatoes, albatrosses' eggs &c. They were dressed in various costumes, & all wore sealskin shoes, and wove worsted stockings; they were a fine healthy looking lot of men, some of them born on the island, & never having once left it. The women had a gipsy looking appearance, several of them were Creoles from the Cape, & S. Helena, but the native born children were very handsome. There are no marriageable couples—or our Captain would have conducted the ceremony, for as many as wished. The chief man went by the name of Mr. Green, he is over 60, & has his mother living with him—aged 91—to whom he wanted our Captain to give a passage to the Cape, at the old lady's request, but he declined fearing that she might die on the way.

One man from each of the seamen's messes was allowed to go on shore & purchase provisions for his mess; nearly all of them have sheep, pigs, & potatoes, many of them Geese, & Fowls. A Bullock weighing 61-½ stones, was procured for the ship's company, & in exchange we gave them—2 casks of Flour, 1 of Vinegar, 1 of Madeira Wine, a case of Sugar, 50 lbs of Chocolate, 20 lbs Tea, 30 lbs Tobacco, some salt, lamp wicks, & Flannel.

They were out of all these articles, but expect a supply shortly from the Cape. The present population is 84; the men all answer to their Christian names—except "Mr. Green". Tristan d'Acunha is the farthest from a continent of any inhabited island in the world;[64] it is 1,750 miles south-west of the Cape, & 1000[65] south-east of Bahia. About 8 months ago, a Portuguese vessel was

Ship's party skinning penguins, Inaccessible Island, Tristan da Cunha, 1873. Reproduced by permission of the Trustees of the National Museums of Scotland. Photo courtesy of the Natural History Museum, London.

wrecked on the other side of the island, with all her crew drowned.
The present season is early spring here, but the snow is plentiful on
the hill tops; the time is almost the same as in England, but we
gain 16 minutes per day when under sail, as our course is due East.
The men here report that on one of the other islands, 22 miles from
here, there are two German seamen, who have lived there two
years. We are just leaving Tristan, to sound &c, round the other 2
islands, & may hear or see something of the "German Crusoes".

October 16th, at Inaccessible Island
We anchored this morning close under the lee of this wild looking
island, and in a small cove about 200 yds in extent, we saw the hut
of the 2 Germans, & themselves walking down to the beach.
Directly our boat landed, they came off, & asked the Captain to
remove them from the island, for they had been living on it for 2
years in great hardship, & with no probability of being removed.
After hearing their story, he consented to give them a passage to
the Cape, & has given them some clothes &c, for their own were
quite worn out, & they had neither boots, nor stockings. I have
heard them relate their adventures, & will give you the rough particulars.—They are brothers, Prussian seamen, named Gustave &
Frederiche Stoltonhoff, speak excellent English, & are very intelligent looking men. On the breaking out of the Franco Prussian war,
they were enrolled & served in the army besieging Metz, the elder
rose to be 2nd Lieutenant. In the early part of [18]71, they were discharged with their regiment, & joined an American Whale ship,
which arrived at this island in October 71.

¶As is not unusual in whalers, they were, with their own consent, victualled & left on the island for 3 months to catch seals,
while their ship went farther South, whaling, but she never
returned, & was in all probability lost. They had the misfortune to
lose their boat soon after landing, & the beach where they had
built their hut, was entirely surrounded by Cliffs, or rather mountains, so that there was no opportunity of getting to the other side
of the island. The patch of land in question was only about 200 yds.
in extent, & on it they had sowed potatoes & vegetables which
thrived very well. The island covers about 3 square miles, & rises
for 2000 feet on all sides—more or less precipitous. On their side it
rose almost perpendicularly, but fortunately there were several
streams of fresh water running down the Clefts in the rocks,
caused by the melting snow & rain. They tried hard to scale the
Cliff but found it impossible, & resolved at last to swim round the

island, in hope of finding an easier ascent to the summit. This they at length found, & managed to reach the top—which was comparatively level, and covered with a rough vegetation; but it abounded with wild boars of such a fierce and ferocious description, that they soon found they should have to return again to their old haunt under the cliff. However they managed to seize 2 tiny sucking pigs, & drag them to their hut through the surf, building them a stye, & feeding them on potatoes &c. They soon increased, but not fast enough to supply them with food, & they had occasionally to swim round & ascend to shoot a Boar; they had a good supply of matches & ammunition. Seals came in the winter & Penguins in such numbers that the beach was covered with their eggs. These they used to eat, & feed their pigs on them partly.

After this manner they have existed during the last 2 years, occasionally seeing a ship at great distance, but never one near enough to attract notice, until last night when they saw a vessel coming under steam, straight for their island home & they never lost sight of its light until she appeared, & anchored this morning, & found her name was the "Challenger"; & this is the conclusion of the "Stoltonhoff's" story.[66]

We have been sounding &c, round the island all day, while the Germans have been picking up their traps, seal skins &c, digging up their vegetables which they gave the ship's company, with the 3 remaining Pigs & some hundreds of Penguin's Eggs, which they had stored up, knowing the Birds would soon be leaving the island for a colder region; their hut they burnt down, & are at present on board with their goods & chattels. They have a sum of money which the Paymaster has taken charge of for them, they kept a Log of their Voyage & adventures, but were a day ahead in their reckoning, not having known it was Leap Year. They are translating their Log into English, & some of the Scientifics & officers, are already concocting a regular romance about it, for the newspapers, which I have no doubt you will see & find far more interesting than my account. We are anchored for the night close to the still smouldering hut, but leave early in the morning to sound &c, round the other islands.

October 17th, Nightingale Island

We steamed over here at 4 AM this morning, & landed the scientifics to explore while we sounded round it. It is 18 miles from Tristan, & 8 from Inaccessible, & is the pleasantest looking of the lot, being covered with vegetation, & rising gradually to the height of about 1000 feet above the sea level. There are 3 or 4 distinct hummocks of land, but the area of the whole is less than Inaccessible, & there are no animals on it whatsoever. The great feature of the island is its caverns, of which there are a great number all round the shores. The Artist & Photographer have taken views of all these scenes. We have several Penguins on board for these islands swarm with them; they are the clumsiest looking creatures you ever saw, being half Bird, & half Seal, having the head & feet of a Duck, with a Seal's body, & habits, except that they lay eggs. They are covered with a fur like feather very pretty to look at; several of our men have skins but they are only good for making muffs. We are laying off from the land, & are to sound between the islands tomorrow; sailing for the Cape at night.

October 27th, 150 miles from the Cape, & expecting to anchor to-morrow
We left the islands at 7 PM on the 18th, obtaining a fine view of the Peak of Tristan as we stood away. We have had a slashing wind thus far, & have sounded four times, depth over 2,500 faths, dredging pretty fair.

We picked up a fine piece of Timber which was covered with fine specimens for the Scientifics; but we have seen no vessel yet. The Germans are getting fat, one of them is reading sketches by "Boz", the other "Nicholas Nickleby".[67]

3   Farthest South
OCTOBER 1873–MARCH 1874

I'm a spar decked Corvette, built of wood, not of iron,
    I am good under steam, under sail.
No Sheffield plate dead weights my topsails environ
    As I ride like a duck thro' a gale.
By my Lords I'm about to be put in commission
    For a cruise of three years, if not four,
And for all I'm short-handed, I carry provision
    Such as Corvette ne'er victualled before.
                —From "The *Challenger*, Her Challenge,"
                Punch, 1872 (reproduced in Swann letter-
                book, pp. 7–10)

After ten months of Atlantic transits, *Challenger* reached the Cape of
Good Hope in preparation for passage into the Indian Ocean. The
crew had weathered sounding, dredging, and trawling operations at no
fewer than 140 stations (not including those first seven experimental,
and expensive, dredgings). But this was only the beginning.

The ship would remain in Simon's Bay for more than a month for
refitting and replenishing. Matkin concluded the previous letter to his
cousin.

October 28th, 8 PM, anchored in Simon's Bay, Cape of Good Hope
At 4 AM this morning we sighted Table mountain, & the adjacent
head lands down as far as the Cape. We were about 60 miles away
then, & hove to for soundings, proceeding again at 9 AM under sail
& steam; at 4 PM we entered False Bay, & thence into Simon's Bay,
which is only a part of the former, but the most secure for vessels
of any in this part of Africa. We found H.M.S. "Rattlesnake", & a
Gun boat at anchor here, as well as a Dutch Frigate that is re-fitting for sea. Directly we had anchored the "Rattlesnake" sent out
letters on board, & I had the satisfaction of receiving yours dated
Sep. 16th which was the only one for me. The mail that brought
yours has been here 2 days, & we have newspapers to Sep. 27th—
The next comes in Nov. 5th.

The "Rattlesnake" Flag Ship on this station has just returned
from Cape Coast Castle, & brings a bad account of the Ashantee

war. Commodore Commerell[1] lies on shore shot through the lungs & thigh, he is in a very bad state, & several of his boat's crew are badly wounded, and some of them I believe were killed.

¶You must have felt rather dull in returning to Barrowden after your London visit, amid so much dissipation & wandering away from your professed faith, among charming preachers, with strange doctrines; but hope you are still sound & unmoved after all this search & curiosity to behold these heterodox religions & preachers. You know what Dryden says of these sects & people— "A rambling crew of dreaming lambs succeed of the true old enthusiastic breed; 'Gainst form and order they their power employ, Nothing to build, & all things to destroy."[2]

The Newspaper paragraph in reference to Joe Pepper was very satisfactory, I also read it in the London News, & expect we shall have him Surgeon to the Queen, or some thing of that sort, before I get back. You have some quite noted cousins; can't you write a novel, or invent something? so as to keep up the family credit; discover a planet, then there'll be an Astronomical Swann, as well as an Anatomical Pepper.

In consequence of our having had the Yellow Fever since leaving Bahia, we were placed in Quarantine for 24 hours, so there is no communication whatever between us and the shore yet. We are to remain here until the 6th of December, so I shall get at least one more letter from you, & I will write again before we sail describing the country in general &c. &c. Cape Town is 22 miles distant from here by land, and about 60 by water; it stands on the shores of Table Bay, and the celebrated Table Mountain, (which we saw at sea) lies just behind the town. I believe we are to go round to Table Bay next week, so as to give leave to the Ship's company in Cape Town; we shall be there about 10 days, & I believe the officers will give a grand Ball during the time. We shall return here to re-fit &c, for the Dockyard Arsenal is here, & we have a great deal to do to the ship preparatory to going down south among the ice. There is a small town here called Simon's Town, where is the Commodore's house, in which he lies wounded at present.

This Bay is about 40 sq. miles in extent, & entirely surrounded by immense mountains, shutting out the sea view inland altogether. The coast has the same barren uninviting appearance it had off Barbary over 5000 miles further north; & from whatever part of the African continent you look at from the sea, the view is everywhere the same, great mountains, & sandy deserts until you get further inland. The Cape of Good Hope lies just to the west of this, and is one of the boldest headlands in the world, but is not, as is generally supposed, the extreme southern point of Africa; Cape Agulhas, 80 miles to the south east, extends nearly 40 miles further to the southward.

Sunday, Nov. 2nd
The mail leaves for England on the 4th, and the one from there is expected on the 3rd, tomorrow. I believe the Captain has decided on not going to Table Bay, so in that case, I don't suppose we shall see Cape Town, for it is very expensive riding up from here, & there is not much to see for the money. At present the ship is stripped for re-fitting, the yards & top-masts being on deck, the fore & mainyards are sprung, & the Dock yard Carpenter will have to make new ones. We were 3 days in Quarantine, & directly the Yellow Flag was hauled down, leave was given to one watch for 4 days, & the greater part of them are up at Cape Town; the ship is very quiet now, for nearly all the officers & scientifics are also on leave. The new Schoolmaster from England is on board, came out in the same mail that brought your letter, he is a young man, & had been married 5 days, when he received 24 hours notice to embark for the "Challenger".

¶It has been very hot here to-day, & I have landed on African soil for the first time. All hands went to Church to day, Catholics, Presbyterians, Wesleyans & "Anglo Catholics" each going to his own Church. There is also a Dutch Reformed Church but we could not muster any for that, of course we have some who profess no religion at all, who have to look after the Boats meanwhile. We all went to the Dockyard Church, & the color of the congregation was even more mixed than at Bermuda; there were the usual soldiers & sailors, with Negroes, Kroomen, Hottentots, Malays, Creoles, & children belonging to the lot mixed. This afternoon we have had lots of visitors on board, several from the "Rattlesnake" describing their exploits at the war on the coast; one of their men was killed on the very day that his 20 year's service expired, which circumstance caused a general laugh, & the remark that he had not been kept a day in the service beyond his time. There are a good many Malays here, who do washing and all sorts of such work; one of them has done mine better than I have had it anywhere since leaving England.

9 PM
The mail has arrived—27 days from England; she came into Table Bay this afternoon 3 days before her time, but there was nothing of

any sort for me, letters or newspapers, so must make up my mind to wait for another 10 days, at least.

November 3rd
I have read all the English news up to the 4th of October, & know more about the Ashantee war, and the Rattlesnake affair from the Newspapers, than I knew from the account of the men who were there. I see Tichborne is not yet played out;[3] and the pilgrimage to Paray-le Monial, "in the face of a scoffing and unbelieving world" has been fully carried out, & should think it has been very beneficial to the shopkeepers & Hotels at Paray.[4] You will get this about Dec. the 2nd my 20th birthday. We are almost going out of the world for the next 4 months. Yours as ever

Joe Matkin[5]

The following day Matkin finished off the letter to his mother, concurrent to the one above. It is here that he refers to his sense of solitude despite the general amicability of *Challenger*'s crew.

I believe we are not going to Table Bay, so as the journey to Cape town costs something like 15/-, & the Town is not much to see when you get there—I sha'nt go—for I have several things to buy here in the way of Boots & warm clothing. Of course I shall go on shore to Simon's town, & there are some fine walks here, also a fine beach for sea bathing which will suit me better than lounging about Cape town. It is not like Melbourne or Sydney, & I invariably go on shore by myself. I have not a single companion in the ship—that I call a companion, though of course we are all sociable & friendly.

To morrow being Charlie's birthday I shall probably go on shore & have a spree, also on the 2nd December[6] take myself out to tea somewhere. At present the ship looks as she did in Sheerness 6 weeks before she was commissioned, & I used to go & look at her every morning & wonder whether I should go away in her: all the yards & topmasts are on deck, the fore & main yards are badly sprung & the Dockyard Carpenters are busy making new ones. The ship looks something like the Dutch ship that is lying here, condemned; she came from the seat of the Dutch war in Sumatra,[7] & has been condemned as unseaworthy, the crew are waiting for another ship to come out here for them.

¶A great many of the people here are Dutch, especially the farmers, for this colony formerly belonged to them, & was taken by the English in 1794. There is a Dutch Reformed Church here at Simon's town, also a Wesleyan, Church of England, Presbyterian & Catholic, although the town here has not 1500 inhabitants. Last Sunday we all landed & went to Church, we mustered some for every church but the Dutch; the few that profess no religion have to mind the boats while we are at Church. The Boats race back to the ship to see who shall be first for dinner, the Catholics are always first as they leave Church at 12 & we were nearly 1 o'clock before we got on board again. We went to the Dockyard Church which was something similar to the one at Bermuda, & the congregation more mixed as to colour. There were soldiers, sailors, negroes, Creoles, Hottentots, Caffres, Kroomen, Mulattos, Malays, & children belonging to the lot mixed; there are some shades of colour among the children. . . .

¶The weather is very hot here especially when there is no sea breeze; it is the beginning of summer here at present & there is no fruit but oranges. There are plenty of Fish in Simon's Bay, & meat & Flour is very cheap; we get fresh meat & Bread now all the time we are here. . . . Several of our men wish to leave the ship here as they don't like her, the work is too hard & the sea time so long—compared to other men of war on foreign stations. Several men also to be invalided home, as they are not able to stand the many sudden changes of climate experienced in this ship. Last night the bush was all on fire on the mountains, about 20 miles distant & it looked like a burning volcano.

¶Early in January you will get another letter containing full instructions as to future letters for Australia. I shall answer the Boys' letters before then, if I get any. Of course you will send this on to them, Charlie first, then Bill, & Fred, & also let Walter Thornton read it, for I dare say I have some from him now on the way. Hoping Father & you are all well, & with best love to all.[8]

From Simon's Bay, Matkin sent birthday greetings to his brother Charlie. The letter describes only briefly a trip ashore but we do learn a good deal about his sojourn in Australia three years earlier.

Dear Charlie,

This being your 21st birthday, I write—wishing you many happy returns of the day.

I expect you have had several letters this morning, containing the usual compliments, small presents, & advice;—of course you are at Bedford still, & by the time you get this will be thinking about going home for X.mas.

We have had 2 mails from England since we came in here & the only letter I have had was from J. T. S. Barr'dn. He was just closing the letter—he said—when you arrived, & he also said you were looking stout,—& courting agreed with you. I fully expected a letter from you, & Willie, by the first mail that came, for it is more than 6 months since you last wrote, & 4 months since Willie did; however I am expecting one next week from you all,—Fred included,—for I don't know whether he is staying at Stamford or not. I shall write to Will & Fred by this same mail if I hear from them, but not to Mother until the following mail, as I have just sent her a long letter describing our voyage from Bahia &c.—which she has of course sent on to you. So I need not repeat any of that news again; the 2 Germans [the Stoltonhoff brothers] have gone to Cape town & will most likely sail from thence to their own country.

I expect you would read long accounts about them in the newspapers & also about the curious island of Tristan d'Acunha &c. My Log-Book[9] will be finished very shortly & I shall send it home from here, if possible; it has lasted exactly 12 months, & I have now got a larger one that will last 2 years,—perhaps the remainder of the commission. My increase of Pay will commence on the 1st December, making it £24-6-4 per annum, & I hope to save £40 by the time we return;[10] of course I have not been able to save anything yet, owing to the small pay & the lot of clothes I have had to buy, for we want Tropical clothes one month, & Arctic clothing the next; besides in these foreign places everything is twice as dear as in England.

We shall soon begin to think about Melbourne now, & of course I am very much interested & curious to see the old familiar places again. There are 3 separate shops in Melbourne—wherein I was employed for more than 6 months; all Furniture Shops & Upholsterers; & I have several friends to hunt up there, some who were very kind to me; 1 independent couple who went out Passengers with me in the "Essex", & who kept me over a week when I was out of work & waiting for a passage to England in the "Agamemnon". They live at a place called Flemington about 2 miles from Melbourne, they are an old-fashioned couple, names Mr. & Mrs. Appleby. Altho' they are independent, they are not clever or 'stuck' up. Mr. Appleby would have me call him William, or Bill, & the old Lady used to make me what Puddings I liked for dinner, she used to say I had eyes just like her boy that was dead a long while ago. Of course I shall go & see if they are still there, but *he* was talking about going to England then; the old Lady didn't want to go for when he was home before, after he had done well there, his poor relations & friends were always borrowing & sponging on him; so much that he couldn't stay & went out to Melbourne again with us in the "Essex". However he was still hankering to go back in spite of the old Lady & I am not at all positive of finding him there now.

¶There are several more middle-class people who were very kind to me there, & to those houses I used to often go of an evening, or Sunday,—that I hope to see again there; & also the Museums, Library, Public Gardens, Theatres &c, will remind me of that strange twelve months I spent in the colony. I had almost forgotten the names of the streets & people there, but as we are getting nearer, I have thought more about it. I wish I had kept a Diary while I was out there, it would be more interesting to me now, than the one I am writing in this Ship. Still it will be a long time & a rough time before we get there, as we are ~~to~~ going to 4 or 5 ice bound uninhabited islands, besides the long journey down among the ice; making it nearly 4 months, & longer by half than we were going all the way from England in the "Essex". We shall get there at the same time I did in the old "Sussex" & shall spend Easter there, I expect; it is the best time of the year there, end of Summer: we shall see the English cricketers in Melbourne or Sydney.

¶The season here is early summer & quite warm enough for bathing, I was on shore at Simon's town yesterday, & had a nice walk along the foot of the mountains, facing the sea, I could see Table mountain quite plain altho' it is over 22 miles from here. I have not been to Cape town, & shall not go unless the ship goes round to Table Bay, it is not a very large place but there is more life there.   *   *   *

Nov. 10th

I was on shore yesterday afternoon, Sunday, & had a stiff walk to the top of the mountains which extend all round False Bay & Simon's Bay. The view from the top was splendid, on one side was the Atlantic, & the coast up as far as Table Bay, & behind was Simon's & False Bay,—with the shipping which was right underneath, so that you could look down on to their decks. A large whale was spouting in the Bay; he will probably be caught before he can get out to sea again. The land all round False Bay is very high & terminates at the Southern portion in the famous promontory called the Cape of Good Hope. Far away to sea—several ships could be seen doubling the Cape, some bound to Australia & others coming to England from China, & Australia. The mountains were covered with great boulders of rock, but the soil is sandy & produces very little, there was a sort of rough brush wood here & there, with shrubs bearing splendid silver coloured flowers—similar to 'Immortelles', these flowers are very rare in other parts of the world & will keep for ever, almost. There were plenty of birds also, good whistlers, but not handsome ones like those in Brazil. In the scrub there are large Baboons which frequently attack people unless they are armed with a gun or stick. I kept a good look out for them but I saw nothing but 2 half wild pigs, which I pelted. On the slopes of the mountains facing the Atlantic there were several Dutch farms with pretty white houses & nice gardens attached. The labourers were chiefly Hottentots, & Kroomen who spoke nothing but Dutch, but the Dutch themselves can all speak English. Not more than 70 years ago all this part of Africa belonged to the Dutch & the principal farmers are still Dutchmen.

¶After my long walk I went to the British Hotel & had tea; in the Hotel there is a piano & when I was in there the other day I played over all my old tunes from 'Hamilton's Instructor'—which was the only music book there. After tea I went to the Dutch Reformed Church, which is the prettiest & neatest of all the Churches in Simon's town. The service is something like the Baptists in England & is conducted in the Dutch language in the morning & in English at night. The Parson was a Dutchman but spoke excellent English & preached a splendid sermon. He is considered the best preacher in the town & I hope to hear him again next Sunday. The English Hymns & Tunes are used & they have a nice organ & a choir of Dutch girls & boys who sing very well, but rather broad English. They provide Hymn books for strangers & the greater part of the seats are free. There was a collection before leaving of course: there were a good many English people there altho' the majority were Dutch. At 8 PM I returned on board with about 150 more who had come from Cape town, where they had been on 4 day's leave, & were rather groggy. Last Saturday our Officers played a Cricket match against the Army at a place called Wynberg —(old Dutch name)—& were beaten by 4 runs.

The "Rattlesnake" has just received orders to be ready for sea at an hour's notice as the Caffres[11] have broken out further to the North. We have drawn 10 more seamen from her in the place of those who have deserted at Bahia & this place. 2 have deserted here & 5 or 6 gone to Hospital ashore.

November 14th

The mail leaves the first thing in the morning & the one from England comes to morrow, but I don't think I shall get my letters before this goes, however I will keep this open till the last minute & see. I am not writing to Mother this mail, I shall write by the mail that leaves here Decr. 6th; to Fred I shall write the next mail Nov. 25th & to Willie by this mail if I have time. However you must send this on to Mother as soon as you have read it. We are going to Table Bay on the 25th so I shall see Cape town, after all. One of our Sub Lieutenants named Havergal[12] is the son of the Vicar of C of E Bedfordshire. With best love to all, & hoping you'll have a merry X.mas.

From your Affectionate Brother,
Joe Matkin

P.S. Give my best love to the charmer.
P.S. Nov. 15th.

English mail just arrived but no letters for me—as usual. Our letters just going for England. Goodbye. J.M.[13]

Despite earlier doubts, *Challenger* did eventually return to Cape Town, as he reported to his cousin in early December. In this letter Matkin took the liberty of educating his cousin on the subject of African tribes and documented the grumblings of the crew about the hard life aboard an oceanographic vessel.

H.M.S. "Challenger"
Table Bay
Dec. 4th, 1873

Dear Tom,

Your letter dated Oct. 21st, reached me in Simon's Bay, Nov. 20th having been only 25 days from Southampton. I see by the Flag Staff up on the Green Mountain another mail is in sight, about 50 miles away, so will be in by 2 PM, she is 1 day before her time having left England on Nov. 5th. It is very hot here to day & we can see the summit of the famous Table mountain quite distinctly; the sides are rugged & steep but the top is as flat as a table, about 2 miles in length, it is 3,600 ft. high, I shall go on shore to morrow, & intend to ascend it.

¶Since I last wrote from Simon's Bay, a good deal has been done to the ship; has been newly rigged throughout, caulked inside &

out, & fresh painted. We have also filled up with 6 months provisions, coaled, &c, & fitted up stoves in all parts of the ship against cold weather. As soon as we get to sea, each man will have issued to him gratuitously—1 Pea Jacket, 1 pr Pilot Trousers, 1 pr Mitts, & 1 pr Flannel Drawers.

¶About 10 days ago news came from Port Natal that the Caffres had broken out, so the "Rattlesnake" was sent round at once with 300 Troops to quell the disturbance. Our Captain had orders to be ready at a day's notice to take more Troops round if any further assistance was required, but as we have heard nothing more I daresay it is all settled. Natal is 700 miles from here, & the "Rattlesnake" is not back yet; Commodore Commerrel did not go in his ship, as he is not better yet; he was at Church last Sunday, & looked very ill indeed.

The Caffres are a more war like race than the Ashantees, & are considered the finest colored race in Africa, there is nothing of the Negro about them, and are more like the North American Indians or Maories than any other. They come down here to buy Rifles, Ammunition &c, & live chiefly by hunting. The Malays here are principally fishermen, & come from the Dutch settlements of Sumatra & Java, in vessels. The Arabs trade along the coast from Egypt & Abyssynia. The Hottentots belong to this part of Africa, & are handsomer than Negroes, & not quite so black, they are employed by the Dutch Farmers, & speak that language. The Kroomen are the blackest race of people on the face of the Earth, they come from the country round the river Gambia, not far from the Ashantees. These we see here belong to Her Majesty's Navy on the coast of Africa, & never leave it; they are lent to any ships that may be in the station to do the duty of Stokers, being able to stand the great heat of the Tropics better than white men. They wear the same dress as our men but Mess by themselves in the ship, one of their own race acts as foreman over them. They are paid about 10/ per week, & have all kinds of fancy names as very few can speak English, so there is any amount to answer to Tom Teapot, Flying Gib, Jem Gridiron, Dixie Land &c.

¶Simon's Town is not much of a place for going on shore, I went though for 4 or 5 times, & Sunday Evenings generally went to the Dutch Reformed Church, where the service is conducted in that language in the morning, & in English at night. Great shoals of Mackerel & Salmon came into Simon's Bay last week, & you could catch 20 lbs in about 5 minutes, rather different to fishing in the Welland.[14]

¶On December 2nd, we steamed away from there to come round here, 60 miles further north; we had a party of Ladies & Gentlemen, friends of our Commander, come round with us, & they were sea sick, just like the party we took for a cruise at Madeira. Our Commander belongs here, (his father is Sir Thos Maclear, the Astronomer Royal at this place;[15]) he was appointed to this ship on purpose for astronomical observations; he has been on leave ever since we came here. The coast all the way round from Simon's bay was wild and mountainous, we anchored in Table bay at 8 PM same night, but the party of Ladies slept in hammocks in the Captain's Cabin. When they were just ready for turning in, they found they were not able to do it, so called the Captain, & he lowered the end of the hammock down until they got in them, & hauled them up, one at a time. They had to be hoisted out of the ship into boats (like cattle) as it was blowing hard. We have been more than a year in commission, & I have only slept out of the ship 1 night, that was in Lisbon. I think when we pay off I must bring my 2 hammock hooks away with me, that I hung up to for 3-½ years. They are fitting a stove close to mine, so I shall be warm in the cold weather.

¶Table Bay is about as large as Simon's Bay, & similarly surrounded by great Mountains; there is an island at the entrance, upon which is a lighthouse. The bay is not so land locked as Simon's bay, & does not afford such good shelter for shipping unless they go into the dock basin. There is a Breakwater under construction being made by convicts & prisoners; 4 or 5 of our Seamen are working on it, who preferred going to Gaol for a month, & afterwards to a fresh ship, instead of going further in the Challenger, 5 or 6 have gone to Hospital, and 3 have deserted to the Diamond Fields, but we shall fill up from the Guard Ship before leaving. Many will desert in Australia, & New Zealand, as the men are very dissatisfied at not getting extra pay for this cold weather trip, & the work is so much harder for every one, than in an ordinary man of war, when the ship is in harbour six months at a time. Men who were in this ship last commission say she used to lie in Sydney for months at a time, & was a very happy ship then, but now when we go in harbour it's to refit, coal, or provision, & the men get scarcely any leave. Several of the Officers are great bullies, the most popular of the lot is Lord Campbell.

¶We are anchored about ½ a mile from the city of Cape town, which has a very clean appearance, & looks something like Penzance in England, only larger; it is about four miles in length but not very broad, is well fortified, & there is a railway which runs up the country; the first I have seen since we left England, though there were some at Lisbon, Halifax & Bahia. We are to lay in our Xmas stores here, & on the 8th the Officers give a grand ball in the Royal Exchange on shore; we shall have any amount of visitors on board here.

Last night the Cable parted with 1 anchor & all to-day the Diver has been at work picking it up, & has now got it all right again. On the 10th we leave here for Simon's bay again to fill up with coal, and also to take in a deck cargo of it. We leave there before the 15th Dec^r & are to inspect a lot of islands in the Southern ocean, to report which would be the most suitable for an expedition to go to from England next Dec^r, and observe the Transit of Venus.[16] Two or three parties are going from England to observe it in different parts of the world, & our party will watch it from some island in the China sea.

¶We call first at Prince Edward island, and probably spend Xmas there, before going on to Marion island, & thence to the Crozet islands; after that to Kerguelen's Land, a large island discovered by Kerguelen, a French navigator in 1763. Captain Cook was there in 1774, & lay a long time in the only harbour—which he called Xmas harbour. The island he called the Land of Desolation, as there was no vegetation on it, neither any animals except seals, sea-elephants &c, but there is ice and snow all the year round. The island is 100 miles in length, & 50 in breadth; it is 2,200 miles south by east of this colony. We are to lie a month in Xmas harbour, and thoroughly survey, & explore the island, which has never yet been done.

¶Leaving there we steer south until we reach the great Antarctic ice barrier, then run up for Melbourne about the end of March, & stay there a week or 10 days before going on to Sydney to stay 5 weeks, re-fitting &c. From there we go to Wellington & Auckland, after which warm weather, & some far different islands. New Guinea is the first, then Timor, Borneo, Amboyna, & Manilla, Phillipine Islands; thence to Hong Kong, China, & on to Yeddo, Japan, after which the Kurile islands, & Petropaulopsky, the cold capital of Kamchatka. Thence Aleutian islands, Behring's Straits, & down to Vancouvers, about May [18]75 after which begin again.

December 9th
We do not leave here for Simon's bay until the 10th—to-morrow, & sail finally on the 16th. The English mail has not yet arrived, but is expected hourly as she is 4 days behind time. She is a very old

boat, the worst the Company has, & is always late. I daresay we shall receive another mail before the 16th & there is one sails for England on that day which takes this, the Captain is going to send his little boy home by it, as he is not strong enough for the next voyage, & has suffered very much from bad eyes since he came in here.

¶The Officer's Ball came off last night on shore, & was a great success, over 400 being invited; our Band was in attendance & came off this morning quite drunk, & could'nt carry their instruments. The inhabitants of Cape Town give a return Ball to-night to our Officers, and we sail to-morrow night. We have been crowded with visitors since we came in, the ship is full of them now (one young lady just asked me to whom I was writing, & I answering to you, my love). There's some very handsome girls in this place, Dutch & English, very extensive about the Bustle & in the latest fashion. We had the Governor, Sir H. Barkly & party on board on Saturday, also Sir T. Maclear & party, & they danced with the Officers on the main deck, while the sailors dance on the opposite side. 2 more have deserted since we came round to this bay.

¶I was on shore yesterday morning until 7 PM, & went to the top of Table mountain. It is 3,500 feet high, & I was 2 hours going up, the ascent was not half so difficult as that up S. Vincent, but the view was far better. Looking south you saw the Dutch town of Wynberg, & a level track of vineyard country until you came to the mountains round Simon's bay, & the city of Cape Town spread out under your feet. It was intensely hot climbing up, but on arrival at the summit you were enveloped in a damp cold mist which kept the sun off you completely, & made it very chill & cold until you descended about 800 feet—where it was as hot as ever—Cape Town is a pretty good place for business, but nothing to look at, not to be compared to Melbourne, there is not a single striking building or church in the place, all after the same heavy Dutch style, there are a great many Malays & Hottentots in it; about every 3rd house is an Hotel. There are Theatres, Music Halls &c here, & plenty of loafers, & that sort of people, flash women &c. Peaches, Apricots, Oranges, Strawberries are the cheapest fruits at present, but are not so nice as those at Madeira and the Azores. Two fine merchant ships have been here & gone since we came in, they were called Emigrant Ships, & had each on board about 500 coolies from Calcutta for Jamaica & Demerara, to work in the sugar plantations there. They go for 5 years & are paid about 2d per day, are supposed to be voluntarily emigrating, but it is really a quiet way of importing slaves, for they are treated like those in the West Indies and very few ever live to get back to India again. The ships lie a long way from the shore, and don't allow their emigrants to land, neither permit any one on board. I was looking at them the other day through a Telescope, they were dirty miserable looking creatures, all wrapped up in blankets, having had cold weather beating round the Cape of Good Hope. They only remained here 4 days to take in Water, fresh provisions &c.

December 15th, in Simon's Bay
We left Table Bay with a party of ladies on board, & steamed round Robin's island with them, returning the same night & landing the

Table Mountain, Simon's Bay, Cape of Good Hope, 1873 (Natural History Museum, London, Photo No. 271). Photo courtesy of the Natural History Museum, London.

visitors, after which we steered for here and anchored on the 11th. We found here the Rattlesnake again from Port Natal, she sails for England shortly, having been out for 4 years. We have filled up with coal & have 70 tons on the main deck, as no ship could be procured to bring us any down to Kerguelen Land. We have 40 Sheep, 5 Bullocks, & a large quantity of Poultry, belonging to the Officers, also several Goats for breeding purposes that are to be left on different islands. The missing mail came in on the 14th & brought me 2 letters, 1 being from you with 2 newspapers, & shall have no more news this side of Melbourne.

¶We sail to-morrow night for the long voyage, & shall be at Kerguelen by the time you get this. Did you see the Illustrated London News for 1st November? It contained a picture of our ship at S. Paul's Rocks. I daresay you will hear of our arrival in Melbourne, about the end of March, as there is Telegraphic communication between there and England. 2 more men have deserted, and 3 are now absent, 1 boy tried to swim ashore last night, & has never been heard of since, the distance was 1-¼ mile, & it was blowing hard, & the boy had his clothes on.

. . . I have laid in my Antarctic provisions & clothing & expect to weather it admirably; so must conclude for this side Melbourne.[17]

While at Table Bay Matkin also began a letter to his mother, repeating much of the foregoing but adding some comments on his personal situation, experiences ashore at Cape Town, and amusing details of the Officers' Ball, possibly the first ever oceanographic ball.

*   *   *   I am in receipt of the increase of pay I told you of, since the 1st December, & shall open a banking account as soon as we get to sea, though I don't expect to save much until we leave New Zealand as we shall be a long time in harbour in Australia & there, & of course spend more. I have laid in my Antarctic provisions & clothing (a pot & a half of Jam, & 1 Box Collars, size 15-½), & I think I shall weather it admirably. I should have made my Will only I have nothing worth leaving, except the Pen Knife Mr. Huntly gave me 3 years ago, & I lost one blade of that in St. Iago, breaking open a cocoa nut. I have used it for cutting my dinner, cutting Tobacco, & all sorts of Fruits &c, & can warrant it to be a good one. Remember me to Mr. Huntly when next he calls. We have Strawberries, Peaches &c here, now, & the weather is fright-

fully hot, they have to throw water on the decks every hour to keep the pitch from melting in the seams.   *   *   *

¶The Officer's Ball came off on the 8th in the Exchange Hall on shore, over 400 were invited & I believe it was a great success. Our Band was in attendance, & came off the next morning at 6 AM too drunk to carry their instruments. The Ball room was decorated with Flags, Swords, Boarding pikes &c, & at one end of the room was the Royal Arms & the Challenger's Dredge, sounding apparatus &c, alongside of which sat 2 of the sailors in white who got so drunk before 12 o'clock that they had to be relieved by 2 others. I send you a Ball programme I had given me, the device on the outside represents the dredge on the sea bottom, & 2 mermaids examining it & taking out a star fish; it will give you an idea of what the dredge is like. Mr. Wyld the Artist drew the device & the Captain had it engraved on plates here, for £7. The appendages at the bottom of the dredge are for catching up aneroids, insects, &c.   *   *   *

Mermaids attend the dredge: program for the *Challenger* Ball at Cape Town. Photo courtesy of the Natural History Museum, London.

¶I was not much struck with Cape town, there was not a good looking building in the place; the houses & Churches &c are after the Dutch style & only look well in villages. . . . The population is 30,000 including many Malays & Hottentots, & there are plenty of loafers & idlers at the street corners & public bars, who came out to go to the Diamond fields & find it wants a little fortune to get there as it is 750 miles further to the north. One swell loafer here calls himself Lord Darnley & says he is the son of the Earl of Cobham, Kent. He is such a nuisance & has been to Gaol for 14 days so many times, that the inhabitants of Cape town are raising a subscription to send him home. He was in company with a lot of our men the other night & made them a speech after this style. "Many of you, gentlemen, may know my estates in Kent, I am only waiting for the death of the Earl to return again to England. You see me here gentlemen an alien from my country, unrecognised & unknown; you have a 1st cousin of mine—Lord Campbell, on board your ship, but he won't recognise me & I don't care for that one penny. I am proud to be in the company of a lot of British sailors, this evening—(Hear, hear, Get yourself another drink Lord, you'll soon want a new pair of boots.)—Yes Gentlemen, my appearance is not all that might be desired, but the man's the man for a' that, & although the people of this town may not think it, I shall always look back with pleasure to my temporary sojourn in this colony, & gentlemen I hope that the breakwater (which I have helped to make) may afford you shelter against the winds & storms in many years to come. (Give us your fin, Lord, & take a mouthful of this.)" When committed for 14 days Lord Darnley has to go & help build the Breakwater. Some of our men say he is quite a gentleman & believe him to be Lord Darnley; I have not seen him myself but think he's one of the George Barrington stamp,—Barrington was a swell pickpocket 70 years ago & robbed the Duke of Norfolk of his (order of the Garter) while talking to him in a crowded Ballroom. He was transported to Botany Bay, & died a rich man out there. When he arrived in Sydney harbour in the convict ship he introduced himself to the Governor & people of Sydney in a Rhyme commencing thus:

> From distant climes, o'er wide spread seas we've come,
> Tho' not with much eclat or beat of drum,
> True patriots we, for it be understood,
> We left our country for our country's good;
> No selfish interests helped our generous zeal

> What urged our travels was our country's weal,
> 'Twas motives such as actuated Cook
> When round the world his way he took
> And fell a martyr to a savage bank
> Far from his home in a far off land.
> &c. &c. &c.
> * * *

The letter closes with the usual reactions to news received in family letters, plus gossip about desertions and Wyville Thomson's servants.

Professor Thompson has engaged a young Caffre for a servant in the Analyzing room, I expect he will be a curious colour in about 3 weeks time. The Professor has also a Bermuda Negro for his private servant, & I have heard that he will get a Maorie at New Zealand, also a couple of New Guinea men, & a Chinaman, a Japanese, & an Esquimaux bye & bye. We have lost 2 more men by desertion, & a boy swam ashore last night from the ship but was brought back this afternoon. The Photographer is also absent somewhere, & the Police are after him as we are to sail finally to morrow night or Wednesday morning.[18]

As *Challenger* was preparing to depart the Cape for the long voyage to Australia via Antarctica, Matkin penned a quick letter to his elder brother. Charlie had been wed in the interval since the previous letter, a marriage that would produce eleven children.

> H.M. Ship "Challenger"
> Simon's Bay
> Decr. 15, [18]73

Dear Charlie,

I have only time for a few lines this time but thought I should like to write a few lines as we are to sail to morrow & it will be some months before I shall be able to write again.

I suppose the marriage is over by this time, Fred said in his letter it was to be the 25th November & I am very glad to hear it, altho' it came rather suddenly.

I hope to hear all about it from yourself when we get to Melbourne, about the 20th March: also tell me how & where you spend your X.mas &c. Don't forget to send my sister's[19] portrait, & tell me if she has any more unmarried sisters as nice looking as herself. You see I know she is nice looking, Miss Wildman told me *that.* I suppose you thought it didn't concern me. I have had a letter from Fred, but none from Will—tho' I have written him twice. If he has written I shall get it at Melbourne. I have written a long letter to Mother by this mail, & will write you a longer & more interesting one from Melbourne or Sydney.

It is blowing hard here now I hope it will be calmer to morrow.

With best love to you & to my new sister[20]

During the first week in the Indian Ocean, Matkin was too busy for correspondence. On Christmas Day 1873, with *Challenger* in the vicinity of Prince Edward's Island, he began letters to his mother and to Swann "compiled from the Log book."[21] Each covered the entire transit to Melbourne (nearly three months) and ran to many pages. The letter to his mother, for example, originally consisted of five folded sheets (one is now missing), written both front and back, with the horizontal text overwritten with a vertical text, thus yielding forty pages.

The following letter to his cousin takes *Challenger* to the Prince Edward, Crozet, and Kerguélen islands before steering south for Antarctica. Additional comments from the concurrent letter to his mother appear in curly brackets ({ }).

H.M.S. "Challenger"
Off Prince Edward Island
Xmas Day, 1873

Dear Tom,

Our 2nd Xmas on board has arrived & finds us in sight of two snow covered isl[ds], but they are still 20 miles away, so we shall not anchor until to-morrow; the weather is just about the same as with you, & the day has passed away far more pleasantly than last X.mas in the Bay of Biscay, O., still the sea is very rough to-day, & the ship rolls a good deal, but we are pretty well used to that now, it was too rough for the regular church, but we had prayers as usual, and are now wearing our warm clothing for the first time, as the Thermometer now stands at 40°, & was down to freezing point last night; rather a sudden change as it stood 70° only 4 days ago. We left the Cape on the 17th for this island, distant 1200 miles, & had a good wind, we sounded 2 or 3 times, depth about 2 miles, & the island hove in sight about 10 o'clock this morning.

All hands were on deck looking out for land till dinner time, 12 o'clock, when the Band struck up—"The Roast Beef of Old England" which sounded very nice, though the dinner was salt Pork, & Pea Soup. Of course we brought private stores with us, and had a fine dinner considering; the Captain walks round just to see the tables laid out, and tastes each messes plum duff, a very ancient custom in the Navy; there are 20 messes, but he only managed about 5; after 1/3 pint of Madeira was issued to each man, and the Band played dance music, which they are still doing, so I being no dancer have got this letter under weigh, with 2 or 3 others. We were considering at dinner time where we are likely to be next X.mas, & find that we shall be somewhere in the neighbourhood of Hong Kong, & X.mas/75, close to Valparaiso. Dec. 21st was the anniversary of our leaving Portsmouth, so the Captain issued Wine to all hands. Just before we left the Cape our Photographer deserted, so at the last moment Professor Thompson had to rush up to Cape Town, & engage a young fellow there; who has been very sea sick since coming across here.

We are now standing between the islands until morning, when we run up to Marion island.

Jan. 3rd, 1874, off the Croizet islands

On the day after X.mas we stood in to Marion, & landed an exploring party while the ship sounded round the island. It is about 4 miles in length, & 3 in breadth, & has mountains on it about 3000 ft high, covered with snow nearly all the way down. There was grass on it but no animals what ever except Penguins, Sea Elephants, & hosts of great Albatrosses, some of them measuring 17 ft from wing to wing. All of these creatures, as well as lots of other sea birds were there for breeding purposes, as the season lasts from November to May. The Albatrosses & Penguins are very friendly, the former generally makes his nest between two Penguins, which make none, but merely a hole in the ground & lays 2 eggs, but the Albatross has a regular nest of dirt & moss, & raises it about 4 ft. from the ground, they also lay 2eggs, which are nearly as large as Ostrich's, after they are laid one of the birds sits on them until they are hatched. The old Man goes to sea and gets his belly full, then returns & sits on the eggs while my lady goes for a feed. A good many of these birds were killed for the sake of the breast of white down, & the web of the foot which our men make tobacco pouches of. They can't fly from the ground on account of the great length of their wings, but must start from some elevation. They are very tame, & would never leave the nest until toppled over with a stick. A large sea-elephant was shot on the beach, but I did not see him. {The scientifics caught as many Albatrosses & Penguins as they liked & brought them on board, also shot a Sea Elephant, an animal like a Walrus, & as large as a cow.[22]}

¶On the 27th we stood across to Prince Edward's Island, which is much smaller & rather higher, but no one landed. We had dredging between the islands and obtained some good specimens, & the same night we made sail for the Crozet islands distant 540 miles. On the 31st of Decr we sighted Hog island, but could not go in very close on account of the fog.

¶New Year's Day was terribly foggy, but on the 2nd we sighted the 12 Apostles' rocks, Penguin island, Possession island, & East island. In the evening we stood in for an anchorage off Possession island, but found it was blowing too hard, with a very rough sea, so we stood out again, & obtained a good view of the island which was something like Marion, but larger. As the sun was setting the mountains in East island came out right above the fog in the most wonderful manner, of which the Artist made a fine sketch; to-day we have been dredging, depth 700 faths. but as it still keeps very foggy we are just going to leave the Crozets, to make sail for Kerguelen Land, distant 700 miles.

January 7th, 1874, anchored in X.mas harbour, Kerguelen's Land

We have only been 3-½ days coming the 700 miles, & sighted this island at 4 AM anchoring at 8 in this magnificent harbour, the best one we have yet been in, it is about a mile in length, & ½ a mile in breadth, & has deep water all round, it is surrounded on 3 sides by high lands, and there are 2 or 3 small islands abreast of the entrance, the land has a very dead appearance, the hills are covered in snow, & there is not a sight of a tree to be seen. There is a very peculiar rock at the entrance to the harbour; it lies in the middle of the passage, & has a tunnel through it large enough for a Train to travel. This has been made by the ceaseless action of the water, & makes it look like a Railway bridge. X.mas harbour is at the extreme north west of the island; it is in Lat. 49° South, Long. 70° East, the same latitude as the Scilly islands, & although this is the middle of summer down here, the hills are all snow covered.

¶Several parties have been on shore during the day exploring & have returned again, bringing with them various kinds of sea birds, sea elephants, Penguins &c. I shall tell you more of the sea elephants after I have been on shore & seen them. There are 4 different sorts of Penguins on this island, the largest and handsomest being the King Penguin, which stands about 3-½ ft high, & has a golden crest on its head. Our men say they look like the Russian sentry's up the Crimea, for they stand as stiff & immovable as a soldier, with their flippers standing out straight. They are of a white & purple black colour, with yellow feet, & are perfectly harmless, & look so helpless when they are being killed. They stand on the beach in droves, & in the evenings go through all sorts of military evolutions. Besides this kind there are the Jackass, Royal & common Penguin here, all breeding on the beach and rocks of this island. Then there are Petrels, Albatrosses, Cape Hens, Tern, sea swallows, rock pigeons, & carrion hawks; the latter 2 kinds are land birds, but migratory and get their living along the sea beach. There are also plenty of wild ducks, but they like the rest are only here for the summer. Specimens of all these creatures, with a couple of sea elephants, & a large seal were shot, and brought on board, & are now stuffed and pickled for the museums. The Seal weighed about 40 lbs., & the sea elephants each 1100 lbs, requiring a good number of men to hoist them on board. Bundles of the Kerguelen Cabbage were also brought, which proves very palatable, & eats something like Spinach though rather bitter. It is the only plant to be seen on the island & grows close to the sea, is in shape something between a lettuce & a palm cabbage. Captain Cook found it very efficacious in curing Scurvy among his crew, so we shall eat plenty of it.

¶We go to sea again to morrow for a cruise round the island, but are to come back here again before we leave it finally. It is snowing now, & very cold.

¶This island was discovered by Kerguelen a French Navigator, who, on his return to France published a grossly exaggerated account of its capabilities & extent. In the year 1776 Captain Cook visited it, & called it the "Land of Desolation," from its dead & inhospitable appearance; it is composed of igneous rocks, which rise into hills from 2 to 3000 ft high. The island is 100 miles in length, & 50 in breadth; there are great many smaller islands round it, and plenty of bays & good harbours. The interior of it has never been visited by man, & perhaps never will, for the ground is frightfully irregular & boggy, so impassable that all progress is debarred inland. Captain Cook spent five weeks round its shores, & also visited Marion island before leaving for the South. He started from the Cape of Good Hope, and after visiting these islands, penetrated to 71° south, then ran up to New Zealand in 110 days. Captain Weddel in 1824, penetrated to Lat. 74° South, & in 1840, Captain Ross got as far as 78°4′ S. the farthest anyone has ever been, or is likely to reach.[23] His ships were the "Erebus" & "Terror", afterwards lost in the Arctic under Sir John Franklin. In Lat. 78° Ross found an impenetrable barrier of ice, & also discovered land with burning volcanoes on it 10,000 ft. high. The French, Russians, & Americans have each succeeded in sailing within the Antarctic Circle but never reached so high as Ross.

Our Captain expects to reach Lat. 65° S. but his orders forbid him from forcing his way through the outer Ice barrier, for he will only have 3 months provisions left then, and dare not risk being frozen in.

February 1st, off Kerguelens Land, bound South

We have finished surveying & exploring the island, so will give you further particulars before leaving it finally. On the 8th of Jan. we left Xmas harbour, and anchored the same night in Betsy Cove, a snug harbour named after some whale ship. It is about 40 miles from X.mas harbour further East, there are several small islands as you steam along the coast, also plenty of rocks & shoals, so that a very careful look out has to be kept, & soundings continually taken. The coast as we came along to Betsy Cove, & indeed, all round the island, presented the same dreary & desolate appearance, in some places the hills were higher, & quaintly shaped, but were always covered in snow; & a heavy fog hangs continually over the land, so that the view inland is completely obscured.

We remained in Betsy Cove 8 days, while the boats went away surveying the adjacent bays & harbours, which were very numerous, & parties of Officers & Scientifics went exploring, shooting & botanizing inland, but they never succeeded in getting more than about 10 miles. Numerous birds, sea elephants &c, were shot, & enough wild duck to supply all hands 2 days, for dinner. I went on shore once with the Paymaster shooting ducks, and we went about 5 miles inland to see some wonderful cascades of water which come down from the mountains. The walking was something frightful, the island is one vast swamp. At every other step you sink up to your knees in the boggy ground. What looked like grass from the ship—turned out to be moss, & it was the mossy ground which was the most treacherous. Not a tree or shrub was to be seen anywhere, no animals of any sort, neither insects on the earth though we looked carefully, except wild ducks & carrion hawks, we saw no birds, so that we may call it truly a "Land of Desolation".

¶Just abreast of where we are anchored, on a prominent hummock of land were 7 or 8 gravestones, we were greatly surprised to see them when first we entered the harbour, they stand upright, & being painted white, are very conspicuous long before a ship is near enough to see what they are. They have been erected by whalers who rendezvous in this harbour, to the memory of their ship-mates who have died or been drowned off the island. They are made of ship's timber, well shaped, painted &, I need not say how solitary & desolate they look in this dismal spot. They are all English & American, some of them dating back to 1840, & nearly all read after this description—In memory of George Eccles, Second Mate of the Barque "Julius Caezar" who was drowned whilst whaling off Desolation Island, May 1863. "And so he bringeth them into the haven where they would be."

Great quantities of Kelp abounded in the sea round this, & other harbours, and on one part of the beach we found an old ship's copper, & a lot of iron hoops that had been left by whalers years ago, after boiling down their whales. The beach was strewn with whale bones, some of enormous size—much larger than an Elephants. A day or two after we came into Betsy Cove, we were surprised to see a small schooner standing into the bay, which anchored the same night, & proved to be the "Imogen" whaler of New London, America. During the rest of the time we lay there the whale men were always on board our ship, telling us all about the island, their line of life, pay &c. The Captain dined with our officers every night, & having been down in this part of the world for a good many years, was able to give us a great deal of information.

Their vessel had just come from Heard is[ld] further south, & was waiting to be joined by another schooner from America, to commence the season's whaling. The crew composed English, Scotch, Americans, Negroes, most of whom came on board before we left. They join the vessel for 4 years, & are paid according to their success, but the average was much less than that of our men, while the hardship is something awful. They would gladly have changed places with ours, for from the time they leave America, until their return they scarcely see a strange face. These men have been here 3 years, & the only inhabited place they have landed on is Tristan d'Acunha. Once a year another vessel comes out to them bringing provisions, clothing, &c, & taking back their oil & skins; as yet they had none, having just sent their stock off. They carry casks, staves, and hoops as cargo, & when they kill a whale, take the blubber to the nearest island, & boil it down, the Cooper goes on shore & sets up as many casks as may be required, after which they are filled with oil, marked & left on the island, until the store vessel comes for them at the end of the year. This one had a store of oil on one of the Crozet isles, & when their annual store vessel was removing it, one of the boats was capsized whereby the mate & 3 men were drowned.

¶This island is their great rendezvous & they generally winter in X.mas harbor. They live very well for they eat Wild Ducks, Albatross, Penguins, rock pigeons, sea elephants, sea leopards, & any amount of cabbage. Some of our men have tried these animals & approve them all, I have tackled an Albatross and liked it, but don't care to try sea elephants. They have told us all about the latter, so I will describe the animal to you. He[24] belongs to the Walrus & Seal family, but is much larger, & not so valuable, though the whalers often kill them for the sake of their oil. They weigh from 12 to 1500 lbs, we killed one weighing 1100, they are covered with a coarse furry skin, which is only of value to themselves. They live in the sea in winter, but come on shore in the summer to breed the same as Seal's, Penguins &c, they live on the same food as Seals & Whales, and during the time they are on shore they eat nothing, so get very thin during their 3 months on land. Their head is exactly like that of a lioness, they have whiskers, but of course have no feet, like animals have large eyes, and 2 large flippers where their fore legs should be, & 2 very strong ones where their tail is situated, which he uses to propel himself through the Water, & by raising himself on them manages to hobble along on land; they come to the island in November, with the Seals & Penguins from the south, for the purpose of breeding. The males come a week or two before the females, & wait on the beach until they arrive, who, on their arrival retire farther up, while the gentlemen have a very long & fierce combat for the possession of the best looking wives; after this little affair they choose their partners, & settle down comfortably. There is a sort of general confinement in May, during which time the ladies are very helpless, & the gentlemen most affectionate & attentive; they are also very savage if anyone attempts to intrude on the mother & infant, & have been known to defend them to the last extremity. I read the other day Anson's Voyages,[25] in which was an account of a sailor who had killed the mother & calf, & while engaged skinning the latter, the male animal came unperceived behind him, & getting his head in his mouth, inflicted on him such frightful injuries, that he died. This was at Juan Fernandez. About a fortnight after the calf is born, the mother seizes it in her mouth & throws him into the sea, for they do not take to the water naturally, but require preparatory lessons in swimming. After a few trials he manages it, & at the latter end of May the whole herd leave the island for the south again. They are quite harmless if let alone, if you poke him with a stick, he raises himself on his tail & makes a sort of helpless jump at you, but of course falls down; if you throw a stone in his mouth, he roars &

shows his great teeth, grinding the stone to powder. An enormous quantity of blood comes from them when they are killed; the Whalers generally spear them, & they tell us, that, like the elephant on land, they often shed tears while dying, & look at you with such a pitiful & accusing gaze. Seals & Penguins have very similar habits.

On the 16th Jan. we left Betsy Cove for another harbour further east; we had a severe gale during the night, which drifted us to the extreme east end of the island. We passed several small islands, & found the east coast very mountainous & wild looking, the mountains generally being sugar-loaf shape.

At 7 PM on the 17th we steamed into the Royal Sound, & after steaming for about 8 miles where the navigation was very intricate (for the land was broken up into hundreds of small islands, rocks & reefs) we came to an anchor between 3 islands, & the boats were got out for surveying the harbour &c.

Here we found another Whaler, called the "Charles Colcott," also from New London, but in opposition to the "Imogen". The crew were mostly living in a hut on shore, where they had their oil stored in casks on the beach. On a small island opposite their hut was another lot of tombstones,—all whalers but not belonging to this vessel.

The crew was very much surprised to see a large man of war steaming into their harbour & fully expected that we should run on shore. Most of the men came on board while we were there, but the Captain, who was a regular churlish Yankee, refused to come, or to give our captain any information about the island or harbour. He [our captain] invited him to come to church on the Sunday morning, & to dinner in the evening, but he wouldn't come. No Capn, he says—I a'nt been to no church for 20 years, & I guess I aint going to begin now.

¶We remained there 3 days, & on the 20th steamed out of Royal Sound, & hove the dredges. Some good specimens were obtained, but no fish have been caught round the island. We anchored off the land that night, and after dredging again on the 21st stood in, & anchored in Greenland Bay, which is at the extreme south east of the island. Several parties landed there, but the island was exactly the same—no animals or trees, & vegetation to be seen. On the 22nd we left & made sail for Xmas harbour, distant 140 miles. On the 23rd we had a tremendous gale, during which we were struck by a very heavy sea which smashed the sounding platform, & went down the funnel into the stoke hole {frightening the engineers & stokers a good deal. They thought we had run into an ice berg.[26]}

¶On the 24th we anchored in Cascade Bay, & on the 25th in Betsy Cove again, where we rode out another heavy gale. We had landed a pair of Goats on the island before we left, hoping that they would find their way into the interior, and increase & multiply; but we found them grazing among the gravestones, & as soon as they saw the ship, they came down to the beach & commenced bleating. However we left them there but whether they will live through the winter is a query—no other domestic animals could I'm sure.

On the 26th we made sail again for Xmas harbour, but were obliged to anchor that night in Hopeful harbour. There we took in another good supply of Kerguelen Cabbage, & on the 27th made another start. During that day we fell in with the "Imogen" & her partner the "Royal King" also from New London. They were out on a whaling cruize, & each vessel had 2 men on the look out for whales. They anchored with us that night in Rhodes Harbour, & the two Captains came on board our ship to dinner. Several of the crews of the vessels also came & we had the band under weigh, & some good dancing.

¶On the 28th we got out the boats & surveyed the harbour, which is a very extensive one. There are plenty of good ones round Kerguelens Land but X.mas harbour is the best. We anchored there again on the 29th & remained there 2 days to get some specimens of coal, which were found in the rocks & cliffs, & on the 31st we made sail from there, intending to leave the island for good. Previous to leaving another good supply of cabbage was taken on board, & a paper recording the ship's Visit, & other particulars was buried on shore. All to-day, Sunday, we have been sailing parallel with the land, and have just cleared the south point of the island. The west side we did not survey at all as Ross examined it pretty well in 1840; the south side is as barren as the rest, I think the mountains are higher, some quite 4000 feet, & shaped like church spires. We have been 25 days round the island, and anchored in most of the principal harbours. The mountains are now almost out of sight, & we are under all plain sail for McDonald & Heard Island, distant 300 miles, due south.

February 6th, anchored in Whisky Bay, Heards Island

We have been 6 days coming this distance, owing to the frightfully foggy weather we have had, & the uncertain information as to the correct position of the islands. We sighted them twice but were obliged to keep away on account of the fog; at 10 this morning we came up to McDonald Is^ld or rocks, which are a group very similar, and as barren as S. Paul's Rocks, the highest point would be about 400 feet.

Heards Is^ld lies 20 miles further south, and is in the same Latitude as the Falkland Isles, it is about as long as Marion Is^d, & Whisky bay is the only anchorage. The island looks like an immense iceberg, being covered with snow & ice glaciers, which are formed all down the mountain slopes. There are several men on it belonging to the "Imogen", who are here for the summer to kill sea elephants for the sake of their oil. The Captain landed & got what information he could from the men, & we leave for the south to-morrow, as any exploring party is impossible, for the snow lies several feet thick on the island. There is a large iceberg aground on the other side of the island, which the whalers say has been there for years.

"Ye gentlemen of England, who live at home at ease,—Oh, little do ye think upon the dangers of the seas." (Poetical Soliloquy.)

February 16th, within the Antarctic Circle,
Latitude 66°39′ S, off the great Ice barrier, & surrounded by icebergs

We left Heard's island on the 7th for the south, with a good breeze, which increased to a regular gale during the night; the ship was knocked about any how by the tremendous seas, 2 of the main deck ports were stove in & men knocked out of their hammocks by the sea, the Carpenters were up all the night repairing the damage. It was considered the worst gale we have yet had, tho' not so dangerous as the one in the Channel {there was 12 inches of snow on the decks & we got up anchor at 5 AM & left Heard island for the south. We got some fine coral up in the dredge just off the island.[27]}

We ran 214 miles during the gale, with storm staysails only, the best part of the time, the gale moderated on the 8th Sunday, & we were in the Latitude of Cape Horn.

On the 9th we were out of the track of navigation being in Lat. 58° S. & on the 10th we were in Lat. 60°15′ S. & no icebergs in sight. The Thermometer stood at 33-½ in the shade, & the temperature of the surface water was exactly the same. On the 11th at 4 AM the first iceberg hove in sight, & we lay to, close to it, & hove the Trawl. This iceberg was about a mile in length, & 400 ft high, it was quite square, & the summit was covered in snow. We found the depth 1,275 faths.; the surface water was 32-½, only half a degree above freezing point for fresh water, & from this depth we brought up living specimens, the best being a large Prawn, & a curious sharp snouted fish about a foot long, & of a sort hitherto unknown.[28] We passed another large iceberg quite close the same night. On the 12th we were in Lat. 62°40′ S. and sighted several more icebergs; one we passed quite near which looked something like an Iron clad Ship with a splendid ram in front, it was a beautiful blue color, & the summit was frosted with snow, giving it the appearance of a Lion's mane. There was an opening in its centre large enough for a ship to sail through, but the two apparently separate masses were joined together beneath the water. According to Commodore Maury[29] only one tenth of a large Iceberg appears above the water, so the weight & total extent of some of those we have seen must be enormous.

Great quantities of small ice was floating in the wake of this Iceberg, which would have been invaluable to us had we wanted any fresh water: for, you doubtless know when salt water is at freezing point, the salt sinks to the bottom, & the water frozen is fresh, & when thawed, fit to drink. Captain Cook obtained several supplies when he was down here. 27°2′ is the freezing point for salt water at rest, but if in motion 25-½° is the average, this varies according to the degree of saltness. A strong under current drifts the icebergs to the north from 5 to 8 miles a day, & the prevailing westerly winds also drives them to the East, so that their course is about N.E. By the time that they arrive at Lat. 50. S. they meet with a very warm under current from the Indian Ocean, which rapidly melts them beneath the water, & as soon as the submerged part is lighter than that above the surface, they topple over, & either sink or break up into small pieces. They seldom reach farther than Lat. 40 S. altho' they have been found as high as the Cape, Lat. 34 S. Those in the Arctic Ocean generally drift down to Lat. 50. N. where they are met off the banks of Newfoundland, by the warm current of the Gulf stream which very soon settles them. Those banks have been formed of earth, rocks, &c, which the icebergs bear away with them from the land, & the sea above them gets shallower every year. The vapour which this icy current causes to rise from the Gulf Stream is the cause of the general fog to be met with off Newfoundland. The same meeting of the warm current from the Indian Ocean with the Antarctic current, cause those divers fogs we found off Prince Edward's, Marions, Crozets & Kerguelen islands.

¶After we were in Lat. 60° S. we lost those brave westerly winds, & are now in a region of light wind and calm; if it were not for the very cold weather we could fancy ourselves in the Tropics. On the 13th, we passed numbers of icebergs, & on the preceding night nearly ran into one during a snow squall; nearly all the sail is taken in for the night on account of the icebergs, the Lat. was 65°15' S. and the weather mild for the region, & we began to think we were going to the South Pole. But during the same night we ran into the great outer ice barrier, for some distance, & had a job to get out again and as soon as we did the ship hove to until morning. Her course was much impeded by the great lumps of floating ice, & the roar they made against the bows of the vessel during the night, I shall never forget. It was like thunder more than anything else, & the grinding noise woke all who were in their hammocks, and nearly every one was on deck. It was a beautiful sight to see the ship ploughing her way thro' it, & the light emitted from the ice made it nearly as light as day. This light is called by polar voyagers, the Ice blink.

¶Twilight in these high latitudes lasts until 12 o'clock, & daylight breaks at 1 AM so there is night in one sense of the word. Of course this is the middle of summer down here, for 7 months in the year the sun is seldom seen above 65° S. The Aurora Australis was visible during the night, and the coruscations of light were much more brilliant than the Aurora Borealis we saw at Halifax. {The Northern Lights are seen frequently but the Southern phenomena very seldom. Those of the Northern hemisphere are generally in the form of a rainbow, but here they are of no particular form, but shoot like distant sheet lightning, from all parts of the heavens, & the light varies in colour, changing white, orange, & purple & scarlet alternately.[30]}

¶On the 14th over 30 large Icebergs were in sight from the deck, and the great ice barrier about 5 miles away looking like low-land covered with snow, & a pale misty light hanging over it. The icebergs were of all shapes & sizes from a Gothic Cathedral to a Haystack, and varied from 100 to 500 feet in height, & the largest were about 4 square miles in extent. Some had spires to them like churches, others bays and harbours running into them, & many were honey combed with caverns extending through them in all directions. These had been formed by the restless action of the waves, & through these caves the sea rushed & roared with tremendous force, the spray often rising quite 100 feet alongside the bergs. These caverns were of a dark blue colour, the iceberg itself being of a starch shade. Several of them looked exactly like the Spithead Forts, but the majority were more like haystacks with sloping snow covered roofs. Many of them had great fissures through them, as if they had been stricken with Lightning, these rents were of a dark blue colour.

¶Close to the Ice barrier we sounded at 1,657 fths., the water temperature was 29° the Lat. 66° S. only 30 miles from the Antarctic Circle. After sounding, sail was made to the West along the barrier, in hopes of finding an entrance, the sea was as calm as a fish pond but covered with cakes of ice, the wind was very light, & not much progress made; a new kind of sea bird called the Snow Petrel was first seen on the 14th; they are white, & about as large as a dove, they breed on the icebergs but are never seen below Lat. 65 S. On some of the icebergs we saw numbers of Penguins, & on one we saw a large rock, & some earth, showing that the bergs are formed in bays & harbours near the south pole. The Artist made sketches of most of the large icebergs. We passed one on the night of the 14th, which looked something like Windsor Castle, & had a large cavern through it which showed the daylight on the other side; nearly all hands were on deck to see it, & one of the men even said, he had often paid a penny in England to see not half such a sight. At midnight it was light enough to see to read, & the sky was still scarlet in the west where the sun had gone down.

H.M.S. *Challenger* among Antarctic icebergs, from John James Wild, *At Anchor: A Narrative of Experiences Afloat and Ashore During the Voyage of H.M.S. "Challenger" from 1872 to 1876* (London: Marcus Ward, 1878), at p. 74.

¶On Sunday, 15th, we had church in the morning, but only sailed 15 miles. Over 30 icebergs were in sight at once, and no two were alike. The Thermometer stood at 27-½, water temperature 29°. During the night we ran into the ice barrier again, but the ship was at once put about, & we sailed out until this morning when we were about 8 miles from it. The barrier extended from east to west as far as the eye could reach, & looked like a marble wall 50 ft high. We saw this morning the largest iceberg we have yet seen, it was about 4 sq. miles in extent, and looked like a piece of ploughed land in winter covered in snow. It was over 500 feet high, & sloped up very gradually; if I can get you some photographs of these bergs will send some. We saw another that was quite blue, it having very recently turned over.

At 9 AM the sails were furled, & steam got up, & we steamed as far south as the barrier would allow us. The Lat. was 66°40' S. Longitude 78° E. Sail was then made to the west in hopes of finding an entrance to the southward, but the wind was very light, & not much progress made. Wine was issued to night to celebrate our entry within the Antarctic Circle.[31] At present the barrier is still in sight, & we are surrounded by icebergs of all sizes & shapes. There are also any amount of great Whales in sight, blowing & having such a game among themselves. It is now 10 PM and the sun is still above the horizon, & the sky like fire, giving all the icebergs to the West a scarlet tinge. A fine breeze has just sprung up, but they will have to shorten sail on account of the icebergs. The Captain would have tried to force his way through the barrier, but his instructions from the Admiralty forbad him, & we have only 3 months provisions, & 80 tons of coal remaining in the ship. I hope we shall reach Lat 70° S.[32] Every one is in good health & spirits, and very few wish to turn back yet; and those who do are reccomended by the others to get out & walk. We are all wearing our polar clothing, large caps, with flaps to them to cover the ears & neck, & great jackets, trousers, boots & mitts. I am considered to look like the Shah of Persia in my cap. A great by-word among the men at present is "Do you think you shall weather it Bill? Yes, I think I shall go round in her Bob."

Sunday, March 1st, bound to Melbourne, blowing hard
On the 17th of Feb. we were in Lat. 65° S. & could find no opening in the ice barrier. We sailed through several fields of floating ice, and had some snow squalls; it was one of the coldest days as the wind came from the south pole. On the 18th it was still colder, the

Thermometer stood at 24° in the shade; salt water froze as it was thrown on the deck, & the surface temperature was 28-½°. We passed through more fields of thick floating ice, which shook the ship most terribly as she forced her way through it; her sides leak in several places, and she will require docking at Sydney. Numbers of immense icebergs were passed on that day, one large one looked exactly like the pictures we see of the Coliseum at Rome; another like the great fort at Sheerness, and while we were gazing at it, one of the natives of that port sang out—"Change here for Sheerness".

The Lat. was 64°30', & the Long. 84° E. on that day. The Captain finding he could not get a passage through the ice barrier, made sail for Termination Land, a discovery of the United States expedition under Captain Wilkes in 1839 & 40. This expedition sailed from Sydney for the South, & consisted of 5 ships, but none of them could find their way through the barrier of ice, Latitude 67° S. was their farthest, about 20 miles beyond us. They discovered a few islands, & this Termination Land.[33] A French expedition came down here at the same time, which consisted of 2 Corvettes,[34] but they could not get beyond Lat. 66°40', where the barrier stopped them. The ships were nearly lost amongst the ice, & only discovered a few islands. Sir James Ross was starting from Tasmania just as the French & Americans were returning; his ships were strongly protected against the ice, and he forced them through the outer barrier which extended for 200 miles, there he found an open sea teeming with whales, & free from icebergs. He reached to Lat 78°4' the first year, & wintered in New Zealand; the second year he reached to 78°11' S. & wintered at the Falkland Islands, the third he reached to 71°10' S. & sailed for England. His ships the "Erebus" & "Terror" were much damaged, but were repaired, & sailed again with Franklin's expedition to the Arctic region. Ross discovered the north magnetic pole, & found the position of the eountry south which is in the interior of a range of mountains discov^d by him in Victoria Land. A Russian expedition under Bellingshauser, penetrated to Lat. 73° S. & discov^d Alexander Land.[35] So we have been as far south as the French & Americans, & have been beaten only by the Russians, & our own countrymen. As some of our men say "we shall be able to talk Antarctic with any one in England when we get back."

¶On the 19th Feb. we passed numbers of icebergs, looking like the Chalk Cliffs of Dover; the weather was very cold and the Barometer stood 28.5 inches, lower than in any other part of world. We sounded that day at 1,650 fths. & at night passed an immense iceberg, shaped like a Castle on the summit of a Cliff, and we had to lay all that night, the icebergs were so numerous. On the 20th we were in Lat. 63°20′ S. Long. 88° E, the Thermometer stood at 27°, the Barometer at 28.7 inches, decks covered with snow.

¶On the 21st we were obliged to furl the sails & steam as there was no wind, & the sea was quite calm, we were surrounded by immense icebergs, 50 could be counted at once, the ship was steamed within 30 yds of one, & a 9 pounder shot fired at it which brought down tons of ice. On the 22nd, Sunday, there was no wind again; it was like being in the Tropics only the weather was freezing.

¶On the 23rd we were on the exact spot marked on the Chart as Termination Land, & no sign of land was to be seen; soundings were taken, & the depth was 1300 faths. which was a certain indication of no land in these high latitudes, for all round Kerguelen & Heards Island, the depth was only about 100 faths. We steamed for 2 days in all directions, but saw nothing but ice, so may conclude that Termination Land has no existence, & the Yankees were deceived by a large iceberg, or well defined cloud. One of our men (guessed) it was low water when they were down here. On several occasions land has been reported from our Mast Head, & had it not been for our having steam, we might have marked new lands down on the Southern Charts; but, altho' we could all have sworn, that we saw land on one occasion, on steaming towards it, it proved to be a peculiar, & remarkably defined vapour cloud. We have passed over several spots marked on the Chart as "Indications of Land" without finding any, & I daresay a steam ship would dispel several of the mythical discoveries in this part of the world.

¶We had a magnificent sunset on the 23rd, and 45 icebergs were sighted at once, besides any amount of Penguins & Whales jumping about all round us. On the 24th we sounded at 1300 faths. & hove the dredge, when some good specimens[36] were obtained. As soon as the dredge came up, it came on to blow, & all hands were called to make sail for Melbourne. The ship was steamed close under the lee of a large iceberg to allow the topsails to be reefed, and some under current drifted her right on to the berg; the engines were turned full speed astern, but to no purpose & she drifted right on to it; but fortunately struck it very gently, though with sufficient force to carry away the Jib boom, dolphin striker, and other head gear. As soon as the ship got free, the wreck was cleared away, & the Carpenters set to work to make fresh spars, which took them 2 days, & delayed us about 300 miles.[37] The wind increased to a gale, & it was the coldest day we have had, snowing & freezing all the time; the Thermometer stood at 24°, & the wind was fearfully cutting. The Lat^de was 64°15′ S. & Long. 95° E. The wind kept increasing, & the ship lay to under bare poles, the fog came on so thickly that nothing could be seen 50 yards ahead, & the snow was blinding.

¶At 3 PM, a large iceberg was discov^d drifting right on to us; the hands were called to make sail, but by the time they got on deck the berg was only about 20 yds from us, & rose right above the ship's masts. The confusion was something fearful; nearly everyone was on deck, it was snowing & blowing hard all the time; one officer was yelling out one order, & another something else. The engines were steaming full speed astern, & by hoisting the topsail, the ship shot past it in safety. A seaman fell from the trysail while they were hoisting it, & was much hurt.[38] The gale still increased, & the fog was still thick, 3 men were kept on the look out all day &

H.M.S. *Challenger* after collision with an iceberg, from a sketch by Lt. Aldrich, in T. H. Tizard, H. N. Moseley, J. Y. Buchanan, and John Murray, *Narrative of the Cruise of H.M.S. Challenger with a General Account of the Scientific Results of the Expedition*, 2 vols. in 3 (London: HMSO, 1885), vol. 1, pt. 1, p. 404. Photo courtesy of Scripps Institution of Oceanography.

night. During that night the ship was put about 9 times to keep her clear of the icebergs. On the 25th there was ½ an inch of ice all over the outside of the ship, & while she was forcing her way through a large field of floating ice, the Artist sketched her from a Boat. It came on to blow again that night, & most of the sails had to be furled as there were such numbers of icebergs about. While furling a boy fell from the lower rigging overboard; the Life Boat was manned, but was not required as he had caught a stray rope & was hauled on board unhurt, but half frozen.

¶On the 26th were in Lat. 63° S. and Long. 94° E. & there were plenty of icebergs to be seen. The new Jib boom &c was got out that day & re-rigged. We also sounded at 1,975 faths. & hove the Trawl, obtaining some rarities.[39] It was the greatest depth since leaving the Cape, & a series of bottom temperatures were also taken which showed a warmer current beneath than at the surface. It came on to blow a regular gale that night, with snow, and a thick fog. The icebergs were so numerous that the Captain thought it would be safer to lie to under the lee of a large one than to drift with the wind; so for 6 hours we did so, until the wind shifted & blew us away from it. The ship was knocked about frightfully during the day by the great rolling seas, & every one was glad when morning appeared. It was considered the worst & most dangerous night we have had, & the Captain never left the deck for hours. Almost all the men's crockery was broken & many messes used Australian meat tins for basins. Altho' we were all eager to see an iceberg, we are just as anxious to lose them now, it is so dangerous sailing these foggy nights with such masses of destruction all round us.

¶On the 27th it cleared up a little, & we passed numbers more, one tabular shaped was 3 miles in length, & was the very largest we have seen. Yesterday it blew very hard & snowed a good deal, & we only ran 60 miles. If you give your finger a slight cut, or knock, in these cold latitudes, it takes weeks to get better.

¶To-day, Sunday, we ran 150 miles, we are in Lat. 58° S. Long. 102°50′ E. weather rather warmer; Thermometer standing at 35°; we have passed several icebergs but expect they will be the last. We shall sail all night after this, & shall soon get into the belt of westerly winds, we are only 2000 miles from Melbourne to-day, & expect to be there by the 15th March, the day I told you some time ago. We had church this morning as usual, and at present 7 PM the ship is sailing 10 knots per hour.[40] I have got much thinner this trip, finding I have lost 12 lbs since leaving England.

Sunday, March 8th, 960 miles from Melbourne, but a head wind

On the 2nd March we ran 118 miles but ran nearly into a large iceberg during the night in a fog. We were going 9 knots per hour at the time & as soon as it was seen, the helm was put hard down, & the ship shot past it close enough to allow a biscuit to be thrown on to it, and had we struck it the ship would have gone down stern first; some of the men got so timid of them, that they used to remain up all night in case the ship ran into one. For my part, altho' I always slept as soundly as in harbour, I don't care if I never see an Iceberg again, except in a picture; as some of our men say— "we have had our whack of them". On that day we caught the brave westerly counter trades—the best winds for sailing to be found anywhere. Australian Clipper ships coming across here in the depth of winter have been known to average 334 knots[41] 40 per day, for 8 days; one famous Clipper the "James Baines" with a gale of wind behind her ran 420 knots, or 462 miles in the 24 hours.

¶On the 3rd we ran 119 miles, and furled sails for Trawling &c. the depth was 2000 faths, the deepest since we left the Cape; it brought up some beautiful specimens, fish of the most brilliant colours & curious forms.[42] On the 4th at 1 AM, we had a splendid display of the Aurora Australis, & turned out of my roost to see it; the flashes of light varied in colour, being white, orange, yellow, & purple. It was a magnificent sight, tho' rather chilly the way I was dressed. We only ran 40 miles that day, but on the 5th 175 miles, & on the 6th 224; on the 7th 177 in 17 hours, when we sounded at 1800 faths, and obtained some fine specimens in the Trawl, notably a transparent star fish, that looked like red silk velvet. The creature's heart could be seen beating & some young ones inside it with the naked eye. Some curious fish with long tails like rats were brought up.[43] We ran 120 miles to-day, but have a head wind, & only 40 tons of coal on board; unless the wind changes the Captain will put into Hobart Town for coal. The last 2 icebergs were passed on the 4th. Today we are 960 miles from Melbourne in Lat. 48° S. and Long. 130° E.[44]

By the middle of March *Challenger* had completed its long trans–Indian Ocean passage via the Antarctic Circle—the longest period the expedition would be out of contact with civilization. An earlier plan to visit Tasmania was scrapped, the ship passing King's Island and Cape Otway in the Bass Strait before putting into Melbourne on 17 March. Matkin could finally close the letters he had begun on Christmas day of the previous year. In spite of the excitement of the long-awaited arrival and

mail call in Australia, he recorded faithfully, if briefly, the oceanographic work and some tidbits about the vicinity. The version to his
mother concludes as follows.

Monday, March 16th, off King's Island, 100 miles from Melbourne

The wind shifted aft again & we did not require to put into
Hobart Town. We have sounded twice this week, average depth 2-
¾ miles. Some very fine specimens have also been obtained in the
Trawl net. The weather is very warm again, thermometer at 70° in
the shade; the days have also shortened considerably, it is dark
now at 7 PM. We had Church yesterday & the Captain returned
Thanks for our safe deliverance from the recent dangers amongst
the ice.

¶At 5 PM we sighted the first ship we have seen for 13 weeks,
except the whale ships. We sighted King's Island at 12 o'clock to
day; it is 25 miles from Cape Otway, & is about 12 miles long & 5
broad; no one lives on it except the Lighthouse men. About 9 years
ago a large passenger ship was lost on this island & the passengers
were 3 or 4 days on the island before they were discovered &
relieved from Melbourne. A child was born on the island the same
night the ship was wrecked.

¶There is no wind at present & we are steaming half speed; we
have just enough coal to take us in, & expect to anchor in Hobson's
bay at 11 AM to morrow 17th, so that we shall be 3 months to the day
since leaving the Cape Dec 17th. They will telegraph up to Melbourne before we arrive, & thence to England so that as you are 10
hours behind Melbourne time you will know of our arrival in Melbourne before they know it there (according to the time of day). I
hope to be reading some letters to morrow, I expect one from you
all. I believe the mail leaves to morrow, so shall be all ready & will
write to all the boys by next mail. You must send them this to sat-
isfy them in the mean-time, & I sha'nt have to repeat all this news
again. Also let Walter Thornton read it as I shan't have time to
write all this to him. I shall write to him by next mail. I have written to Barrowden & Swadlincote, & as the letters are compiled
from the Log book, of course they are very similar to yours. I hope
to hear that Father has improved & spent a better winter this year.
Tell Miss Wildman I found the cuffs she sent me very comfortable
down south. I have written 50 pages of my new Log book & it is
much larger than the old one; I shall know to morrow if you
received it all right from the Cape.

¶We shall be in Melbourne in time to see the last cricket match with the All England. We are to stay in Melbourne 10 days I hear, & at Sydney 6 weeks. Cape Otway is now in sight, it is 90 miles from Melbourne. By 8 AM tomorrow we shall be off Barwon Heads where the poor old "Sussex" lies at her last anchor.

March 17th
We anchored at half past one PM off Sandridge pier, & expect our letters from Melbourne every minute. We have already heard that Tichborne has got 14 years, & Bazaine's trial & sentence,[45] also that the Ashantee War is over. I will now conclude with very best love to you all, & hoping you are all well.

From Your Affectionate Son,
Joseph Matkin

P.S. March 18th
   Our letters came on board at 4 PM yesterday. I received 8 letters & 8 newspapers. One was from Mr. Daddo—in deep mourning envelope—which was sufficient to indicate its contents. Just as I was about to open the letter, I was called to see 2 gentlemen who had just come on board to see me. One was Edward Daddo, the other a friend of his. Edward wore crepe round his hat, & told me of Mrs. Daddo's death & his elder brothers' before I have read Mr. Daddo's letter. It must have been a great blow to them here, & in England also. Edward was very pleased to see me, I showed him over the ship & am to spend the day with him to morrow in Melbourne. The mail leaves for England on the 26th so will write again before it goes & answer your letters. I was very pleased with Miss W[ildman]'s letter & hope to answer it from New Zealand. 2 of my letters were from Barrowden, & the rest from yourself, none from the boys as yet. The registered letter I received safely, so without taxing myself shall be able to send Father the last loan I had & also Willie & Fred 5/- each.[46]

Melbourne and the Pacific at last!

4 Australia and the South Pacific

MARCH 1874–NOVEMBER 1874

> In these 40 or 50 years the Wesleyans have instructed &
> civilised the natives of these & many other groups of
> islands far more than the Roman Catholics have done the
> Negroes of Brazil, and other countries, in the course of 200
> years. . . . The Wesleyan Miss^{ry} Society deserves the sup-
> port of all denominations in England, & I should think
> their faith is the simplest & best for these primitive
> people.
>
> —Joseph Matkin in Tonga[1]

Having been twice before to Australia, Matkin felt a sense of "coming
home" at Melbourne that was absent at all the other ports of call.
Friends arrived to greet him almost as soon as the mail from home. But
after three months without communication from his family, Matkin was
anxious to catch up on and respond to the news from Rutland.

> H.M.S. "Challenger"
> Melbourne
> March 23rd, [18]74

Dear Mother,

    Besides Mr. Daddo's & J. T. Swann's letters, I received 5 from
you: 2 were addressed to the Cape of Good Hope, one containing
this notepaper which I did not require as I have plenty by me & can
buy it anywhere much cheaper than you can send it out. The other
Cape letter was written the day before my Birthday when you were
having such stormy weather, & we were lying at anchor all the
time.

    You certainly did not tell me much about the wedding,[2] but I
suppose you had nothing much to say. Like you, I hope it will be
for the best—but I don't see why it shouldn't—both of them bringing grist to the mill. You were not much older when you were married. Had it been me, with no trade at my finger ends, you might
have feared & prognosticated (as all Mothers do where the sons
don't marry for money) but Charlie, as long as he has health &

strength—will always be able to live.[3] I hope to hear from Charlie himself by next mail. I am writing a short letter to him now, & will write to the other boys as soon as I hear from them. I was much amused with Fred's letters you sent; he writes nicely and will improve his spelling. I don't take much interest in the Cattle Show now I am away for there is such a lot more interesting news to read out of 3 months English papers, but I hope it brings you more custom[4] every year. I have read the 8 newspapers & lent them round. Thank Mr/s Wildman for the Cornish papers. The registered letter I received all secure & think I need not look too far to find the sender. I always intended to send Father the last 10/- I borrowed at Portsmouth from Melbourne, & now I can do it & also send Will & Fred 5/- each without stinting myself. I have laid out over 3£ in clothes here for we shall not be going to such another cheap place all the voyage. I enclose P.O.O. [Postal Office Order] for the £1 payable to Father & he can send Will & Fred their pocket money when you think they need it.

Mr/s Jones wrote me a nice letter which I shall answer from New Zealand. I don't like to answer it just at once, & of course it will not be a long letter, so if you think Mr/s J. would care to read about Kerguelens & icebergs you can forward your Antarctic letter. I was very sorry to read of Mr. Blackwood's death, he was a good friend to me & a thorough gentleman. I have his last letter to me & think I will send it you home to keep for me as I should not wish to lose it. It might even be of use to me some day. I am glad you spent a pleasant X.mas, your weather was something like ours. Fred had a good holiday, I suppose Will likes his in the summer best. I hope J.T.S. will spend Easter Sunday with you, I expect we shall be at Sydney. I was not *much* surprised to hear of Mrs. Daddo's death, but it is a very sad affair, such a young family too. I am writing Mr. Daddo Sr. this mail, but of course not a newsy letter, chiefly about Australian friends. I am sure he would like to read about the trip down south &c, so I have told him you will send this letter for him to read in a week or so, & after the boys have overhauled it.

Edward Daddo was very kind to me on shore here & very much pressed me to go home with him, but they live 200 miles up the country by railway, & I could not get leave for any length of time; besides I know pretty well what Australia is like (up the country). I shall tell you more about Melbourne in my next letter. I can read your letters in the day almost, you will have to get specs to see to read my other letters. You never need send any of my letters to Swadlincote & Barrowden as I always write there. I have had my likeness taken & enclose you one to see if the cold has altered me at all. I also sent one to Charlie.

¶I also enclose you 3 aboriginal ladies. The natives of this great country are slowly dwindling away, like the North American Indians & the Maoris of New Zealand. The Negro is the only coloured race that makes any stand against the white man; when Captain Cook discovered Australia 100 years ago the natives might be counted by tens of thousands, & to day you could not muster a thousand in all Australia. The island of Tasmania had from 30 to 40,000 natives on it 100 years ago; the last native King "Billy" died when I was out here before. The Government of Victoria has collected what few natives there are left in the colony & formed them into a sort of settlement where they are partly civilized & encouraged to get on, but every year lessens their number & before many years have passed the aboriginals of Australia will belong only to the past. The Maoris of New Zealand are a handsome & quite distinct race from the natives of this country; and not being of such a desponding temperament seem inclined to make a stand against this declining tendency. The missionaries have done a good deal for them & will do more, but there is a sensible decrease in their number every year & unless they intermarry with the white race, their enduring existence & progress is doubtful. These natives here are the ugliest race of people there are almost; I saw a few of them when I was out here before (where I worked in the vineyard); the women were smoking & asked me for tobacco. There are few more degraded or inferior races than the natives of this country; unlike the North American Indians & Maoris, these natives gave the early settlers very little more trouble or concern than the kangaroos. They killed & stole the sheep but never attempted to dispute their land with the whites.

On Thursday last I went out to Brunswick to see Mr. & Mrs. Appleby, the best friends I ever had while I was out here. You know I told you they came out with us on the "Essex" & that I lived with them a fortnight. I used to have what I liked for dinner every day, & when I went on board the "Agamemnon" they gave me some clothes & pocket money. They have been out here 20 years, but went home once for 18 months. They were very poor when first they came out, but they have worked hard & made a lot of money. They have land & houses now, & live in a nice cottage of their own which I helped to build when I was out here. He is a sort of small contractor, carpenter & builder, & he still works. Earns £6 & £7 a week. In 2 years time they are going to retire & come to England &

enjoy themselves. He is no scholar & they are both rough & ready sort; they talk of going to France & Italy & all parts of England when they come back. They belong some where near Ely; they have no children & not many relations, & they say that if they find themselves near Oakham any time they should call & enquire how I was getting on. Of course I told them that Father & you would be glad to see them, but she says "I don't mean to visit them any day, only to have a little chat with your Mother for half an hour." I had to take all my family likenesses out for them to see, & she said we were a fine healthy looking family. Mr. A. thought Miss Wildman the finest looking young woman he ever saw (except his wife) & he thought I was courting her. Mrs. A. thought Will looked older than me, & she thought J. T. Swann a wild looking youth, a mischievous sort of young fellow. Olivers where I used to work in Melbourne are gone, turned bankrupt.

March 25th

I went out to Brunswick again yesterday & slept there, so that it does not cost me much going on shore. As we don't go to sea before Monday, for Sydney, I am to go out again on Sunday for dinner & Tea; & Mrs. A. is going to make me a plum & apricot pudding because I used to be so fond of them. There were hundreds of visitors on board yesterday, & Mrs. A. came on board to see me, but I had gone to see her, so some of the sailors took her all over the ship; she was pleased & they are never tired of hearing about the voyage &c. She has one of my likenesses of course, she says they are like me, but too dark. I don't think them as good as the others. The mail goes in the morning, so will conclude with very best love to all, & hoping Father gets better.

From your Affec'te Son,
Joseph

P.S. After you get this until the middle of July, address Singapore, China, via Southampton; they will be sent on to us about the end of August.[5]

The concurrent letter to his cousin, much shorter, told of the same events, but as usual the tone was more playful.

Dear Tom,

The mail came on board the same evening we anchored, and brought me 7 letters, 8 newspapers &c. I enclose you a splendid specimen of the aboriginal ladies of this country, who neither wears chignon or bustle as you'll see, still she has good pi'nts about her, & in summer her dress is even less expensive than at present, as it is the fall of the year. However I should like the ladies opinion before I commit myself, & its just possible, I may be able to find a handsomer sort at New Zealand or the South Sea Islands, and one that a'int so fond of dress. The first news we heard on arrival here, was that Tichborne had 14 years, & think he has been dealt to[o] leniently with, many better men than he have been hung for much smaller offences.   *   *   *

The English cricketers[6] went home the day before we arrived, they were beaten on one or two occasions, & I believe quarrelled between themselves on money matters. I was very sorry to hear of Mr. Blackwood's death, my last letter from him I received at Lisbon. Soon after we anchored 2 gentlemen came on board to see me; one an old schoolfellow at Oakham, he is a Draper, & lives 200 miles up the country, had come down to buy goods, & wanted me to return with him for a few days, which I could not manage, but spent 2 days with him, and a friend of his at Melbourne, who was a Danish gentleman, & had a great dislike to Germans; but admired the English, because they were beaten at Copenhagen, under Nelson, and on other occasions. We went about Melbourne a good deal, & I visited all my old favorite haunts & walks; it has increased wonderfully but must tell you more about it in my next letter.   *   *   *[7]

With the opening letter of this chapter, Matkin had enclosed a £1 postal money order for his father. The elder Charles Matkin returned this favor with a letter of thanks, which Matkin retained. It is the only known surviving correspondence from his father, the only surviving letter from anyone *to* Matkin during the *Challenger* period, and probably the last communication from his father prior to the latter's death. It is reproduced here, although Matkin would have received it several months later.

Oakham
May 29, 1874

Dear Joe,

I beg with thanks to acknowledge the P.O. Order for £1 which came to hand all right. We sent Fred his 5/ and Willie 5/. Your mother had 5/ and myself 5/. I expect Uncle Joe from Stamford to-morrow (Saturday) for a bit of fishing at the Canal and spend Sunday with us. We have been busy getting our bedroom painted and papered and also Ann's bedroom. Also 2 new bedsteads from Royces for your room, and sold the old ones to Smith the broker.

I saw accounts of the Challenger in different papers very similar in their description to yours. D^r. Wood has had your long letter to read, and very pleased and interested with it. It took me a long time to get through it, as my eyesight is not very good, and [the] letter being crossed, rather bothered me; but I managed to get through it at last, and a very interesting letter it was.

I don't get rid of my pain at the chest; sometimes better than other times.

I saw by this morning's paper that some of the Astrologers have started for different stations to view the *"Transit of Venus."* Xmas Harbour, Kerguelan is maned [named], where you have been.

Your Mother unites with me in love to you, & hope you are well, & will have a prosperous voyage.

> I remain,
> Your affect. Father,
> C. Matkin

P.S. My birthday to day.[8]

Matkin was not alone in finding Melbourne to be one of the most agreeable ports of call of the voyage. Quite a few of his mates fell to the attractions of Victoria and failed to return to the ship at the end of the two-week visit. Departing on 1 April, *Challenger* sailed up the coast to Sydney, a passage of six days including the Easter holiday. The ship would remain in Sydney for nearly two months. The crew (but not the officers) judged Sydney more English yet less hospitable than Melbourne.

> H.M.S "Challenger"
> Sydney
> April 10th, 1874

Dear Tom,

The English Mail arrived here to-day from Melbourne, & brought me only one letter from yourself, dated Feb^{ry} 6th. A mail leaves to morrow for England, via San Francisco, & the Pacific Railway, and think it will reach England within a week of the last, posted a fortnight ago.

¶We remained in Melbourne until April 1st, & arrived here on the 6th, the distance is 580 miles, but we were sounding & dredging as we came along. I enjoyed myself very much in Melbourne, spent 2 or 3 days on shore with my old friends at Brunswick. We had such numbers of visitors on board before we left; the Officers gave a dance party on the 28th; over a hundred ladies & gentlemen came off. Our men say Melbourne is the finest place in the world, they were so well treated by the people; the railway authorities gave them free passes for travelling up the country. Many went to Ballarat, the great gold mining metropolis,[9] & 6 of them never came back. I spent a fortnight there, when out here before, it is a fine town, over 90,000 inhabitants and is 100 miles from Melbourne. We read an account of the University Boat race in the Monday papers after the event.

¶On the 30th 2 seamen tried to escape from the ship to the shore in one of the boats, but they were re-captured & are now in irons waiting court martial. On the 31st, I had to go up with the Paymaster to Melbourne to fetch £800 from one of the Banks to pay the Ship's company.

¶We left on the 1st of April, & sailed & steamed round; we sighted the coast nearly all the way and passed several small islands. The coast is moderately high, & wholly covered in forest, very unlike the African coast. We passed several ships & sounded once within 15 miles of the land, at a depth of 2,375 fths; 5 miles nearer the depth was 100 fths. This was our first sounding in the Pacific Ocean. We spent our Good Friday & Easter Sunday at sea, so you may be sure we were not over comfortable. The German Frigate "Arcona" came round with us, & arrived a few hours earlier.

¶We entered Sydney Heads at 1 PM Easter Monday, & at 2 PM dropped anchor in Farm Cove, close to the "Arcona", and H.M.S. "Dido". It is 9 miles from the Heads to Sydney, & the scenery down the bay is magnificent, & the harbour of Sydney is justly considered one of the finest in the world; there are so many Bays and Creeks branching off from Botany Bay, which afford fine shelter for shipping. There are several small islands close to the town, in one of them there is a large suburb like New York & Brooklyn, & steamers cross every few minutes. We are lying nearly a mile from Sydney & can't see much of it, but are within a stone's throw of the public park in a fine bay reserved for the anchorage of Men of War only. When we came in on Easter Monday the Park was crowded with people holiday making, picnicing &c. The ladies saluted us bravely with their handkerchiefs, & we were soon surrounded with pleasure boats. The houses here are older than at Melbourne, & the place has a more English appearance, some parts of the harbour are very similar to that at Plymouth. This bay is not one fourth as large as Hobson's Bay, but much better sheltered being perfectly land locked, we are unable to see the sea from where we are lying, but there is a fine view of the Pacific from the high lands of the town. The shores of the bay are covered with forest & park, with residences of the gentry peeping pleasantly out between the trees.

¶There are some splendid merchant ships here; the famous China Clipper "Thermopylea" is at anchor, she ran from London to Melbourne in 58 days, and in one 24 hours ran 420 miles; but the finest and largest merchant vessel we can see is a new one, the "Samuel Plimsole" from London. I have not been on shore yet, but am going for 2 or 3 days next week, all our men who have been say that it is inferior to Melbourne. One watch is now on 4 days leave, and returns on Monday next, & on Tuesday the ship goes in dry dock at Cockatoo island for some time, & expect we shall lose more men during our stay. The weather is very hot although winter is coming on, the Thermometer standing at 80° in the shade, but in the summer months Sydney is often cooler than Melbourne & Adelaide. I am anxiously looking out for English papers containing the termination of the Tichborne trial; the people here always thought him an imposter & rogue.

You will receive this by the end of May, & if you write at once, and address Sydney, it will be forwarded to us at the Fiji islands. Put (elsewhere) on it, & its bound to reach us sometime. One of our men has his letters addressed "Challenger" voyage round the world (or elsewhere). After the middle of June until the latter end of July, address Singapore, & it will reach us at Somerset, Cape York, by the end of August. After leaving New Zealand, the Fijis, New Caledonia &c, &c., we stand west towards Torres Straits, and New Guinea, calling at Cape York, the most northerly point of Australia; further particulars in my next.[10]

The two-month layover in Sydney gave Matkin ample time for correspondence. The first of the surviving letters to his youngest brother, Fred, now seventeen years old, was begun on 13 April. It lays out (as Matkin understood it) the ship's itinerary for the coming months in greater detail and with a tone of positive anticipation.

Dear Fred,

I wrote to Willie last Saturday to tell him I had not heard from him for a very long time but the same afternoon I received a letter from him & also one from Charlie that had been delayed at the

Post Office. As for you I know you have not had time to write to me before owing to the great amount of business responsibility on your shoulders, but you must manage to write occasionally. I suppose you are nearly 5 feet high now, you must send me your portrait when you have it taken, then I shall have the family complete —the first & second generations. I enclose you one of mine taken in Melbourne, but they are not as good as the Hull ones.

¶Of course Mother sent you the cold letter to read so I need not tell you any more about icebergs or sea-elephants; we are in rather different weather now. Thermometer at 80° in the shade & we shall have no more cold weather until we leave Japan in March 1875 for Kamchatka. This next year's cruise will be the pleasantest of the lot I fancy for we are going to nearly all the large groups of the South sea islands besides New Guinea, Borneo, Timor, Moluccas, & Phillipine islands before going to Hong Kong & Japan. Every week will see us at different islands & among different people so that I shall be able to fill up my "Journal" round those pretty well.

¶I hope to pick up a few curiosities among the natives. I should have liked to have got a Seal skin at Kerguelen's island but the Officers always took care to go ashore first & kill what few there were. The Paymaster has 6 but he got 4 of them from the Whale Ships, gave them Rum & Tobacco for them, & now they are properly dressed he says they are worth 20 Guineas apiece. Perhaps I may get hold of one or two up Siberia way, the Russian Tartars will give any thing for Spirit.

¶We are going to the French penal settlement of New Caledonia[11] after leaving the Fi Ji islands &~Solomon~islands. Some of the French political prisoners escaped from there a few days ago & landed here. We are also going to New Britain & New Ireland— islands concerning which very little is known. The natives are half savage, & something like the Maories of New Zealand; our Photographer takes portraits of the natives of all the islands, & I believe we are to bring a few natives of New Guinea home with us. There is also a rumour that that island is to be taken possession of by our Captain & declared an English possession,[12] as Gold has recently been discovered there. We go from here to Newcastle for coal on the 20th May. Newcastle is 90 miles further north. From there we go to Wellington & Auckland, both in the north island of New Zealand; & we shall be there until the end of June, after which we go to Norfolk island, New Caledonia, Fijis, New Hebrides, Santa Cruz islands, Solomon islands, New Britain & New Ireland. Thence thro' Torres straits to New Guinea, calling at Somerset,

Cape Yorke the most northerly point of Australia. We go to many smaller islands that I don't know the names of; & nearly all are inhabited & fertile. You can trace our course on the map to New Guinea, & before we get there you will hear from me again—most likely from Kandavau—Fiji islands.

How would you like some piebald Matkins if I get married & settle at any of the South sea islands, I'll send you one home in a band box, addressed to *Uncle Cash*. I need not tell you much about Melbourne for you have heard me talk of it many times. It is the finest built town in the world & increasing fast, there are nearly 300,000 inhabitants there; it is twice as large as Hull & of course better built & laid out as it is only 37 years old. Sydney is 90 years old, but not half the place Melbourne is, tho' the Sydney people think it is. The discovery of Gold caused Melbourne to shoot up so; there are no important diggings in New South Wales. Sydney harbour is the

Challenger's steam pinnace in Sydney Harbor, 1874. Reproduced by permission of the Master and Fellows of Christ's College, Cambridge. Photo courtesy of the Natural History Museum, London.

best we have yet anchored in, & is one of the finest & prettiest harbours in the world.

April 15th

I was on shore yesterday & had a good look round the place; altho' it is not such a fine built town as Melbourne it is nearly as large & more like England; the streets are narrower & dirtier but there are some very fine buildings in Sydney, the Town Hall & Cathedral are the best & equal to any in Melbourne. The Cathedral is Protestant & the prettiest I have seen out of England, the bells were ringing last night & I thought they sounded more like Oakham bells than any I have heard elsewhere. I intend going to the Cathedral on Sunday for I am going on 3 days' leave on Friday morning. Friday I am going to spend with Mr. Sleath's brother out at Cook's river, 6 miles from Sydney. Saturday I am going to Parramatta 15 miles down the river; it is one of the prettiest places in Australia & where all the oranges grow; they are just coming in season here. Sunday I spend in Sydney, & I am to call & see Mr. Sleath's eldest daughter who is in business in Sydney, she is about 18 & nice looking.

¶I went out to West Botany when I was on shore, to see Mr. Sleath. He is a farmer & keeps a hotel & has been out here 20 years. He was very pleased to see me for during the whole 20 years he has lived here he has never met any one from Oakham. He is very well off but does not care to go back to England again, his first wife has been dead a long time, & left him 3 or 4 children, he is married again & his second wife has children I think. He is going to show me round Sydney & the neighbourhood on Friday, & he is coming on board next week. From his house you have a fine view of Botany Bay & the Pacific Ocean. Botany Bay is where the convicts used to be sent & is a fine piece of water, but it is shallow & has a bar of sand running across it. Had it not been for this Botany bay would have been the port & harbour of Sydney instead of Port Jackson, 10 miles farther north. Captain Cook landed in Botany Bay 100 years ago, when he first landed in Australia, a stone monument marks the supposed landing place; he called it Botany Bay on account of the fine botanical specimens he found on its shores. The Dutch, Spaniards & French discovered Australia, Tasmania, & New Zealand before we did but Cook was the first to explore them; a boat's crew belonging to him were eaten by the natives of New Zealand, in the bay of Wellington. There are no convicts here now, they are all at Freemantle, West Australia, & no more will be sent from England. Australia is nearly as large as-large as all Europe, & the 3 islands of New Zealand are as large as Great Britain.

¶This letter leaves on Friday but will not reach you before about the 9th June, you can write before the 15th July & address Singapore, or elsewhere, & it will reach me at Cape Yorke, North Australia, the end of Aug. I am writing to Mother, but you can send this to Charlie & Willie so that they may know I received theirs all serene & I shall write to them either from here or New Zealand. I hope Uncle Joe's shoulder is better again, tell him I have taken all my old Mercuries[13] out to Mr. Sleath's brother at Cook's river. Also give my love to him, & to Aunt & Cousin Lucy, & accept the same.

From Your Affec^{te} Brother,
Joe

P.S. Remember me to Lowell & to Seaton when you next see him. Tell me how you are getting on, & don't forget your music.[14]

Three weeks later and still in Sydney, Matkin began another round of correspondence in response to four letters received via San Francisco. To his mother he focused on the themes of shipboard living and working conditions, morale, desertion, and his future career. The accommodations aboard *Challenger* were not up to the standard of his previous ships, and certain personalities among the officers and crew only made matters worse. Crew members were less than enthusiastic, it seems, about Professor Wyville Thomson's popularization efforts in the Edinburgh magazine *Good Words*.

. . . I shall write to Miss W[ildman] by next mail from here, if I have not departed on the Wallaby track again. We have lost nearly 30 since we came to Australia, & I am certain that if this had been any other ship than the Challenger, I should have gone too. But as it is, I intend to go round—as our men say—for I shall never have such another opportunity of seeing the world. Still I am quite sure I shall never remain 20 years in the Navy, but after a few more years & a trip up the Mediterranean I shall settle down as Will's & Fred's foreman.[15] Another mail via Suez comes in to morrow & will bring me some newspapers & perhaps a letter from Walter Thornton.

I was much interested in what you told me of Mrs. Blackwood, if she should think of the book you must save it for me, I shall prize it very much indeed. I am writing to Mr. Sleath by this mail, concerning his brother &c. I am going out to Cook's river to morrow & I expect Mr. S. on board next week. They are very kind & pleased to see me always. Miss Augusta Sleath, the second daughter I am quite partial to, the eldest I don't care much about; she is not as good looking & inclined to affectation having had a year or so at a boarding school; her sister is a regular little housekeeper & her Father's favorite. I am going to take her to morrow a fire screen made of gold & silver everlasting flowers & grasses—gathered from the mountains at the Cape of Good Hope. I should have brought you one, but they take up a lot of room, & I have nowhere to keep it for 2 years. I hear that after we leave New Zealand until we arrive at Hong Kong in Nov'r they will give us articles to barter with the natives of the Polynesian islands, instead of money; so of course I shall pick up some things. I have spent a lot of money in clothes here & am afraid that instead of the £40 I talked of bringing home the sum will be nearer 4. Those men who had money in the Bank have taken it all out here & talk of starting afresh after leaving New Zealand. On your birthday we were in the thickest of the ice. You can't call yourself an old woman until you are as old as Grannie & have as many grandchildren.

The Queen's birthday is a great holiday out here & the Captain will issue Wine to all hands; I shall drink Father's health before the Queen's. I wrote to Walter T[hornton] last mail & will write again shortly. I sent Charlie a Sydney Illustrated Paper last mail with instructions outside to forward it to Oakham. If he has done so let Mr. Sleath see it. This week's paper contains 2 pictures of our ship at Kerguelen's Land, if I can get you one I will. I must write to Charlie next mail for my last was a very short one. I was pleased to hear from Fred, he improves in his writing, but he tells me very little about Uncle or Aunt Lucy. I hope Uncle has got all right again.

¶We have been in dry dock for a week but the ship's bottom wanted very little doing to it. We have taken in 230 tons of Australian coal & at present are busy taking in 6 months provisions to last us to Hong Kong next Nov. We leave here on the 27th May for New Zealand, we shall lose more men before then & at New Zealand we shall lose some. Two men have gone to Gaol for 3 months for trying to desert at Melbourne & taking with them one of the ship's boats. Two more are waiting court martial for deserting here, & 2 have been invalided. They are not the worst hands either that have gone since we came here & the Captain is in a great way about it. They are treating the men better now in hopes of deterring others from going but the men only laugh. We have some great bullies & snobs amongst the officers & the work is much harder for everyone than it is in an ordinary man of war whilst the pay is the same. There is not half the comfort in this ship that there was in the "Invincible" or "Audacious". Our Issuing Room[16] is a dogs hole compared to what it was there, it is down below the water line & I have to do all my writing by candle light, going up to breathe about every 2 hours. The Steward & I get on very well together but we don't like the Paymaster, he is not to be compared to my last one. He is a Plymouth man & all his family are in the Navy & regularly eaten up with it. He is all there in such places as Portsmouth, Plymouth, Gibraltar, Bermuda &c, but he don't care for Melbourne or Sydney where the people take no more notice of him than they do of any one else. At present the officers are making preparations for a grand Ball they are to give on shore this week to the people of Sydney. Two or three of our Bandsmen are among the deserters so the band is rather demoralized just at present.

¶This is the rainy season here just now & it comes down about every day. It rained in torrents the whole of Saturday evening & Sunday, 36 hours without ceasing one minute. There is a fine Free Public Library here where I generally spend my evenings when on shore. I went to the Cathedral on Sunday morning; it is a very nice one, all the seats are free but there are collections every Sunday. The organ is a fair one, cost £1500, but the singing is not as good as it is in England, or at the Catholic Churches out here. A party of the scientific gents with some officers & seamen went to Brisbane the other day for a three week's tour into the interior of Queensland. Brisbane is 500 miles farther north, & Cape Yorke is 1200 miles farther north. There are plenty of Alligators, snakes &c up there.

Going across to New Zealand we are to make a line of soundings for a cable to be laid from Sydney to Wellington. Our ship is full of mosquitos just now, some of the men can't get a wink of sleep for them & are covered with great lumps, but they don't fancy me at all like they did when I was out here before. I have just read the conclusion of the Tichborne trial; do you read Professor Thompson's letters in Good Words[17]—I don't think much to them, neither does any one amongst the ship's company. This mail leaves on the 9th May via San Francisco & another via Suez will leave in a week or so. Remember me to all enquirers & with best love to everyone, & hoping Father will improve this summer.

From Your Affec'te Son,
Joseph Matkin

P.S. May 9th

The English mail via Suez came in yesterday but brought me no letters, only 2 newspapers, one from J.T.S. & a Grantham announcing Mr. Palmer's death. I am much surprised & hope to hear more about it next mail. Good bye.[18]

The following week Matkin posted letters, the last from Sydney, to his brother Charlie and to his cousin. The comparison of Sydney with Melbourne continues, and the forthcoming movements of *Challenger* are updated.

H.M. Ship "Challenger"

Sydney, May 15th, [18]74

Dear Charlie,

A mail leaves for England this morning, & it will be the last homeward mail while we are here. Your letter came to hand all right but it was posted a day too late for the mail & only reached me at Sydney. I am very pleased with Harriett's figure head,[19] & must get a nice Album at Japan. Tell her I shall bring her a wedding present from Japan, if you had not kept the courting &c so quiet, I might have sent something from the Cape.

¶I sent you an Illustrated Sydney News, with instructions on the outside cover to forward it on to Oakham for them to see, but of course they would send it back to you again. It will give you some idea of what Sydney is & also of the beautiful harbor. Melbourne is half as large again as Sydney & a much finer place, more business, traffic &c. Here the streets are narrower, dirtier, & more like England. There is a splendid Cathedral here tho', & I often go on a Sunday morning. The Town Hall, new Post Office, Museum &c are all fine buildings. The climate is not so healthy as it is in Victoria, it is hotter & damper. Work of all sorts is plentiful & wages good, but I don't think Printers are so well paid as most other trades; they generally work piece work.

¶I was very sorry to hear of Mr. Blackwood's death, & by the mail that arrived here last Friday I heard of Mr. Palmer's sudden death—from the Grantham Journal. I had 4 letters last week, 2 from Oakham, 1 from Swadlincote, & 1 from Fred, Stamford. I never hear from George Barton now; is he still at Hull? Fred never tells me anything about Uncle or Aunt in his letters from Stamford. Just tell him next time you write to him.

¶I have been on shore here several times & often go out to Cook's river—6 miles from Sydney to see Mr. Sleath's brother who came from Oakham 20 years ago. He has 4 children & has just married again.

It is the rainy season or winter here at present & it often rains 24 hours without ceasing. Oranges are just ripe here now, they grow down at Parramatta, 16 miles up the harbour. I am going there next week, it is a very pretty place. It is as warm here now as it is in England in summer, but the evenings & early mornings are damp & chilly. We have any amount of visitors on board here & the officers are invited to balls parties &c, everywhere. Our ship's company are invited to a Tea meeting & Temperance Lecture this evening on shore. I expect it will be rare fun, any amount will sign the pledge & come off drunk tomorrow morning.

Two of our men were tried by court-martial yesterday for deserting & sentenced to 12 month's imprisonment. We have 5 men in gaol & have lost 22 by desertion since we arrived here, & I dare say we shall lose more at New Zealand. They have nearly all drawn their money out of the Ship's Saving's Bank & talk of making a fresh start after we leave New Zealand. There are 4 men of war here now besides several little gun-boats. The "Pearl" is the Flag ship, there is also the "Dido", our ship & the French man of war "Cher" from New Caledonia—about some communist prisoners[20] who escaped from there & landed here a few weeks ago. One of the French sailors was drowned last night,—was drunk & fell off the pier.

Wigram's fine steam ship Northumberland arrived at Melbourne last week in 51 days 18 hours from Plymouth. This is faster than the mail boats do it thro' the Suez canal. We have taken in 230 tons of coal & filled up with provisions &c to last us to Hong Kong next November. We expect to leave here on the 28th May[21] & shall be 14 days going to New Zealand as we are to make a line of soundings across for a Cable to be laid this year between Sydney & Auckland. The distance is 1100 miles. We shall not stay long at Wellington or Auckland, probably leave Auckland for the Fiji Islands about 5th July. We shall be all among the south sea islands by the time you get this. I hear they are going to give us articles to barter with the natives—instead of wages, from New Zealand to Manilla. We have taken in a lot of old iron hoops &c to trade with the natives of New Guinea. We are to spend some considerable time round New Guinea, the Solomon islands, New Britain, New Ireland &c. We shall be some time at Hong Kong & 3 months round Japan, where we are to arrive March 1875. We also call at Cape Yorke the most northerly point of this country—1200 miles north of Brisbane & we are to receive letters there from Singapore, about the middle of September. Letters should be posted to Singapore via Southtn by the mail leaving England on the 30th July, or via Brindisi to the 7th August. Via Brindisi requires a fourpenny stamp for us. Of course you can write before that time if you like as mails leave England for Singapore every 14 days. You can tell Will & Fred this. I wrote to

Mother last Saturday—via San Francisco but I think you will get this first.

Write soon & tell me how you are getting on &c. With best love to Harriett & yourself

From Your affecte Brother
Joseph Matkin[22]

Concluding observations on Australia, ship movements in Sydney harbor, and anticipations of the coming voyage dominate the concurrent letter to his cousin.

* * *   There are 3 men of war here now: our ship, the "Dido" & the "Pearl",—Flag ship of Commodore Goodenough, which has just come from the Fiji islands, now under British protection. The French Man of War, the "Cher", is also close to us, she came in on Sunday morning from New Caledonia to make enquiries concerning Rochefort,[23] & the other 3 communist prisoners who escaped from there a few weeks ago, and landed here. Our Officers gave a grand Ball on shore the other night to the nobility &c of Sydney, & to-morrow our ship's company are invited to a Tea meeting, and Temperance Lecture, at the Masonic Hall on shore. The Tea meeting will be a great success, but the Lecture wont allure many towards the Teetotal brethren. * * *

The children of Australia are not such strong healthy looking children as the English, but the colony of Victoria is healthier than New South Wales; they are very fast children generally, much like the American juveniles, but speak better English here, altho' they call a cow, a Keow.

¶We were only 5 days in dry dock at Cockatoo island, as the ship's bottom was in very good condition; the ice did us but little damage, and when a ship sails through water of such varying temperatures no *barnacles* or other secretions can accumulate on her bottom. An English barque arrived here the other day with the Captain's wife lying dead on board; having died the day before the ship came in—in childbirth.

Professor Thompson, & some of the scientific party are gone to Brisbane for a three weeks tour into the interior of Queensland. The rainy season or winter has commenced here, but it is quite as hot during the day, as summer in England. We have had some magnificent sunsets here for the last month, and for rainbows I never saw a place like Sydney. A fine iron vessel called the "City of Benares," sailed hence yesterday for London, via Cape Town, but should not care to have taken the voyage in her, as she was laden deeper than any ship I ever saw, & it is the worst time of the year for doubling the Horn. The fine ship "Samuel Plimsole", sailed for London a fortnight ago; she is supposed to be a faster vessel than the "Themopyla".

¶ * * * Our Officers have been fêted a great deal since we came here, but the men have not been so well treated as they were at Melbourne. We have had the Governor, & any amount of visitors on board; but we shall all be glad to get to sea again for a change. We shall have been nearly 8 weeks here before we leave for New Zealand, where we do not stay long.

The time slips away very quickly with us; it seems as if we shall be in England again, in almost less than no time. It takes us a rare lot of cash, which one is unable to see the wonders of the world without spending, & some more will have to fly in purchasing curiosities at Japan. Most of our men have drawn their money out of the ship's saving bank since we arrived here, & talk of commencing afresh after leaving New Zealand.[24] * * *

The crossing to New Zealand was a particularly stormy one, making oceanographic operations in the Tasman Sea alternately difficult, impossible, or fruitless—and costly to both life and line. While under way Matkin began a letter to his cousin, later adding details of New Zealand's history, geography, demography, and ethnography.

> H.M.S. "Challenger"
> South Pacific
> June 21st, 1874

Dear Tom,

* * * We left Sydney at 4 PM on the 8th June, for Wellington distant 1180 miles, S. by E. & we were to take a careful line of soundings across; to sound every 2 hours in the shallow water, for the cable that is to be laid from Sydney to New Zealand next year. We are still 25 men short of complement, altho' we shipped 8 new hands as stokers, cooks, &c, and took those out of prison that were sentenced to 3 months for deserting. There was quite a crowd of people on the wharves & shores of the harbour to see us depart, & as we steamed past the "Dido", & "Pearl", their crews mounted the rigging & cheered us while their Bands played "Auld Lang Syne." Our men returned the cheers, & our Band gave 'em "the Girl I left behind me." We had been 9 weeks in Sydney, & were getting tired of it.

Until March [18]75, our address will be Singapore, the letters to be forwarded from there to wherever we are; Mails leave England every fortnight, so you can't make much mistake. We do not expect to be home before July [18]76, as we are a month behind time already. The Melbourne Illustrated paper for June 13th had a picture of our ship at Kerguelen's Land, & also amongst the ice. They were taken from our Photographs; but have not been able to get you any of them, but daresay you will see the same pictures in the London papers.

¶On the 9th of June we sounded in sight of Sydney Head at 50 faths. & 75 faths. but a gale sprang up that night which stayed any further sounding & obliged us to run back & anchor in Watson's Bay for 2 days until the weather moderated. On the 12th we started again & sounded several times from 70 to 900 faths.; on the 13th sounded from 400 to 700, & ran back to the land to verify the line of soundings.

¶On the 14th (Sunday) it blew very hard, & was too rough for Church, but not for sounding the Captain thought; so he sounded at 2,275 faths., about 2-¾ miles, the sea had deepened nearly 2 miles in that run of 40 miles. In hauling in the line 1,500 faths. were carried away, which the men attributed to sounding on the Sabbath, & not having had any surveying Wine lately. The same night it came on a regular gale, the upper yards were sent down & everything prepared for it; the cutter was stove in, & the patent life buoy carried away by a heavy sea in the night, & a good deal of the men's crockery broken.

This life buoy is a new patent, there is a place for the man to put his feet in, if he understands it & can reach it, & a receptacle for 7 days provisions, & on being let go by means of a common bell handle & wire, a magnesium light ignites & burns for 20 minutes, water not affecting it. We had only that one, though we have several old fashioned buoys.

¶On the 15th it was rough, but on the 16th we sounded at 2,550 faths. & carried away 2,500 in hauling in; which line costs 9d per fath. On the 17th the depth was 2,700 faths.—over 3 miles, and we obtained a few living specimens in the dredge. On the 18th & 19th it was a dead calm, & we had to steam—depth 2,600 faths. To day, Sunday, we had no church again, as it was too rough, but we sounded at 1,975 faths—nearly a mile shallower since yesterday. It is now blowing hard, & the Barometer falling.

Saturday night, June 27th, anchored under Long Island, Cook's Straits

Last Sunday night the gale increased, and the fore top mast staysail was carried away. The ship stood it first rate, it was a frightful night, every now & then a heavy sea would strike the ship, & make her quiver again. On the 22nd it cleared a little, & we sounded at 1100 faths, a difference of 2 miles, in a run of about 80. On the 23rd we sailed 141 miles, and sounded at 275 faths. only, (another mile shallower). We were then 240 miles from the nearest land, Cape Farewell, N. Zealand. From there to C. Farewell, the depth varied from 450 to 45 faths. At 1 PM 24th, we sighted Mount Egmont, 65 miles distant. It is 8,200 feet high, and of course was covered in snow, as it is mid-winter here; still there is no snow on hills under 3000 feet high. The wind is frightfully cutting, & we are all wearing our winter clothing.

There are several mountains in N. Zealand over 8000 ft high; the highest Mount Cook, is 13,200 feet above the sea level, & is a fine land mark for ships. The 2 large, & 5 or 6 small islands of New Zealand are nearly as large as Great Britain & Ireland together, & most of the land is cultivable. There are only about 8 towns larger than Oakham, the 4 largest are Dunedin & Christchurch on the south island, & Wellington & Auckland on the North. Dunedin is the youngest & most go ahead place, there are some rich gold diggings there about. Christchurch was formerly the Capital, and a nice English sort of town I am told; it gave place to Auckland, which although the largest place was superseded by Wellington on account of its more central position, which is at the south point of the North Island; and Auckland is 580 miles north of it by water, & Dunedin 630 miles.

¶There are under 300,000 white people on N.Z. as yet so there is plenty of room for emigrants; there are 40,000 Maories partly civilized left now, and are all on the North island except 1000, who are scattered about the South. They speak the same language as the South sea islanders, but were intermixed ages ago with the Papuans of New Guinea, who are supposed to have come from some of the Polynesian islands about the year 1550. This supposition is borne out by their own traditions and war songs, which relate how they sailed here in their canoes, & conquered & exterminated the original Natives, with the names of their Chiefs. The original natives were probably of the same race as the aboriginals of Australia. The Maories are the finest & bravest race of savages in the world, & are good seamen; altho' they were cannibals they lived in houses, & tilled the ground when Cook visited & explored the island in 1776.

On the 25th June, at 1 PM we were close into the land, & as it came on to blow, & we could not get thro' Cook's Straits, the ship was anchored in Annesly Bay[25] on the South island. The coast was mountainous, & something like Kerguelen's, especially coming through the Straits, except the hills being covered in forest. We had to remain all day on the 26th, caught lots of fish, & got plenty of fine shell fish from the rocks.

This morning at 3 AM the anchor was got up, & we steamed out of the Bay for Wellington, distant 45 miles. We found the Wind & sea as rough as ever on getting outside & the wind was right in our teeth. In 8 hours we only steamed 20 miles, and were obliged to run under the lee of Long island, & anchor until the weather moderated. We passed lots of islets & rocks to day, & the coast looks frightfully bleak & wild; the mountains on each side are covered in snow. The Straits are about 20 miles wide just here, & there is a fearful sea running. The Captain is in a great hurry to reach Well.ᵗⁿ, but the men are not while this weather lasts, and until further orders would rather lie here nice & snug catching fish. When we left the Cape on our long voyage we had 280 tons of coal, & had 10 tons left when we reached Melbourne. On leaving Sydney we had 220 tons, & have burnt it all but 10, so what with the coal, & the sounding line carried away, this will be a very expensive trip.

Sunday night, June 28th, at Wellington

The passage is over at last, & so is the long voyage for one poor fellow who was drowned at 12 o'clock to-day, while every one else was dining in comparative comfort; drowned in broad daylight, whilst performing his duty; & not one of us to see him go, or to throw him a life-buoy; the ship went steaming on for Wellington, & he to his last long home.

Altho' it was blowing as hard as ever this morning, & a frightful sea running, the Captain had the anchor got up, & started full speed on the last 25 miles of the journey. We had no church, as it was too rough; the ship was cutting an awful caper but still riding like a duck on the water; the fore & aft sails were set, & as the water is shallow in the straits, a leadsman was in the chains sounding. At 10 o'clock we passed a dismantled ship at anchor under the lee of the land, and at 12 o'clock they piped dinner. The Officers then go to Lunch, & all the rest of the ship's company to dinner, except the Lieut. in charge of her, the 4 men steering, the leadsman, & the boatswain's Mate. The leadsman stands on a platform outside the ship to heave the lead. Just after the crew went to dinner, the man sounding got his lead & line fouled round the anchor, and was seen to climb up to clear it by a marine. Just at that time a frightful sea struck & broke over the fore part of the ship, shaking her from stem to stern. It capsized lots of plates & basins on the mess deck, but only caused a general laugh; one facetious fellow said, "Who's that knocking at the door," & another told the sea to "Come in", that was all that occurred there, but on the upper deck the sea had come in, & gone out again, taking with it the poor leadsman from off the anchor, in all his thick winter clothing, sea boots, & oilskins. Then the ship righted herself again, & went ploughing away for some 10 minutes; at the expiration of that time, the Marine who saw the man go out to clear his line, happened to look out of the port, & saw the line still round the anchor, but no leadsman there, when he ran & told the Lieut. of the Watch, who at once informed the Captain & Commander; the Captain telegraphed the engine room to stop, & all hands were called to put the ship about, and in 2 minutes she was steaming fast back over the original track. Some of the Officers went aloft with glasses, & nearly every soul in the ship was in the rigging looking out for some sign of him. After a few minutes some dark object was seen right ahead, & about a dozen sang out, "there he is, right ahead", but on steaming up it proved to be only seaweed, & nothing more of the poor fellow was seen. The ship then stood on again for Wellington, & we dropped anchor half an hour ago, about half a mile from the town. The deceased "Edwd Winton" was one of the finest & steadiest seamen in the ship, was about 25 years old, & married just before we left England. He was an excellent swimmer, but had on an enormous lot of clothes, & no one knows whether he was stunned against the ship's side by the sea, or he might have been struck by the fans of the screw as he drifted under the stern. No boat could have lived in the sea that was running, but the ship could have steamed up close enough to throw him a rope, had he been seen. No blame attaches to any one.

June 30th
We have just made a subscription on the lower deck for the widow of the poor fellow, & £20 were obtained from the crew alone, & I daresay the Officers will raise another £30. There were no letters here for us.

The bay of Well.^{tn} is something like Table bay, & the place is not unlike Cape Town. It is about as large as Stamford, but the houses are mostly of wood, & it is as dirty as Halifax. Meat is 3d per lb. and is better than the Australian; provisions are generally cheap, but rent & clothes are very dear. We are now coaling to last us to the

Fiji islands, & I hear we are not going to Auckland being so much behind time.

July 1st

The English mail has just arrived, & brought me 3 letters, two from yourself, the last dated May 4th & addressed Auckland, it has not been long on the way, & is the latest news I have from England.

We shall be at the Fiji islands about the 1st August, & shall write you from there; they are in the Tropics, the distance is 1250 miles, so we shall soon be in warm weather again. Any later letters you have written will be kept there until we arrive, or if after we have gone, they will be forwarded to Cape Yorke, Australia.

I have not been on shore yet, but am going on Friday; we *expect* to leave here on Monday the 6th; & will tell you more about the place in my next letter.

I have not yet told you of the loss of the "British Admiral" on King's Island close to Melbourne;[26] it has happened since we left there & 80 lives were lost.

Meat, potatoes, flour, & bread are cheap here but clothes, house rent, & drink are very dear. Wages are even better than in Australia, but are not so steady; a jobbing man wont make under 8/ per day, & mechanics &c. get 12/; still there are plenty of idlers, & loafers from England, America, & Australia, some are continually travelling about to the different Australian colonies, & never work for a week together.

After we leave the Fijis you will not get a letter for 2 months. We shall be dredging round the Malayan islands, and Polynesia, until March [18]76, when we go to Japan. The letters must be addressed Singapore, & the Consul or other authorities there, will forward them by private ships to wherever we are.

Yours faithfully,
Joe Matkin[27]

Challenger remained in Wellington barely more than a week. Matkin apparently wrote his mother, but the letter has not survived; hence the record of his New Zealand observations is slight. Upon reaching Tonga (Friendly Islands) in late July he was able to write at length to both his brother Will and his cousin, concluding both at Fiji. The former letter is reproduced here; substantive differences in the latter are again interpolated within curly brackets ({ }). Tonga seems to have made quite a good impression, Fiji only slightly less so. Amongst these exotic landfalls, Matkin all but forgot oceanography. Even the capture and observation of a live pearly nautilus near Fiji, mentioned in most other accounts, is passed over. The activities of the Wesleyan missionaries at Tonga he found especially creditable, however.

> H.M. Ship "Challenger"
> Tongatabu, Friendly Islands
> July 23d, 1874

Dear Willie,

I think it is your turn for a letter again, & as I shall not have time to write to Mother this time, you must send this on to Oakham as soon as you have read it. I am writing to Charlie & Fred also, so you need not send it to them. I suppose Mother would send you on my Wellington letter & I need not say anything about New Zealand. I had a run on shore at Wellington & saw some of the Maories, they are much like the natives of these islands. Owing to the many desertions we have had since we arrived in Australia we are 34 men short of our complement, but this number will be sent out to us at Hong Kong from England.

¶On the 6th July we left Welltn for these islands—distant 1250 miles north. We are in the tropics again now & it is very hot; we shall be in the tropics for the next 5 months; & we shall find it hotter still at New Guinea. We kept close in to the New Zealand coast for 3 days, sighting Poverty Bay & the East Cape, but as we were so much behind time we did not go to Auckland. Our greatest depth in the Pacific Ocean thus far, has been 2,925 fathoms,[28] & we obtained some fine specimens in the dredge & trawl net, large sponges and &c.

¶On the 10th we passed the 180th degree of longitude, & were exactly 12 hours ahead of you in England. If we had been homeward bound & gone round by Cape Horn we should have had 8 days in that week, but as we shall be steering west again shortly we shall lose 8 or 10 minutes every day for some time. We shall have our 8th day in the week between Japan & Vancouver Island. On the 14th we passed thro' the Kermadec group of islands & sounded off Sunday island, but did not land. They had a fertile appearance but I don't think there are any inhabitants.

¶Last Sunday morning, 19th, we sighted Eoa[29] the first of the Friendly islands & anchored at 1 PM at Tongatabu, [Tongatapu], 10 miles further on. This group of islands was discovered in 1642 by Tasman the Dutch navigator, but received their name from Captain Cook, who was the first to communicate with the natives, & called them the Friendly Islanders. There are about 160 islands of all sizes but only about 20 are inhabited. The total population is 25,000. Tongatabu is the largest island & residence of the King; it is

CAYORK TO HONGKONG

touching at the

ARROE Pt KT Pt BANDA Pt

AMBOINA I. TERNATE Pt SAMBOANGAN

ILO ILO AND MANILA

Sept.-Oct., Nov? 1874

also

HONGKONG TO YOKOHAMA

touching at

MANILA, ZEBU, SAMBOANGAN

HUMBOLDT BAY AND THE ADMIRALTY Pt

Jan? Feb? March, April 1875

For explanation of abbreviations for see Appendix I.

CHINA

CHINA SEA

Map of *Challenger's* track, New Zealand to Japan; from T. H. Tizard, H. N. Moseley, J. Y. Buchanan, and John Murray, *Narrative of the Cruise of H.M.S. Challenger with a General Account of the Scientific Results of the Expedition*, 2 vols. in 3 (London: HMSO, 1885), vol. 1, pt. 2, p. 542.

about 10 miles long & 5 broad & has 5000 inhabitants, besides about 40 white people, carpenters, coopers, small export traders, and missionaries. It is about as large as St Thomas, West Indies, but much more level. There are about 50 small islands in sight from here; some of them are not as large as a small flower garden & scarcely out of the water, but they are all covered with cocoa nut-palms, & vegetation down to the very waters edge. They look like little gardens afloat, & as no one lives on these small ones it was suggested by one of our philosophic seamen, that the Captain should serve out an island to each man, & after bringing him a Wife, some Rum, and Tobacco, leave him for the remainder of his days in Paradise.

We found 3 large American whaling barques at anchor here; they are waiting there a week or two until the hump-backed whales come in & they can commence their fishing season. They are after the sperm or warm water whale & never undergo the hardships of those whale-men we saw down at Kerguelen's Land.

Several boats of natives came out to meet us & we towed them back with us. They were fine able-bodied looking men of a light copper colour & black curly hair which stands up straight from the head & is dyed brown with a sort of lime.[30] They came alongside naked but before they came on board they each put on a linen sheet round their waist with great dignity & satisfaction. They had very pleasing faces & could speak a little English; they looked over the ship & seemed particularly struck with the engines as we steamed in, for they very seldom see a steam ship at these islands. Our men took them down on to the lower deck & gave them some dinner, generally calling them by the name of Mungo Jumbo.

¶There was an Englishman in the boat with them who does a sort of trade with shipping, & he informed the Captain that [Tongan] King George [Tabu] had an Englishman imprisoned on one of the small islands for attempting to kill one of the natives whilst drunk, & that he was waiting for a man of war to arrive and try the prisoner. Our Captain had the matter investigated & decided that the man should be sent away from the island, to Australia or New Zealand at the first opportunity. The man had been several years on the island, & came from Yorkshire.

¶The natives told us they would bring off fruit &c on Monday, but they never traded on the Sabbath day. They go to Church 5 times on a Sunday. Some of our men told them they ought to trade whenever they got the chance, & were answered by the natives— "You too much devil, you no missionary". On Monday morning swarms of canoes came alongside to trade &c. The canoe is made out of a hollow tree & has a long frame-work to balance it on the lee side. There are 2 natives in it—one in the bow, & one in the stern, who paddles. In the canoe there is a large bunch of bananas, a basket of cocoa-nuts, one of melons, one of oranges & limes, a few fowls, a basket of eggs, some sea-shells & in the centre of the lot a small black, & very solemn pig which stands upright & looks as if he were having a cruise in his own trough, for exercise. They asked a shilling for each article & basket of fruit upon hearing which the sailors cursed the missionaries, & said that they would soon spoil the whole world. Some of our men were expecting to pass buttons for money & they expect to get to some islands bye & bye where they can exchange beads for pearls, & where other natives always keep a cold missionary on the table. They were some what mollified on finding that the natives did'nt know a 3d piece from half a crown & small silver coins soon became very scarce, upon which some of them had the cheek to ask for 10 3d pieces for half a crown off the natives, but the heathens did not see it.

I bought a large bunch of bananas, some shells, & a fish-hook made out of a shark's tooth for 3d, a paper collar, & an old Tie. About 50 native priests & Instructors came on board on Monday & they were all very partial to a collar & tie. Some of them were dressed in the most fantastic rigs, old black suits a mile too large, & college capes—relics of departed missionaries. Some had a coat & a shirt, some a pair of trousers & a collar, none any shoes or stockings, & they all had some little article to trade with.

¶Our Photographer went on shore & took portraits of the native men & women for we are to take portraits of the natives of all the groups of islands we visit. The ship's Brass Band also landed on Monday & played selections outside the King's palace, to the great delight of the whole island, assembled round. Last night the chief Wesleyan missionary the Rev.d Baker, his wife & 2 daughters, the 3 American whaling Captains & their wives, & the King's 2nd son, came on board & dined with our officers. King George is nearly 80 years old, & the Queen [Charlotte] weighs over 30 stone, so they do'nt go about much; we took their portraits. His eldest son governs the northern island. The other son was a fine looking man over 6 feet, he was dressed in European costume, & could speak good English. He can also play the violin & harmonium & played some sacred hymns.

¶The missionary seems to have more power than the King; his

cottage is next door to the Royal Palace, & is the prettiest on the whole island. The palace is a long wooden building one story high with a thatched roof. It is surrounded by a wooden palisading & looks very much like a large stable. I went on shore yesterday afternoon & had a long walk into the interior of the island. I had to be carried out of the boat for there is no landing place. The island is one vast garden, covered with cocoa nut palms, & overrun with a thick jungle of creepers & tropical vegetation. The roads are like the avenues in the woods at home only shaded with palm trees. There are no rocks or stones, only the ground is covered with seashells,—shewing that these islands have not existed above the ocean for many hundred years.

¶It was a very hot afternoon & I had on my straw hat & oldest clothes; I also took with me some small silver & tobacco to trade, but I saw nothing worth buying & I made presents to the ladies, of my tobacco. Water is very scarce & is brackish; the natives drink cocoa-nut milk instead. A green cocoa-nut holds over a pint of cool milk, & you could have as many as you liked by climbing the trees or asking at any of the huts—which is the easiest plan, for it is rather a difficult matter to get the outer skin off a cocoa-nut.

Melons, bananas, cocoa-nuts, & yams flourish all the year round; they are the principal food of the natives, & as they grow spontaneously the natives don't cultivate anything to speak of. Cocoa-nut oil is the only export: they won't take copper money.

Many kinds of fruit grow here in the season, but it is mid-winter at present. The natives won't do any regular work but they have to pay a small tax to the King & the missionary. There are some Roman Catholic missionaries on some of the islands, but the two sects don't fraternize together. There are lots of flies, & mosquitos on the island, some fine birds, & splendid butterflies & lizards, but there are no wild animals, or snakes. There are any amount of pigs, cats, dogs, horses, & fowls, that have been originally imported & greatly increased. There are not many sheep or cattle. We took in our days fresh beef, but the men threw it overboard, it was so fat & oily. The cow had been fed on cocoa-nuts.

¶On the highest point of land facing the sea is the principal church. It is built of wood, painted, and thatched, is well seated, & has a very neat interior. It would hold about 800, I should think; there is a nice organ & the singing is first rate. There are other places of worship & several schools &c, where the native teachers instruct, under the missionary. Nearly all of the younger natives can read and many can write. Some of the young women are very handsome & have long black hair, but they get fat as they grow old; they have fine teeth and good figures, & they are very fond of dress —on a Sunday some of them wear silk gowns. On a Sunday in one of the huts I entered the girls were ironing their Sunday muslins. On week days they wear nothing but a linen sheet tied round the waist & reaching to the knees, & this they take off if a shower of rain comes on. The children wear a palm-leaf & a smile—as the Yankees say. They all smell disagreeably of cocoa nut oil; they anoint themselves with this stuff & after they have been out in the sun a short time, it smells very strong. They are very pleased if you will come & sit down in their huts & are very kind and hospitable. The huts are framed of bamboo, & thatched with Palm leaves &c.; there is a hole in the roof to allow the smoke to escape & a hole in the side to enter. There is no furniture to speak of, a few rush mats & rough cooking utensils. You generally find the whole family squatted down on their hams as you enter, & grinning a welcome. The hostess tells the girls to bring some cocoa nut & yams roasted in plantain leaves. A yam is a large root like an old tree stump to look at; it cooks like a potato & tastes something between an artichoke & a potato. We use them instead of potatoes, for English potatoes won't grow in the tropics. Bananas are the most plentiful fruit & the principal food of all these islanders. I like them better than any fruit we get here. The ladies are very fond of tobacco, & I made presents of it to all those who entertained me.

¶The natives invariably ask you if you are a missionary—meaning a Christian; they bring you their Bibles & Hymn books to look at. They also have Almanacs for 1874. All those books &c are printed in London, in the native Tongan language, by the Wesleyan Missionary Society.[31] In these 40 or 50 years the Wesleyans have instructed & civilized the natives of this, & many other islands, far more than the Roman Catholics have done the Negroes of Brazil in the course of 200 years. I consider this island & people a credit to the missionaries. They do their work thoroughly & I doubt whether the missionaries of our Established Church would have succeeded so well {but shall be better able to judge when we get to some of the islands under the Evangelizm of the Established Church[32]}. The Wesleyan Missionary Society deserves the support of all classes in England {& I should think their faith is the simplest & best for these primitive people[33]}. They have made great progress with the Maories in New Zealand.

I don't think there is any more I can tell you to night about the Friendly Islands. It's a little more than the midshipman sent home to his friends when they asked him to write and describe some of these islands—the savages, and their manners and customs. He wrote—"Manners they've none, and their customs are disgusting".

We are now under weigh for Kandavu, Fiji Islands,—distant 450 miles, & the island of Tongatabu is just disappearing. {There are plenty of flying fish, & several hump backed whales to be seen.[34]} The sun-sets & sun-rises in this part of the world are splendid; the sun rises out of the sea like a ball of fire & sets in the same manner. At night the ship looks as if she were ploughing her way thro' fire, owing to the myriads of phosphorescent creatures in the water.

August 6th at Kandavu, Fiji Islands

The English mail from San Francisco is due here tomorrow, & I am expecting letters from some of you; the mail leaves again the next day & will take this; you will get it about Septr 20th, I think. On the 23d July we passed Turtle Island & on the 24th visited the island of Matuku belonging to this group. {A party of Officers & men landed, & explored the island.[35]} It was about as large as Fernando Norhona [Noronha], & had 600 inhabitants. These natives were black and ugly as are most of the Fiji Islanders. They have a fierce morose look & are not to be compared to the Friendly Islanders on strength, or disposition. There is more Papuan blood in their veins, & I expect the nearer we get to New Guinea the blacker we shall find them owing to the mixture of the two races. The genuine South sea Islanders are only to be found now at the Sandwich Islands in the North Pacific; we are to call there coming home.

On the 25th we arrived at this island which is about 29 miles long and 6 broad, it is mountainous & fertile. There are natives on it & a few white people but no town to speak of. There is a good harbour formed by an enormous line of coral reefs which extends across its entrance & acts as a breakwater. There is a narrow entrance for ships & the navigation is very difficult; the mail boat ran on the reef a few months ago & was got off by H.M. Ship "Pearl".

These reefs abound all round these islands of the Pacific, & on the calmest day there is a tremendous surf breaks over them. Sharks abound round these reefs in any quantity, we caught several when we were surveying off the reef the other day. The French man of war "Cher" that was at Sydney with us ran on a reef about 300 miles from here a few weeks ago, & has become a total wreck. She was taking Roman Catholic missionaries round to the different islands. I expect the mail boats will eventually make Levuka their place of call, as there is a good sized town there.

¶We remained at this island 2 days & sailed on the 27th for the island of Ovalau distant 120 miles. We arrived there the next day & anchored off Levuka the white capital of the Fiji islands, receiving some old letters from Sydney. I got one from Seaton, London.

¶Ovalau island is about as large as Kandavu, & mountainous. The Fiji group numbers 220 islands of all sizes. There are several larger than Tongatabu; the 2 largest Vanu Levu, and Vita Levu[36] are each about 300 miles in circumference. They are in the Tropics & vegetation is very luxuriant. {The Bread fruit, banana, plantain, arrowroot, nutmeg, capsicum & tea plant, & sugar cane will all grow here, but the first three, with the cocoa-nut are the principal productions.[37]} Cotton grows wild but is now an object of cultivation by many white colonists on the banks of the river Reiwi on the largest island of Vanu Levu.[38]

¶The whole native population is not less than 200,000; there are several different tribes & great chiefs but Thakombau is the supreme King over the lot. He lives on the large island but has a house at Levuka. In the interior of the large island the natives are still very wild, & uncivilized, & fighting frequently goes on. There are several missionaries of all denominations on the different islands, but they don't find the natives so tractable as those on the Friendly Islands. Some years ago, one of the chiefs offered to sell all the missionaries on the islands for a Keg of Tobacco.

¶I dare say you read in the newspapers of the annexation of these islands to Great Britain.[39] Old King Thakombau has managed within the last few years to contract a small National Debt of about £80,000, & finding himself unable to pay it, handed over a few weeks ago, his Kingdom & his liabilities to England, on the understanding that our Government should be responsible for his debts, & pay him a pension of £3000 per annum, as well as smaller pensions to several minor chiefs. He is to govern the natives &c., but all financial matters will be managed by English officials. The treaty is not concluded yet, but it is only a matter of time, and there is no doubt that the transaction will ultimately prove very profitable to us, & beneficial to the natives.

¶We passed several large islands before we sighted Ovalau, & they are all pretty high out of the water. Levuka stands at the foot of a range of hills; the houses are all of wood & the place is not unlike Wellington. There are no natives living near the town, so I did not go on shore there. The white population numbers about 500; they are from all the colonies & America & are not of the honestest description. They are mostly hotel keepers and store keepers & they call themselves the Pioneers of the Fiji's. (This word is pronounced Fee jee.) There is a post office and a newspaper office, a German, French, American, & British Consul there, I expect there will soon be a British Governor. We had Fresh Meat and Bread all the time we were at Levuka, & we also filled up with 150 tons of coal that was sent from Sydney on purpose for us. H.M. Ship "Dido" came in on the 31st from Sydney.

¶King Thakombau was not at Levuka or we should have had him on board. I saw one of his war canoes; it was a wooden tower built on a double canoe. There was a mast & a large angular sail made of cocoa fibre, & a lot of oars to pull it. There were about 60 natives in it & they were singing a war song as they passed the ship, & beating their tom toms or native drums; it steered with an oar at the stern of each canoe & was about 70 feet long.

¶On the 31st July the ship's barge with a crew of 7 men and 7 officers {& scientifics[40]} left the ship under the charge of the 1st Lieut. for a week's cruise round the larger island & they are going up one of the rivers into the interior of Vanu Levu. They have a tent, rifles, cutlasses &c., & a week's provisions, & they were to join us at this island of Kandavu to day, but they have not arrived yet.

¶On the 1st August leave was given to all who wished to go for 4 hours. More than half of those who went came back so helplessly drunk that they had to be hoisted on board like cattle. The liquor sold to them by the "Pioneers of the Fiji's" was such poisonous muck that some of them were raving mad for some hours. {We shipped at Sydney as a Chartroom servant, a young man of about 24 years of age, who had scarcely a rag to his back, & the Captain took him out of Charity; it appears that his Father is a Church of England Clergyman, & this son was such a wild scamp, that they were glad to get rid of him, and he came out to Australia with £400 some 5 years ago, & has spent the lot. He has been the round of all the colonies, tried the gold fields, & lots of speculations; he is an inveterate drunkard, and has had the delirium tremens several times, so much, that, when he is perfectly sober, his hands are all of a shake, and his gait that of a dicrepit old man. He is a good scholar, & when sober has the manners of a gentleman, is not at all proud & very liberal in standing "Glasses round" when on shore, so is rather popular with the sailors. He says he wants to save enough money by the time he gets back, as will enable him to clothe himself decently before showing up at home. Well! he went on shore at Wellington, & came on board beastly drunk; he went again at Levuka yesterday, & was the worst & most incapable of the lot; broke 17/ worth of glass &c, on shore, that the Paymaster had to pay out of his wages, & he smuggled 2 pints of Rum into the ship by means of a bladder tied round his waist, it was taken away from him, & he got so excited that he went mad for several hours; tried to stab one of the ship's police, broke 3 large lamps, & did much other damage until he was dragged down to the cockpit & put in irons just outside my door; his ravings were of the most blasphemous kind (worse than any drunken sailor) I ever heard, & he said that he had once been in irons for 6 months in Australia, & that they always treated him well in gaol.

This morning he is pitiful to look at, his one suit of clothes all torn to rags, his limbs too weak to keep him up, & one of the worst black eyes I ever saw. He has no recollection of any thing, is very willing to apologize to all, & he says he will drink no more, which resolve I am afraid will be only very temporary, & it is easy to see that his years are numbered for this world. He calls himself "Dawson" but I doubt whether that is his real name.[41]}

¶We left Levuka this same night & arrived at Kandavu again on Monday 3rd. We are thoroughly surveying the harbour and adjacent reefs and expect to leave here next Tuesday 11th, for the New Hebrides islands distant 500 miles West. We brought Mr. Layard the British Consul round here with us, he is a fine looking jolly old man & puts me in mind of Mr. Huntley. He has been out here many years & speaks the native language. We have also on board 10 natives of the New Hebrides that we are to take back to those islands. They have been working for 3 years at these islands for the white people. Instead of wages they each have a chest of clothes, a rifle, tomahawk, axes &c, & I expect they will be great chiefs when they get back. They are something like the Fijian natives but not to be compared to the Friendly Islanders. {Probably they were kidnapped in the first place by some British vessel.   *   *   *[42]}

Owing to the scarcity of labour here & in Queensland a system of kidnapping natives from the different South sea Islands has been much resorted to during the last 10 years. Our Government fitted out at Sydney several small schooners a year or so ago, to put a stop to this labour traffic, for it was worse than slavery, & had become so common that no peaceful vessel dare go near many of the islands for fear of being murdered. The natives of the islands retaliated by murdering any white people they came across. At the Solomon Islands, a ship's crew was massacred a short time ago & this cruel system was the cause of the murder of Bishop Patteson[43] some time ago.

¶We had several Fijian Chiefs on board yesterday & took their portraits; some of them were fine looking men, & so are some of the women—tho' not equal to those of the Friendly Islands; they wear scarcely anything at these islands. We had some enormous canoes alongside the other day with fruit &c, some of them were regular family canoes, & had the whole family thereon. I bought some fruit, shells, coral &c for some small silver. It is very hot here at present. One of our sheep had a lamb this morning, & at meal times you see all hands round it; it is considered the greatest curiosity on board. You will know where to write to for the next 8 months;—Singapore, or elsewhere. I hope you get on allright with your business,—your music, dancing, courting &c.—"Ye gentlemen of England, Who live at home at ease, Oh! little do ye think upon The dangers of the seas." Of course you'll give my love to your young lady; I think you ought to send me her likeness, to take round the world.

Don't forget to send this to Mother. With best love to you & all at home.

From Your Affec[te] Brother,
Joseph Matkin,

I received your Papers all serene—Write soon[44]

The account of Fiji continues in Matkin's next letter to his cousin, begun ten days later in the New Hebrides (now Vanuatu), and concluding with one of his most exciting experiences ashore: getting lost in the bush at Cape York.

We also encounter, in both this letter and the one that follows it, rare references to Matkin's own physical condition—he had become quite thin and pale—and to the circumstances of his work. Storeroom duties apparently occupied his time so completely that he was seldom called to work on deck, even during difficult oceanographic operations. This makes all the more surprising his knowledge of the voyage's scientific accomplishments.

H.M.S. "Challenger"
New Hebrides
Aug. 18th, [18]74

Dear Tom,

As we are just about to make sail for Cape York, I will get a letter under weigh which I hope to send from there, via Torres Straits, & Singapore. My last letter I had to close at a moments notice, & the San Francisco mail never came in with ours, so shall not receive them until we reach Hong Kong or Manilla. On Sunday, Aug. 9th, I went on shore at Kandavu, Fiji Islands, & had a long walk, but not such a pleasant one as Tongat[a]bu, the roads were not so good, nor the island so fertile, neither were the natives as handsome, & very mercenary; there were Missionaries, but they have not made such progress there as at Tongat[a]bu. We heard from Sydney that our Government had refused to annex the Fiji Islands,[45] if so, I believe one of the Colonial Governments will see what can be done.

¶We left Kandavu on the 10th of Aug. for these islands, distant 450 miles west, & sighted them last night. The depth across was about 1500 faths. We were caught in a heavy squall the other night, & lost a spar, & some small sails, before the sail could be shortened, but it was all over in half an hour.

I told you we had 10 natives of this group of islands on board, brought from Levuka, this morning we took their portraits, & landed them on their native shore of the island of Api. This group of islands was discovered in 1606, by Queros, the Spanish navigator; it consists of about 20, most of them of good size, for their area is almost as great as the Fiji group, like which they are of volcanic origin, but are more mountainous, & there are several active Volcanoes. The largest islands are Espirita Santa, (200 miles in circumference) Tanna, Erromango, Api, Aurora, Mallicolo, Amboyna, Penticost, Santa Maria, & Vanua Lava,[46] & are about 300 miles north of New Caledonia. The inhabitants are something like the Fijians but more slender, are broken up into many different tribes, & are always fighting, their weapons being clubs, spears, bows and poisoned arrows, & they have a very bad name.

¶We sighted several peaks on the different islands last night, and lay to until this morning, when we steamed into Api island to land the natives. It is about as large as Tongat[a]bu, and has a beautiful appearance being covered with trees to the very top of the mountains, which are about 2000 feet high. There are not many Palm Trees and of course the island is not so fertile as Tongat[a]bu. There are several reefs in the vicinity, and the ship did not go in very close. We sent 3 Boats in containing the 10 natives, their arms, clothes, &c, & several Officers & Scientifics with fancy articles to make presents, & barter with the natives. The Boat's Crew, Officers &c, were all armed in case of an emergency. The natives came down to meet them, & were all armed, but as soon as they saw their returning countrymen, they yelled with delight. Our 10 natives belonged to several different tribes, & each claimed their own men, with their arms &c, & marched away into the bush. Only one of them could speak English and the sailors called him Toby. When his tribe came down to claim him, he said to our men "All those men belong to me", & he & the Chief stood on the beach, and laughed at each other for about 10 minutes. The Boats brought back Yams, Plantains, Cocoa Nuts, curiosities &c, and are all hoisted on board again, and we are now under sail for Cape York, distant 1500 miles, north-west. The natives of Api wore no clothing, merely a nominal wreath of leaves,—both sexes. There are no missionaries for they ate one at Erromango a short time since.

Sunday, Aug. 30th
Anchored in the Coral Sea, off Raine Island
Our greatest depth on this trip has been 2,450 faths,[47] & we have had good Trade Winds, & pleasant weather. We entered the Coral Sea to day, and are about 120 miles from the Australian continent.

The water was 1400 fams yesterday, and to night we are at anchor owing to the great number of reefs all round us. The great barrier of Coral reefs extends for more than a thousand miles along the east coast of Australia, and runs for more than 100 miles out from the land. We sighted Raine island this morning, & can just see it from where we are anchored; it is not 2 sq. acres in extent, & destitute of trees, being scarcely out of the water; of course there are no people on it, but there is a large stone beacon, 60 feet high on it, that has been erected as a landmark by the Board of Trade. Like S. Paul's Rocks, there are thousands of birds living on the island. We are 8 miles from it now, and anchored off a long reef; it is something wonderful to see the number of Reefs all round us; the Captain was up at the Foretop with a glass as we came in. I believe Captain Cook ran on one, and was repaired at Batavia.

¶There are enormous quantities of sharks here about, as well as other fish; all hands are busy catching them now, and over 20 have been already caught, some of them 8 ft. long. The men cut off the tail, slice open his belly & let him go again, he generally swims about them for half an hour, then dies, & is eaten by his messmates. Other good fish have been caught, one was pulled up, & just as he was level with the water's edge, a Shark came & took half of him. We had no church on board to day owing to the great number of reefs we were among. To-morrow we are going to land a party on Raine island, & shall anchor again at night in the open sea.

September 4th, Somerset, Cape York

On Aug. 31st, we stood in to Raine island, & landed a party to overhaul it, &c. there was nothing on it but Sea birds, and they were thicker than at S. Paul's Rocks, the ground was literally covered with them, their eggs and young. The soil was sandy but there was a little vegetation to be seen, & in the course of time a soil will be formed from the deposit of the birds, and cocoa nuts &c. will wash on shore, & spring up as they have done at other Pacific Islands. The island is of coral formation, & will probably rise higher out of the water, and enlarge every year. The stone tower was sadly in want of repair, on the top there was a cistern for catching fresh water, so that if any one was wrecked there, they would find a supply & be able to live on birds, eggs &c, until succoured; but this cistern, and the whole roof of the Tower had fallen in. On the doors were records of the visits of different ships, but there had been none there for the last six years. A record of our visit was added, & the party returned, after which the boats were hoisted in, & we made sail for Cape York; but owing to the reefs we anchored again the same night in sight of 3 other small islands. Lots more Sharks were caught, I should think there is no sea in the world where they are more abundant.

¶It came on to blow that night, and over 120 faths of cable were paid out to ease the strain on the anchor; this took three hours getting up in the morning, when we made sail again for Cape Yorke. We sighted the Australian Continent again at 11 AM, and anchored the same night, Sep. 1st, off the settlement of Somerset, and the coast here presented the same flat, sterile appearance that it did towards Melbourne, more than 2,200 miles further south. There is not a hill to be seen over 600 feet high, and yet on the opposite side of Torres Straits within 120 miles there are mountains on the island of New Guinea over 13000 feet high, in sight half way across the Strait; & in the interior are some whose peaks are over 16000 feet high above the sea level. This mountain system may be traced across the Pacific from the Rocky Mountains in N. America. It rises at the Sandwich Islands over 13000 feet, again in Japan higher still,—still higher in New Guinea. It does not touch the continent of Australia but stretches more to the East; it appears moderately

high at the New Hebrides, & Fiji Islands, & extends through the
New Zealands Islands rising over 13000 feet on the South Island, as
was found by Ross in the Antarctic Circle in Lat. 76° S. where
burning mountains were seen over 13,000 feet high. Every where
along its course Earthquakes are prevalent & owing to its near
vicinity they have occasional shocks here that are felt in no other
part of Australia.[48]

¶The land here is covered with bush and trees, and on the night
we came in there was a large bush fire inland, that could be seen by
us out at sea before the land was visible. The Colonists say these
fires are caused by the great heat of the sun, & the excessive dry-
ness of the grass &c. A glass bottle has been known to act as a mag-
nifier sufficient to set fire to the grass in Victoria; & here it is
much hotter for it is only 10° S. of the equator. The aboriginals of
Australia are thicker here than in any other portion of the Continent, & are very treacherous & troublesome.

This place belongs to the colony of Queensland, which is a tropical colony and one fourth the size of Europe, & West Australia a little larger. The whole colony is only one fifth less than Europe. As
soon as the Queensland Mail was started via Torres Straits, Somerset sprung into existence. There are not many white people even
now, but still they are all working men. The land is being cleared
for cultivation & there is the nucleus of a town in the shape of a
dozen substantial white verandah'd cottages, a barracks for the
mounted Police, a Post Office & some missionaries to keep the
natives in order, & no doubt Somerset will improve during the
next 20 years, for it is in one of the main thoroughfares of ocean
traffic between America, Queensland & the East Indies.[49]

We took in a Bullock to-day and shall [have] fresh meat while we
are here. The two missionaries are on board now dining with the
Captain, they call themselves Cameronians.[50]

¶I believe we go from here to Timor, thence thro' some small
islands into the Indian Ocean, after which we return thro' the
Lombok and Bali Channel to Amboyna, the Dutch Capital of the
Molucca Islands; thence to Celebes & Borneo to Manilla, after
which Hong Kong, where I have heard we shall be detained re-fitting until after Xmas, & that owing to this delay we shall leave out
the trip to Vancouver's Island, & go from Japan direct to the Sandwich Islands, but this is not decided on yet. After Jan. 25th our
address is Yokohama, Japan, via Southampton. From Hong Kong
we go to New Guinea, New Britain, New Ireland & Caroline
Islands before going on to Japan, where we are to spend three

months surveying in the inland seas &c. It is also decided that we go to Monte Video instead of Rio Janeiro on our way home, for which I am sorry, particularly wishing to see what Rio was like.

Soon after we anchored here our letters came on board, I receiving some from home, with one of yours, and was sorry you did not appreciate my mahogany colored charmer;[51] you must please to remember she a'nt dressed, which makes a considerable difference, but she is not a circumstance to the females of Tongat[a]bu; & wish I could send you the portrait of one I have in my "mind's eyes" now. You would find me much altered since I left England,

The "mahogany coloured charmer"? Natives of Tonga-tabu, Friendly Islands, with Professor Thomson, 1874 (Natural History Museum, London, Photo No. 358). Photo courtesy of the Natural History Museum, London.

am much thinner & as pale as Hamlet's Ghost, with hair getting quite grey. You see I get very little exercise, & my work lies below the water line, where fresh air and daylight never penetrate, having to do all my writing by candle light, the "Challenger's" accomodation is very poor for all except Officers.

September 7th

We expect to leave here to-morrow, but this will not go before the 18th. We had a party of native men & women on board the other day; they were similar to those of Victoria and New South Wales, & as thin as rats, the ugliest race we have yet seen, the Ashantees are handsome compared with them. We had one of the Missionaries on board yesterday to conduct our Church Service.

¶In the afternoon I went on shore with 5 mates & got lost in the bush, for the land is one great forest, & there are no roads, or animals, & but little fresh water. Parrots, Cockatoos &c were very plentiful. There were some magnificent butterflies, also lots of snakes, centipedes, lizards, mosquitos, flies & ants, the latter are the pests of the settler. We got entangled in a mangrove swamp, up [to our] knees in mud, it came on dark, with the tide rising into the swamp in which lizards & even alligators abound. I separated & struck off into the bush, arriving on board at 8 PM, the others coming in this morning by way of the shore, having slept in a tree all night until the tide went down.

Yours Sincerely,
Joe Matkin[52]

For the fascinating passage between Cape York and Hong Kong— "thro' the Lombok and Bali Channel to Amboyna . . . thence to Celebes & Borneo to Manilla"—only one letter, though a long one, to cousin Tom Swann, survives. Matkin later relates (see below) that a major typhoon struck Hong Kong before *Challenger's* arrival, resulting in the loss of a mailboat carrying *Challenger* mail. In the surviving letter he reports that previous delays caused this leg of the voyage to be abbreviated. Several exotic ports of call in the East Indies were thus cancelled.

H.M.S. "Challenger"
Arra Islands,[53] Sea of Arafura
September 15th, 1874

Dear Tom,

We are now within 5 degrees of the Equator again, & are undergoing a sort of mild roasting, which is aggravated by the ship being continually under steam, & there being no wind to speak of, I need not say that this does not improve my health, being shut up all day between decks, and getting no exercise. Did you see the pictures in the "London News" of the ship at Kerguelen & amongst the ice?

Great alterations have been made in our cruise since last I wrote, we are proceeding to Manilla by an entirely different route to that originally laid down, we do not go to Timor or Borneo, Celebes or any of those islands, & after leaving Hong Kong do not again visit New Guinea or New Britain, Solomon Island &c, but go direct from the Caroline Islands to Japan. This alteration is owing to our being so much behind time, & to the refusal of the Admiralty to extend the voyage beyond April 1876, so that you may expect me home about May. After leaving this group of Islands now in sight, we go to New Guinea, thence to Amboina, Mollucca Islands, & on to Manilla, reaching there the 1st of November; from thence to Hong Kong, arriving about the 15th, to refit, & do not leave before the 27th Dec^r, so that we shall have a Xmas in China. From Japan in end of June we proceed direct to the Sandwich Islands, and do not go to Vancouver's at all. The remainder of the voyage is the same except one part we go to Monte Video instead of Rio Janeiro, but this arrangement may possibly be altered again.

¶We left Somerset on the 8th and steamed out through Torres Straits, for these islands distant 500 miles; the straits are about 100 miles across in their narrowest part; we did not sight New Guinea but called at two small islands for a few hours, there are several situated in the straits, some are inhabited, but those we called at were not. As soon as we cleared the Straits we were in the Sea of Arafura, as this part of the Pacific is called. The depth is only about 40 faths., and we have dredged & trawled every day with great success. We passed a Singapore Mail boat the other morning, bound to Brisbane but she had no letters for us. The same day we passed a large Dutch Merchant ship under full sail for Batavia; the Dutch have many valuable possessions in this part of the World. Thunder storms & vivid flashes of lightning are very prevalent after sunset in this part of the world. The southern island of this group[54] is now in sight but our place of call is farther north. The Doctor is now serving out Quinine to the Boat's Crews who have to land to morrow as a preventive of Fever, Ague, &c.

October 2nd, at Banda, Nutmeg Islands

We were a week round the Arra Islands, they are a group of small extent, lying about 90 miles west of New Guinea; the population is about 60,000, & is a mixture of the Malay with the Papuan or Austral Negro Family; they are good seamen and were formerly great Pirates,—the terror of the old China Tea Clippers, & other merchant ships that came this way, but our Steam Gun Boats have stopped all that sort of thing.

¶On the night of the 14th Septr. we anchored about 7 miles from the island of Trangan, the largest of the Arra group, & the next evening we steamed in for the island of Dobba,[55] and anchored off the town of that name. It was a beautiful island, but rather low, but covered with a dense vegetation and forest abounding in Parrots, Cockatoos, Birds of Paradise, & enormous Butterflies & Lizards. Soon after we anchored 3 large Proas or native Canoes came alongside with the principal Chief of the island. These canoes & several Flag Staffs on shore flew the Dutch Flag. All the groups of islands hereabouts either belong to the Dutch, or are under their protection. Our Men of War seldom come this way, they go thro' the Straits of Malacca to China. This route to China is called the Molucca passage, & is little frequented by any but Dutch ships, so here they reign supreme & have it all their own way.

¶There was not a sign of a European on the island, but lots of Chinese, trading & scraping a fortune together. All the rest were natives. Those in the canoes came alongside in great style, singing their war songs & beating their tom-toms, or native drums. The singing was something frightful, all in one key, and sounds very doleful & monotonous. The Chief with a few Chinese came on board, and were received by the Captain & Professor Thompson in the cabin, leaving again after having refreshment & cigars. The Chiefs were dressed in Chinese costume, but the other natives had only a cloth round the loins, & were the ugliest race of people we have yet seen, except the Aboriginals of Australia. They were slender, ill shaped, with irregular features & horrible mouths, this latter deformity is caused by their chewing a root called Betel nut, which rots their teeth away altogether, destroys their voices, & makes their mouths look like a piece of raw liver. They are Mahommedans, & live chiefly on fish, eggs, & vegetables; they are very chary about letting you see their wives, or females of any sort. Cocoa nut trees & bananas grow there, but not abundantly, & fruit was very scarce.

¶The town consists of a lot of Bamboo huts, two storys high, thatched, & open at the gable ends to admit a current of air right through. I went on shore one afternoon, & had a fine swim in a salt water lake, the water of which was hotter than any I ever before bathed in. The people & houses were very dirty, some of the Malay children were pretty, those who had never chewed the betel root, or been scorched in the sun. The Betel root is a shrub common to the East Indies & China, & is a species of pepper vine. The lower story of the huts is used by the pigs & fowls, a bamboo ladder leads into the upper loft, and as soon as the women and girls saw us coming, they bolted up this ladder like a shot, while the men & boys stood below & laughed. Those women we did see were very ugly.

¶The Chinese do all the trade, keep stores, and sell rice, rum, tobacco, &c, and they buy all the Birds of Paradise, & Pearls the natives can get. I saw 2 or 3 Chinese women, but they were very shy about shewing themselves, though they were very fair looking. In the middle of the streets were wells of fresh water, and the Chinese community bathe at the mouth of these wells by drawing buckets of water & pouring over themselves, so that the water finds its way back into the well again, this way of bathing is economical, but not our style. The Chinese brought off lots of Birds of Paradise, ready stuffed & cured, & sold them each at 10/. I did not buy any but may do if we go to New Guinea, where they are natives of, but a few of them find their way to these islands; they are about as large as a thrush, but have golden tails over 2 feet long, they look beautiful while flying, but are very timid & difficult to shoot. Our Officers went to several of the other Arroo Islands in the Steam Pinnace, & shot a few Birds of Paradise, as well as other splendid Parrots, &c. They also shot some large Flying Foxes, caught a young Cassowary or Emu, and several Snakes, butterflies, lizards, &c. Some of the butterflies were as large as English sparrows.

¶The native Canoes of the larger sort are thatched over, and look like floating hay-stacks. When they go to sea the whole family, fowls, pigs, &c, go also, and they make voyages of 11 & 1200 miles at a stretch, for the sea about here is as calm as a mill pond generally, & gales of wind are unheard of. On arrival in port they draw the canoe upon the beach, & live in it the same as at sea. These are the migratory Malays; their sail are made of Cocoa nut Fibre, & also their ropes, and they live on Fish when at sea. The natives gamble & play at cards to a great extent. The Chinese have taught them to gamble, & the Dutch to play cards. The real natives of the island are Papuans; the Malays & mixed races being emigrants.

¶On the 21st Sept.r the sun crossed to the [this] side of the Equator, & the Southern summer commenced. On the 23rd we left Dobba for another group called the Ki Islands,⁵⁶ about 80 miles, farther west. We sighted the largest of the group, called Great Ki, the next day; it was about as large as S. Thomas, & much higher than any we had seen since leaving the New Hebrides. Off these the sea deepened to 800 faths. Lots of natives came off there, in their canoes, drum beating, & singing the same old tunes. They were very similar in appearance to those of the Arroo islands, but not quite so ugly. All that day we steamed through the Ki group, which is very numerous though of small extent, & anchored at night in a bay surrounded by beautiful islands. The sunsets in this part of the world are splendid & nearly every night, lightning, without either rain or thunder. We had a whole tribe of natives off that night, & they were allowed to climb the rigging, & ramble all over the ship. After dark they were had on the quarter deck to dance & sing their war songs; all hands stood round them, & the Captain burned a Blue light to illuminate the scene; after which they had a feed of Biscuits & departed.

We left on the 25th, & anchored again at night at another beautiful Island of the Ki group. On the 25th [26th] we left for the Nutmeg Islands distant about 330 miles & passed numbers of others on the passage. On the 28th we sighted Banda, & took soundings at the enormous depth of 2,875 fathoms; the next day we anchored there, & is the largest island & capital of the group called Banda or Nutmeg Islands, which consist of nine, all belonging to the Dutch.

¶Banda is a very clean little town, & is used as a convict settlement for the natives of all the Dutch possessions in the East Indies. There were many women convicts, & they were employed in making roads &c. There is a large Fort and Barracks built by the Portuguese before the year 1700, after which the islands were taken by the Dutch. Banda & nearly all the Dutch possessions in the East, were taken by the British in 1775, but they were nearly all given back again at the conclusion of the war, except Ceylon, the richest island in this part of the globe. The natives of Banda are Malays, Papuans, and all sorts of mixed races, & a few Chinese merchants; but there are not many Dutch or white, except soldiers & officials. We saluted the Dutch Flag with 21 guns, and are to do so at all their possessions we visit. The Governor dined on board one day. Fruit, fowls, and eggs were plentiful, but other provisions were dear.

¶The sight of the town is the Volcanic mountain of Goomongapi, 2,200 ft. high, and still active. There is a book on board written by an American—Professor Beckford—describing his visit to these islands, & his perilous ascent of the burning mountain. He relates how he had native guides, & all sorts of appliances for making the ascent, & that it took him 2 days, there is also a picture of him hanging on to [by] his eyebrows to the side of the mountain with all the blocks of lava cinders rolling from under his feet. The mountain rises very abruptly, & the ascent is certainly difficult & dangerous; several people have lost their lives in the attempt at different times, and not long ago an Englishman was suffocated while looking over into the Crater. On the day after we arrived several of our officers ascended, with guides, &c. and were only five hours absent from the ship, & had had a rest in the town. On the following day, myself & a messmate went on shore, and after having a look over the town, started to go up the mountain, without guides or anything. We were 2-¼ hours going up, & 50 minutes coming down; the last half of the ascent is very dangerous, as the surface of the mountain is nothing but a mass of loose cinders, which rolled down with us 2 ft for every 3 we ascended. On arriving at the summit we saw a Flag-Staff erected by our Officers at the mouth of the Crater, which was about 100 feet deep, & a mass of sulphur inside; the ground is hot enough on the surface to cook eggs, & dense volumes of smoke & vapour rose from the crater & numerous cracks in the earth. It is always smoking, & we were obliged to cover our faces whilst we looked over into the crater; for the whole air round the summit is suffocating. We made the ascent in the heat of the day, & forgot to take any water with us, but we soon got some from a Malay hut at the foot of the mountain. We descended at a rattling pace, for there was no walking down the first half of the way. It is nearly 100 years since the last great eruption which was accompanied by an earthquake that divided the island asunder, & separated the mountain from the town, & the sea is now a ¼ of a mile wide between the two islands & very deep; we had to cross in a native canoe. The Dutch Fleet was left high & dry for some hours during the earthquake, by the waters retiring from the bay, but as the sea returned again to its former position, they received very little damage.

¶I had a fine bathe at the foot of the mountain, then we visited the Spice plantations. The Clove tree is very pretty and the buds that grow in clusters at the top of it, form, when dried, the spice called Cloves. The Nutmeg tree is very like a pear tree, and the fruit when ripe resembles an Apricot, & nearly as large; the outer shell is yellow, & is allowed to rot off as a Walnut, leaving the Nutmeg enveloped in a thin membranous, sweet scented substance, which when dried, forms the spice called Mace.

Yesterday we went out to sea to look for a disabled Dutch mail boat, but found nothing of her, so returned at night. We are not going to New Guinea until we return from Hong Kong, as this is the rainy & sickly season. This evening we leave Banda for Amboina, distant 150 miles, north. We are only about 200 miles from the Line, & the weather is roasting.

Sunday, Oct. 25th, at Zamboanga

We anchored off the town of Amboina on Sunday, Oct 4th, and remained until the 10th coaling &c. Amboina is the capital of the Molucca Islands, & has about 30,000 inhabitants, the town has 16,000, mostly Malays, Papuans, mixed races, Chinese, Arabs, & Europeans. There was a large Fort, & lots of soldiers there, & we had the Dutch governor, & numerous visitors on board. I went on shore several times, & also played Cricket, the first time for 2 years.

¶The island is mountainous, but fertile, is a large size, & has a good bay. It was intensely hot there, 140° in the shade sun, & over 90° in the bottom of the ship. Cloves, & spices are the principal export, & fruit of all kinds was cheap and plentiful. The colored people are all Mohammedans, I looked through the windows into one of their Mosques, which was built of wood, and shaped like a Chinese pagoda; the interior was quite bare, and the floor of stone; the natives stood up to pray, and went though all sorts of antics. The Malay people do washing beautifully, they use no soap, but beat the clothes on the stones in the river beds. They are very fond of wearing white garments when they do have any clothing; and some of the children who had not been scorched in the sun, were very pretty, had splendid eyes & teeth, & long black hair.

¶They paid us a month's money at Amboina in Dutch dollars at 4/2d each. We had fresh meat, vegetables, and bread there for two days, for our health's sake, the meat was 2/5 per lb, vegetables 1/6, & bread 1/ per lb. The Admiralty allow 2/6 per lb. to be paid for fresh meat, & vegetables when a ship has been long at sea, & the doctor recommends it. I was reading the other day the voyages of Cook, & other English navigators amongst these islands over 100 years ago; they used to bury 2 & 3 men each day at the Arroo Islands. In those days Lime Juices & Quinine were unknown in the Navy.

¶We left Amboina on the 10th for Ternati [Ternate], another of the Molucca Islands, 50 miles north of the Line; we passed several others of the group, & crossed the Equator on the 13th at 10 PM in Long. 127°30′, East.

¶On the 13th we anchored off the town and island of Ternate, on which was another volcano 3000 ft high, & very active, on one of the adjacent islands there is another 5000 feet high. The town of Ternate was not so large as Amboina, but much cleaner & prettier. There was a Fort & Barracks, & a Dutch man of war at anchor off the town. I went on shore there, & spent a pleasant evening. Fruit was plentiful, especially Pine Apples, which sold 1-½ each. Fowls were 6d each, and eggs ½.

¶On the 17th we left there for the Phillipine Islands, passing lots more on which were volcanoes, and sighting the large Dutch island of Celebes, on which were 3 active volcanoes. The Dutch islands and possessions are very quiet, dead & alive sort of places, there is no enterprise or progress to be seen anywhere; no railways or improvements of any kind, & the people are the same as they were 100 years ago. I wonder what the Spaniards, Dutch, or any other people think when they visit Melbourne, Sydney, or any English colony.

¶We sounded in the Celebes sea, at 2,375 faths. & 2,675 faths. & on the 22nd sighted the large island of Mindanae [Mindanao], Phillipine group, which group is large & very valuable, they are in the North Pacific, and belong to the Spaniards, they were discovered in 1521 by Magellan, and taken possession of in 1570, in the reign of Philip the 2nd, after whom they were named. They number nearly 1000, but only about 150 are inhabited. Luzon is the largest island, & is only one fifth smaller than England, Mindanai island is only a trifle less. They are rich, abounding in all sorts of fruits & minerals, even gold & diamonds have been found, and there are also Coal mines in Mindanai; they are mountainous, volcanic, & subject to earthquakes; they export rice, tobacco, cigars, hemp, &c, and Manilla is the Capital. The whole population is over 8 millions, & about 2 millions are either Spanish, or of Spanish descent, that portion are Roman Catholics, the remainder Mahommedans, or Idolaters. There are two tribes of natives, the Aboriginals, called Karasoras, of Papuan origin, and the Malays, & mixed races. The Karasoras are nearly yellow, small, ugly, with long black hair; something between the Maories and Papuans; there are also great quantities of Chinese.

¶On the 23rd we anchored off the town of Zamboanga, which is the Capital of the large island of Mindanae. The town stands at the foot of a range of high mountains, which are clothed to their summit in dense forests of ebony and sapan woods. There is a large Spanish man of war at anchor here, & we have had most of the Officers, as well as the Spanish governor on board. The town looks something like Amboina, but is much larger, stronger Fort, more soldiers, and is much dirtier. There are great quantities of Cocoa nut trees here, & fruit is very plentiful, especially oranges, guavas, pine apples, shaddocks, & bananas, as well as the bread fruit. There are lots of wild creatures, serpents, &c, wild buffalos, & small half wild horses, called Mustangs, inland. The Buffaloes have been tamed, & crossed with the domestic Ox, and are used for all sorts of heavy draught labor. The small horses are very fleet & pretty, and were to be hired, with a native attendant, for a dollar per day; lots of our men had them and astonished everyone, as well as themselves by their riding.

¶This morning we had Church, & after the heat of the day was over I went on shore, & had a good walk and look about. The Spanish half-caste women are very handsome, and have beautiful black hair, almost reaching the ground. The natives here are far more industrious than at the Dutch islands, in cultivating the soil, and any manual work. The town trade appears to be in the hands of the Chinese, but the principal export trade in the management of the British & American Merchants. I looked into one of the Roman Catholic Churches during service, there was an organ, & the singing good. Just opposite the Church was the Spanish Officers quarters; they invited all our Officers this evening to come and see a native dance; the dancers were natives with Chinese girls, women, and some men, & were dressed in Chinese Fancy Silks & all sorts of costumes; the music was horrible, and the dancing peculiar. They build the houses two story's high, with open verandahs, so from the street we could see all that was going on. The native houses are built of bamboo on piles, and were similar to those at the Arroo islands.

¶We leave here to-morrow for another large town further north, and shall pass very close to the island of Borneo, but shall not touch it. The weather is still roasting, and we are almost always under steam, the decks have to be kept constantly wetted to keep the pitch from melting. Mindanao is very rich in minerals & coal, but the Spaniards are too lazy to work the mines, and all the coal used by steamers here, comes either from Sydney, or the British Island of Labuan, to the north of Borneo. We hope to get some letters & English newspapers at Manilla next week.

November 11th,[57] off Luzon, under steam for Manilla
We left Zamboanga on the 26th and on the 27th sounded at three miles in depth, and that night there was Lightning from every

quarter of the heavens, such flashes as I never saw before, & neither rain or thunder. On the 28th we sighted the island of Panay in this group, and anchored at night off the Capital town of Iloili (pronouced Elo-Eli). It is much larger than Zamboanga, having 60,000 inhabitants, & is the second city of importance in the Phillipine Islands. It stands on a river, in a wide marshy plain, at the foot of a chain of mountains; is notoriously unhealthy, & very dirty my messmate said, so I did not go on shore, there being nothing to see different to Zamboanga. We had the Spanish Governor on board, and also took in Fresh Meat, Vegetables, & bread; the meat was only 6d per lb. We also took in 50 tons of coal, English coal. We heard news from the English merchants of Bazaine's escape,[58] and of a terrible Typhoon near Hong Kong, and that some of our letters had gone down in the mail boat there. We left on the 31st for Manilla, distant 350 miles, and on our passage, saw scores of fine fertile isles, all mountains. Are now off the large island of Luzon, which is nearly the size of England, and to-morrow expect to be at anchor at Manilla, by 3 PM hoping to receive some welcome letters.

Sunday, Nov. 15th, in the China Sea

We anchored at Manilla at 4 PM on the 4th and received our letters at 9 PM, mine numbering nine, besides newspapers. We anchored nearly 2 miles from the city of Manilla, which stands on a flat peninsula at the bottom of the bay, which is as large as Table Bay, and surrounded by high lands. Nothing but the spires and domes of the Churches could be seen from where we lay, but the city had a very pleasant appearance especially at sunrise and sunset, when the great Chains of mountains situated some 15 & 20 miles inland came out very distinctly. Luzon is the seat of government for all the Phillipine Islands, its capital, Manilla, stands on a river, which is navigable up as far as the town, and was full of shipping, I think even more than at Bahia, which town it somewhat resembles, though a much busier and gayer place, and quite equals it in dirt.

We saluted the Spanish Flag with 21 guns, and the same number was returned from Fort S. Philip, which defends the city. Manilla is the oldest European city in this part of the world, it was founded in 1571, and was formerly much more important than it is now. The Spanish treasure ships used to sail from Acapulco in Mexico across the Pacific to Manilla, & home to Spain by the Cape of Good Hope. Commodore Anson took one of these galleons worth over one million pounds sterling. Its present population is about 230,000, of all colors, but the majority are Malays, Papuans, and Chinese. Of course there are Spaniards, and there is a Spanish squadron always stationed at Manilla, as well as lots of Troops, but the principal Foreign trade is in the hands of the British and American Merchants; the chief exports being cigars, rice, indigo, tobacco, hemp, coffee, and sugar.

The British Consul gets £1100 a year, he, with his wife and family were often on board during our stay. Fruit, eggs, &c. were plentiful, and we had Fresh Meat, Bread, and Vegetables all the time we were there. A month's money in Spanish dollars at 4/ each was paid to the ship's company, and leave was given every night from 4 till 10 PM.

The Steam Pinnace did all the boating, as it was a long pull for a rowing boat. The weather was much cooler at Manilla, and we had plenty of rain. We had the Spanish Governor, and Admiral, with lots of military and naval Officers and men, Catholic priests, ladies, &c, on board during our stay, and the day before we left, the Captain gave a Lunch & Dance party. The ladies were tolerably handsome & well-dressed. I like their style of dancing better than the English.

¶I went on shore one evening, but it so soon got dark, that I did not see much of the town, which is lighted by oil lamps. There is a fine Cathedral, and several large Churches, and also some enormous government cigar manufactories, employing thousands of native women. Most of our men have laid in a stock of, but being no great smoker, I did not go in for a bargain. The price of the very best is 8d per hundred, and you could buy the same quantity for 3d. The houses are built of stone two storys high, with flat roof, and open balconies, but there is nothing pretty about the best streets, and the back ones, where the Malay and Chinese reside, were something horrible. There was a bit of a Park, full of well dressed people, for it was Sunday night, and the Military Bands were all playing. The ladies wear no bonnets only lace veils, and many of them smoked cigars in their carriages. Horses were very plentiful, and vehicles of all descriptions were to be seen driving about the fashionable streets; Carriage exercise is the principal amusement. The natives were about the same as at the other islands; they, and the Chinese seemed to be the only hard workers in the place. Altogether I think Manilla is the finest place we have visited except Lisbon, Melbourne, and Sydney, but everything is very dear there.

¶At 2 PM on the 11th we got up anchor and left for Hong Kong, distant 700 miles. We were two days getting clear of the Phillipine

Islands, and are now in the China Sea, sailing with the northeast monsoons very strong, and 2 days and nights we have had very rough weather, the worst since we left Wellington. We were only 170 miles from Hong Kong to day at 12 o'clock, and expect to reach there to-morrow night. We have been two years in commission today. Are expecting the weather to be comfortably cool at Hong Kong, as it is in Lat. 22° N.; from there I shall write you by the next mail, which of course will be all about China & the children of the Sun.[59]

Steaming for Hong Kong, *Challenger* was now two years into the circumnavigation and well past the midpoint of the expedition. The ship had spent nearly a year in the Pacific, and it would be yet another year before it would again see the Atlantic. Despite the discomfort of his cramped working space, Matkin's spirits, to this point, seem to have remained high and his curiosity about foreign lands and seas unflagging. Ahead lay further experiences of the exotic, but also sad news and sailor's malaise.

5   Two Tastes of Asia
NOVEMBER 1874–JUNE 1875

> It takes a good deal to make me laugh in a foreign country,
> for I generally wear my features screwed up into an uncon-
> cerned indifferent sort of focus; but on this occasion had
> to run for it although I stood it better than my friend.
> —Joseph Matkin in Osaka[1]

Challenger was five days en route from Manila to Hong Kong, where it would remain from 16 November until the new year. Just recovering from a severe typhoon, Hong Kong was a source of continuing fascination to the ship's steward's assistant. Matkin sent off no fewer than six letters, nearly all lengthy, during this port stay. First he concluded the letter to his cousin begun at Aru two months earlier.

Nov. 17th, at Hong Kong
 We anchored yesterday at 2 PM & received our letters. Shall not tell you anything of the place or people this time, but am delighted with both; this does not go until the 29th. Expect if we stay here 6 weeks, I shall be able to tell the women from the men, but can't at present, as they dress exactly alike. Miss Kong Whop has just taken my washing, & I was telling her "let's have it off sharp young fellow" when one of our old sea-dogs said "Why he's a gal, you donkey".

> Yours Faithfully,
> Joe Matkin[2]

Although there was no hint in the foregoing letter, Matkin had received dismaying news from home while in the Philippines. Brother Will, younger than Joseph by just a year, had somehow dishonored himself. This may have been connected with the "young lady" referred to in JM's letter to Will a few months before; or with the "business" that he had begun with family financial support—an apprenticeship in the town of March, Cambridgeshire. In any case he had left this situation to join the army, and we later learn that he was posted to Edinburgh and eventually became a fencing instructor. The week after arrival in Hong Kong, Matkin wrote to his mother about the difficulty. This is among

202

the most personal of the letters and shows Matkin in his most serious mood.

> H.M. Ship "Challenger"
> Hong Kong
> Nov 25th, [18]74

Dear Mother,

I received the news of Willie's discreditable affair at Manilla 3 weeks ago, & I need not say how surprised, grieved, & disappointed I was to think that all Father's money had been wasted for nothing, & I can scarcely believe it yet.

When I think of his last letter to me, about his joining the "Glee Club", & his courting &c., I can hardly credit that he has gone from March in such a disgraceful manner.

He talked about my coming to see him when I get back & I thought he was going on so quietly & well, & settled at March for the next 3 years. I consider the bills for clothes &c., disgracefully selfish, & almost as bad as the dishonesty & now he is hooked for the next 10 years, I suppose, on 6d a day, & if Father or you ever wanted a pound he could no more help you than he could if he were a helpless cripple.

However I hope you have spent the last money over him & that he will bear any future trouble on his own shoulders.

I have not heard from him, neither have I written to him yet, for I have not had time this mail to write to any of the boys, & you must send Father's long letter on to them to read. I read Will's letter to W. Thornton & I thought it was written in a very swaggering, impenitent spirit; & I have heard that he wrote to young Tippin of Barrowden advising him to enlist in the same regiment.

I sha'nt say anything to him about it when I write for there is no undoing the job, & I hope he will like his new employment & will get on. His education ought to prove of great service to him in getting him promotion & he will have ample time & means of improving himself, for they have splendid reading rooms attached to those regiments. The mails leave here every fortnight, so perhaps I may hear from him before the next goes.

I have £4 in the Savings Bank which I will send to Father if he wishes it, next March from Japan. I suppose the money had to be borrowed to settle this affair & the sooner it is paid the better.

Let me know as soon as you get this—address Yokohama, Japan. You must tell me how you enjoyed yourselves at Bedford. I was very pleased to hear Miss W's let version of the baby.[3] You did not say whether it was like its Father or its Mother. I received Miss W's letter all serene as well as 2 from Aunt Lizzy, 2 from J.T.S., 2 from W. Thornton & one each from Fred & Charlie which I shall answer next mail. I have written to Walter Thornton this mail & also to J.T.S. and Mr Daddo. I can't tell you all news this time as this mail is just closing. I am not sure whether I shall have time to write home by next mail but Charlie or Fred will send theirs on if I don't write.

With best love to all,

From your affec^te Son,
Joseph Matkin

P.S. Remember me to all Enquirers.[4]

The extant letters resume a month later, after Matkin's twenty-first birthday (which passes unmentioned) and as his third Christmas aboard *Challenger* approaches. However dire Will's problem may have seemed, we do not hear of it again except for a passing reference in a letter to Charlie;[5] the "normalcy" of Matkin's voyage account has returned. The letter to his mother appears below, with interpolations from the contemporary letters to his cousin and brother Charlie. Matkin was greatly affected by the seven-week sojourn in Hong Kong, making several trips ashore. The letters demonstrate that he was eager to expand his knowledge through reading, but also that he did not escape absorbing some of the Western stereotypic features of the Chinese that plague even more modern minds.

H.M. Ship "Challenger"
Hong Kong
Dec. 19th, 1874

Dear Mother,
Your letter of Oct. 15th reached me Dec. 10th, & was the only one by that mail, but I received a "Graphic" from W[alter] T[hornton]. I wrote to Willie last week. I see the letters of Professor Thompson in "Good Words" are finished;[6] they are not as interesting as his "Lecture", I think.

Since we arrived here on the 16th we have taken in all our coal, all Provisions to last us to Yokohama April 15th, & they have nearly finished refitting. We are not going in dock, & shall be ready for sea the 1st week in January, tho' we do not know when we are to leave.

This island of Hong Kong belongs to the "Ladrone,"[7] or

"Thieves" Islands; it is barren & mountainous & only 29 square miles in extent. Hong Kong means in English "Fragrant Streams", & it is just in the Tropics & is about 15 miles from the "Pi-Kiang" or "River of Pearls", generally called the Canton River. The city of Canton is 85 miles from Hong Kong, & there is a Portuguese settlement called Macao 40 miles up the river before you reach Canton.[8] Macao has belonged to the Portuguese for the last 200 years. The island is separated from the mainland of China by the strait called Ly-ce-moon Pass which is about a mile wide in its narrowest part. It is shaped like a half moon the two ends of it approaching the opposite peninsula of Kow-loon & forming one of the most magnificent harbours in the world. In the hollow of the harbour stands the capital town of Victoria at the foot of a range of hills terminating in Victoria Peak, which is 1975 feet above the level of the sea. The town has at least 200,000 inhabitants, besides an immense floating & migratory population numbering many thousands.

From the harbour of Hong Kong nothing of the sea can been seen; it is surrounded by mountains both on the island and main land; these mountains have a very wild, sterile appearance, but make a beautiful picture. Looking at the splendid land-locked harbour you would wonder how the recent terrible Typhoon could cause the damage it did to shipping, over 1000 lives were lost in this little harbor. This you would not wonder at if you saw the

Hong Kong Harbor, 1874 (Natural History Museum, London, Photo No. 450). Photo courtesy of the Natural History Museum, London.

myriads of Chinese junks & sampans in which the floating population live & die. Lives count for nothing in this country, it was the
"Dollars" absorbed by the typhoon that made it such a terrible
calamity here. The sea rose many feet into the lower parts of the
town, & the water was nearly up to the 2nd story in the houses facing the bay: ships were driven from their anchors & thrown almost
into the town. Two steamers are now lying one on top of the other
close to the wharf, with their masts only above water. The Captain
of one of them, a Spaniard, with his wife, were drowned & their
bodies washed up into the 2nd street from the bay.

The typhoon extended as far as Macao up the river, & no one
knows how many Chinese lives were lost up there. There was not
a single man-of-war damaged, altho there were several lying here:
they were moored most securely. Hong Kong was ceded to the British in 1842 & the opposite peninsula of Kowloon in 1851. Of course
it was not taken for its richness or fertility, for nothing grows on it,
but as a military & naval station for the protection of our commerce, & as the centre of an immense export trade its value cannot
be over-rated. The amount of shipping that comes into Hong Kong
on their way to Japan & the Seaports of China in the course of a
year is something enormous. Hong Kong was the base of operations during the long Chinese war between 1848 & 1859,[9] & from its
position at the mouth of the great Canton River its possession by
us is of the last importance. We have a Dock yard & Victualling
Yard here to accomodate our enormous Fleet on the China Station,
& its fine harbour is the admiration & envy of all other European
naval powers. At present there are at least 300 vessels here of all
nations & flags, & myriads of Chinese coasting junks &c. There
are 10 British men of war lying here, 2 American, 2 Portuguese, 1
Russian, 1 Siamese, & 3 Chinese. Every vessel & junk carries a
light at night, & the bay looks like a town. Victoria is well lighted
with gas, & has a very brilliant appearance at night, for the streets
rise one above another for 800 feet.

There is a Governor here, a General, a Commodore in charge of
the Dock yard, & the Admiral in command of the Fleet. There are
not many European merchants, or people of any sort except soldiers; the white civilian population is under 1000. They are either
merchants, hotel keepers, chemists, Doctors &c; all the minor
trades & all mechanical trades are monopolized by the Chinese, for
no European could compete with them in prices &c. Seen from the
water Hong Kong looks something between Lisbon, Wellington &
Levuka as it stands at the base of a range of hills. The climate is

generally healthy & dry; we have had no rain since we have been here; we had two cold days & nights, all the rest of the time it has been as warm as an average English summer.

¶There are 3 old wooden ships here, 2 of them Line of Battleships, and one 3 decker. The 3 decker "Princess Charlotte" was Flag Ship of Admiral Codrington at the battle of Navarino,[10] & is the only 3 decker that ever crossed the Equator. We generally go on board there to Church on a Sunday; she is a roomy old ship used to carry 1000 men & 110 guns {but is now used as a supernumerary ship & prison[11]}. One of the Line of Battleships the "Victor Emmanuel" was Hospital ship at the Ashantee War & is just out from England, & brought out all our new hands. There are 8 large Gunboats for going up the river, & chasing the numerous Chinese Pirates; & there are 2 wooden corvettes, one at Shanghai, & one at Yokohama for the summer. The Flag Ship "Iron Duke" Admiral Shadwell, is a sister ship to the Invincible & Audacious, she has been out here 4 years & is the largest & heaviest ship that ever came thro' the Suez Canal. She was towed thro' the canal by a smaller man of war & grounded on the mud 17 times.

The Audacious is now on the way out from England to relieve her.

Soon after we anchored here scores of small Chinese junks & sampans swarmed round the ship, containing tailors, shoe-makers, washer women, artists, dealers in curiosities, provision boats &c— one of the latter was nominated to supply the ship's company all the time we remain here.

These boats are called in the Navy "bum-boats," & supply the men with all sorts of provisions from the shore.

The Chinese bum boats had bread, all sorts of fruit, foreign butter & cheese, fried fish & prawns, & boiled & fried eggs—the eggs are very large being all Turkey & Ducks, the price is 6d a doz. Very sweet Mandarin oranges, & bananas are the principal fruit. The oranges are 3 a penny, other fruits are dearer for it is the winter season just now. All these provisions & those sold on shore come from the main land of China, & are brought in the coasting junks.

Every meal hour these boats come alongside, so also do the other boats containing washerwomen, tailors &c. I should think there were 500 people round our ship when we came in & the other ships were similarly surrounded. These sampans, as well as the largest junks are built in the shape of a Chinese slipper. The reason for this is reported by tradition to be as follows—About 3000 years ago the people of China began to build their vessels after all sorts of patterns & shapes, so the reigning Emperor was consulted concerning these new fangled notions & asked to furnish a model for all future generations of naval architecture, so he threw down his slipper for a design & it has been adhered to to this day, even their steam war junks are built in this fashion, & looking at their vessels from the stern you see an exact likeness to a Chinese slipper. The smallest boats are decked over & there is a sort of bamboo hood built over the stern of the boat, after the style of a gypsy's caravan. Between the deck & the bottom of the boat is the sleeping apartment where they sleep like bees in a hive. The smaller boats generally carry one man, two women & any amount of children. The large boats carry several families. They live chiefly on rice & fish, but nothing comes amiss to them. At meal hours they come alongside & send all the children on board with bags to pick up the scraps & broken victuals.

About 10 boats attend the ship day & night to take any one on shore who wishes to go, the fare is 10 cents at night, & 5 in the day time.

The money here is in Dollar bank notes & American silver Dollars—value £3 each, & 100 cents go to a dollar. There is also Chinese & Japanese copper money. The floating population of China on the sea & the inland rivers & numerous canals is estimated at over 15 million, & the total population 400 millions, nearly half the whole population of the world. Before coming here I always considered this a gross exaggeration, but can quite believe it now.

Canton has a million & a half of people within its walls, Pekin has more, & there are dozens of cities in China with over a million inhabitants. Still their immense number counts for nothing, for they can scarcely be called a united people. There are distinct languages & religions amongst them, & not the slightest fellowship or national feeling whatever.

If the people here were to hear that Canton had been destroyed by an earthquake or by fire they would send no relief or assistance to the sufferers, & evince no concern whatever, unless it touched their own pockets. Human life in this country is considered of no account. During the recent typhoon scores of people were drowned in sight of their countrymen without the slightest notice being taken of it.

The Chinese have a very numerous Army & Navy, & their Gov.' is more despotic than that of Russia, but to say that our Empire in India is in danger from them, is all moonshine. They are well able to take care of themselves, but as an offensive Power they count for nothing: they have no ambition, no curiosity, & very few of them know that there is such a country as India. I mention this because I read an article the other day in one of the leading magazines, wherein the design of the Chinese & also the Japanese, on India, in the future, was distinctly foretold.

The Chinese do not recognize such a division of time as a week, & have no sabbath whatever: they appear to be even busier on a Sunday if anything. They are idolaters & each household & junk has its own peculiar deity.

Notwithstanding this is a British colony, they have not adopted a single custom from us, nor introduced a single article of European dress. They live in exactly the same style as their fore-fathers 2000 years ago, & they use exactly the same mechanical tools. Their motto is to "adhere to all that is established, & to reject all that is new." Such things as railways, machinery &c &c, find no advocates among the Celestials. They have condescended to erect English sign boards over their shops, as well as signs in Chinese characters, & this is the only innovation I could detect when I have been on shore. There is no alphabet in their language, each character represents a distinct word, & this is what makes the Chinese such a difficult language to learn. They write with a small brush, & do it very neat & quick too. I need not describe their dress—the long cotton blouse & enormous wide trousers, white gaiters & clumsy looking shoes. Their dress is generally blue, & costs very little; their long pig-tail reaches nearly to the ground. The women dress & look very similar to the men; they generally wear ear-rings & all their hair {done in a chignon[12]}, & this is the only means of distinguishing the 2 sexes. The men & women living afloat wear no shoes or stockings; those on shore do. I saw scores of women on shore with the proverbial small feet: their feet are bound up, & compressed into shoes about 3 inches long, & they amble along with great difficulty, like a cat on hot bricks. This cruel custom is falling into disuse I think, for I saw no children being sacrificed to this queer national custom, & imagine it is now confined to the upper ten thousand. The women work terribly hard, & on the water they invariably do all the rowing & pulling, it is nothing unusual to see a boat full of men being rowed by 2 women, & they can navigate a boat better than the men. They carry their babies on their backs in a sort of gipsy's sling, & nurse & work at the same time; the baby's nose is compressed close to the mother's back, & this is the reason the Chinese all have such flat noses. They are a dirty, but very industrious, &, in many things, a clever people.

¶I have been more struck with this country & people than with any other we have visited, & it is the only one that has exceeded my expectations. Provisions, clothing &c are cheaper here than in any other country in the world: any sort of manual labour can be performed in this country at about half the European price, & in many things they can compete with machinery. Rice is their principal food, & they can exist on a halfpenny a day: a whole family of the middle classes will subsist on 3 shillings a week. Their vegetables are the best I ever tasted: they don't eat much animal food, tho' fish & game are very plentiful; they drink nothing but water, & their only vice is opium smoking & eating, & this is too expensive for the majority to indulge in, but they smoke tobacco & cigars. All our carpentering, plumbing &c, repairs, have been effected by Chinese mechanics, & nearly all the workmen in the Gov^t Dockyard are Chinese. They are good citizens, & give very little trouble in the town, but they are a treacherous & a mercenary people, & would do anything for money. During the China war, they attempted to poison all the European community by putting arsenic in the bread, but they put in such a large quantity that its presence was detected & they were discovered. They do not scruple to rob a drunken white man in the streets, or even murder him; & piracy is still very prevalent on these coasts altho' the penalty is Death.

¶We get good Meat, Vegetables, & Bread here & also English beer in the canteen at night. All sorts of provisions are cheap, especially poultry & eggs. We often have Geese for dinner in my mess & sometimes a Chinese dinner from the shore of Curried Cats, dogs & Sundries; cats & dogs are great delicacies in this country & my mess always go in for delicacies. At Kerguelen's Land our greatest delicacy was Carrion Hawks. You can get a good suit of clothes for 30/-, the cost of making is only 6/-, & a pair of English style boots or shoes from 3 to 6 shillings, made of kid or cat skin, & they will wear a long time if not wetted. There are scores of Artists come on board every time our money is paid: they do oil paintings from photographs very tolerably for 8 shillings; also ships, landscapes &c. &c. I have had yourself & Father done in oil for 17 shillings, & framed. They are about 18 inches long (Bust only) & done very creditably, & they are worth a better frame. There are one or two little mistakes if you stand close to them, but I don't stand close to them when I look at them, & I had them done for myself & children.

¶The Chinese invariably ask twice as much for an article as they expect to get, & are wonderful fellows at driving a bargain. Their English is very peculiar, they say "Can do" & "no Can do" for Yes & No, & instead of saying "this is good" they say "this number are nicey", & they always pronounce the letter R as a L: for American they say "Mellican". They are very skilful in carving ivory or Jewelry, & can imitate anything to a nicety. If you go to a Tailor & order a suit of clothes, he will scan your figure for a minute & say "Can do", & without measuring you at all, fit you to a T. They do it very quick too for they are so eager to get the money. Washing is 1/6d per dozen, & they will take it ashore one day & bring it on board again the next. Any rubbish, ashes, &c, thrown overboard is eagerly collected by the sampans, & nothing in the shape of food is thrown away as refuse {& you should see some of the Chinese cook-shops: all the offal, entrails &c of a beast they eat with curry . . .[13]}. The first day our money was paid, the ship was swarmed with Traders of all sorts, & a good deal of money was spent. For an article worth a Dollar they will ask 2-½ & say "No have got 2 tongues"—then if you offer them a Dollar & show them the money they will sing out "Can do". It's just the same in their Bazaars on shore; if you only shew them the money they will sell you something. I have spent a lot of money here in Curiosities, Oil Paintings, Clothes &c. &c. I have bought a large Camphor Wood Chest to keep my clothes in, for 3 Dollars (that is 12/9d) & it would fetch over £2 in England. The wood smells beautiful & no moth or insect will go near it. You could spend a fortune here, in a day, in silks, fans, ivory goods &c. & when you visit one of their large Canton Bazaars you can't come away with money in your pocket. I did intend going to Canton but I have spent so much money that I can't afford it. Lots of our officers & men are up there now, for it is the Race week. I must get a description of that wonderful city from some of them. You can get a box of good Tooth powder here for a penny, & the box would cost as much as that in England, for it is made of wood like a small slate pencil box.

¶Our Photographer that was engaged at Cape Town left the ship here, making the 3rd we have had; however we have got another. The Carpenter of the ship has deserted, & also one of the new boys; but we have taken in over 60 new hands from England to fill up vacancies. We have invalided 2 or 3 here, & several have exchanged into other ships. I do'nt think we shall have 60 of our original 240 hands left, when we pay off. We have 40 boys from Plymouth, just out from England, & they can eat Bananas & oranges.

¶But the greatest alteration of the lot happened a few days ago

when Captain Nares received a Telegram from England to say that
he was appointed to the command of the new Arctic Expedition
that leaves England next April, & was to proceed home at once.
Lieutenant Aldrich (our 1st Lieut) was also to accompany him. The
Captain would have preferred remaining in this ship until the
cruise was finished & then going up [to] the Arctic, but he could
not refuse it. Professor Thompson was in a great way about it, &
talked of throwing up the whole affair & coming home, but the
captain persuaded him not; however he will go home before we get
back.[14] The officers gave a grand farewell dinner & made the Captain a handsome present. The Captain made a short speech & said
how sorry he was to go & how he should often be thinking of his
old ship mates & that he hoped to be back from the Arctic almost
as soon as we get back. They will be away one winter & 2 summers
& return about Sept'r [18]76. The recent Austrian expedition is I
think, the cause of the present hurry in getting away.[15] Captain
Nares said he owed his command to the zeal of his officers & men,
& to their great success; & he thought the new voyage was an off-
shoot of the "Challenger" expedition. You will read all about it in
the Papers.

I don't think Captain Nares is quite strong enough for such a
voyage, he suffered from "Rheumatics" on the Antarctic trip, & he
is rather a timid man I think—not enterprise enough for such a
command.[16] He was up in the "Arctic" 16 years ago, in the "Resolute", & another ship, in search of Sir John Franklin: he was a
Lieut. at that time. The "Resolute" was frozen in so hard that they
had to abandon her & make their way in sledges over the ice until
they reached a settlement. Four years after the "Resolute" was
picked up in the Atlantic by an American Whale ship, having
floated with the ice nearly 2000 miles south of where she was
frozen in. She was taken to New York & put in a thorough state of
repair by the American Government, & presented to the Queen at
Portsmouth.

¶On the 10th Decr the Capt'n & 1st Lieut. left for England in the
mail boat, & will be home long before you get this: the Captain
travels from Alexandria overland. He took with him his Steward &
his coxswain, one of the finest & most popular seaman we had in
the ship. The Captain was rowed on board the mail boat by 8 of the
junior officers; all hands mustered in the rigging & gave him 3 good
cheers, & the Band played "Auld Lang Syne". We are all sorry to
lose him for he was a very kind & good man.

¶A new Captain has been appointed & will be here in 2 days

from Shanghai, where he is in command of the "Modesty". His name is Thompson,[17] & he bears a bad name for tyranny on this station. He plays the fiddle & preaches his own sermons, I believe, but will tell you more about him bye & bye. His coming will probably make our commission longer. We have a new Lieutenant named Carpenter from the "Iron Duke". The "Transit of Venus"[18] was distinctly visible here, but there was no party of observation here abouts. I wonder how they would get on down at Kerguelen's Land.

¶I have been on shore here several times, but the first time was the greatest novelty. The town of Victoria extends from east to west for 4 miles, but from north to south it is of no great extent on account of the steep hills that rise above it. The houses are mostly of stone 3 & 4 storys high with open balconies; & they are wonderfully thick & over crowded with inhabitants. Fires are very frequent here. The population is estimated at 110,000, but I should think 300,000 would be nearer the mark. There are no cabs or omnibuses, so the traffic is not confined to the road sides. Instead of cabs there are hundreds of sedan chairs carried on bamboo poles by 4 Chinamen. You can hire them for a dollar a day, or for 5d an hour. Very few white people ever walk here, & soldiers & sailors invariably use these chairs; it is a very common occurrence to see 20 or 30 sedan chairs coming along the street, each containing a sailor with a cigar in his mouth, & one foot out of each of the small windows. When they reach a public house they sing out "Shorten sail" & "Heave to", & out they all get, keeping the chair waiting until they have had a good "soaking", as they call it {until they have got pretty well "three parts seven eighths"[19]}. Very often they will meet an opposition party of American seamen when there is generally a squabble until the patrol comes & separates them. The Russian & Yankee seamen invariably fight when they meet. Last X.mas day there was a regular fight between the "Iron Duke's" men & the Yankees, & several were killed & hurt. When the "Iron Duke" gives leave, they always send an armed patrol party to try to prevent street rows. The Police of Hong Kong are Sikhs from India, they wear blue uniforms & white linen turbans & are a fine body of men. The Chinese never mingle in these street rows, they walk on about their business as solemn & staid as can be, & seldom give the police any trouble except for stealing & gambling in the gambling houses.

¶The people dry their clothes on long bamboo poles thrust out from the balconies, & if you look down a street you see nothing but these flags of blue clothes. Besides being lighted with gas, each house & shop burns a large Chinese oil lamp made of coloured Paper, & the streets look very brilliant at night, & thronged with people. On each side of the road are fruit stalls, & movable Chinese Restaurants where you can get a good dinner of Rice & "Curried sundries" for a half penny. The streets & houses are very dirty & smell very disagreeably. There are very few European shops or bazaars, as the Chinese can undersell most of the trade. There are lots of bazaars where they sell silk, ivory goods, jewelry, fans, cabinets &c &c & these shops you could spend a small fortune in. There are scores of artists, printers, book-binders, carpenters, coopers, cabinet makers, stone-masons, &, above all barbers; about every 10th shop is a barber's, & they do gossip while they are getting their heads shaved & their pig tails dressed. The undertakers make a fine show of coffins: these are made something like a slipper in shape & generally of camphor wood. The manufactures are chiefly silk, cotton, paper, & porcelain, & the chief export is Tea. There are some immense stone quarries here, worked by a small army of Chinamen, & there are several dry docks round the other side of the island. The Gov^t docks are over at Kowloon, & there is also the Prison. There are 3 British regiments here, & one Lascar regiment from Bombay. There are several fine Public Buildings, & Merchants' Houses here, & a fine Sailor's Home for the benefit of the numerous merchant seamen of all nations. There is a Roman Catholic Church, & a Protestant Cathedral for Hong Kong with Canton & Shanghai comprises a Bishopric: I was at St. John's Cathedral the other Sunday afternoon: it is very pretty & neat & there is a nice organ & good singing! In the Cathedral grounds is a large granite monument to the memory of the brave Captain Thornton Bate, who was killed at the storming of Canton by the British & French in 1859. There is a fine Public Botanical Garden in which stands the Bishop's Palace. No other people could have made such a garden on the slope of a steep mountain 800 feet high.

The women work at many manual trades as hard as the men. I must finish my account of Hong Kong & China in my next letter after X.mas. There was an awful wreck here last week of the steam ship "Mongol" just off this island & 20 lives lost. {She left here in the morning for Japan & ran on a rock whilst steaming 12 knots per hour, & she went down in 3 minutes.[20]} Today we have just received news of the total loss of the American mail steamer "Japan" by fire, about 80 miles from here. She runs between here,

Japan & San Francisco. Several hundred lives are supposed to be lost, & the 2 American Men of War & a British Gun-boat have just gone out to look for any survivors. I must tell you about it in my next letter. {Only 7 people have arrived here yet, out of over 700 passengers & crew, & it is known that she had not half boats enough for such a number of people. The 2 American men of war here got up steam at once & started for the scene of the wreck; they have just returned but I don't know with what result. I can see several bales & packages on their decks.[21]}

I must now conclude with best love to all, hoping we shall all have a merry X.mas & a happy New Year.

From Your Affec^{te} Son,
Joseph Matkin

P.S. December 22nd

Our new Captain has arrived from Shanghai with his servants, his fiddles, & his Piano, looks like a decent man {but it is too early yet to form much of an opinion, as the old saying is "New brooms sweep clean," but if he proves equal to his predecessor we shall have no reason to complain[22]}. I am writing to Charlie, but not to Fred this mail: you must send this on to Fred & to Willie; I shall write to Fred & Miss W. next mail in January. You need not send this on to Walter Thornton as I shall be writing to him in a day or two, & also to Mr. Daddo.

Dec. 23d

The English mail has arrived & brought me 2 letters, one from J.T.S. & one from Fred dated Nov. 4th. Fred says he never received my letter from Tongatabu, so it must have been lost. He did not say whether you had received the Cape York letter, you should have done so early in Nov. Fred's was a very amusing letter, & he writes very well. I shall write to him in a fortnight's time, tell him. I shall write to Will again as soon as I hear from him. J.T.S. tells me that Father was better again; I think we shall receive one more mail here before we leave for the South. I have just received an Illustrated Paper from W.T. of Oct. 31st. The Prussian Frigate "Arcona" that was at Melb^{ne} & Sydney with us, has just arrived. The mail leaves to morrow & we leave here Jany 8th, I hear. We are just going to make our X.mas pudding.

Remember me to all friends.

Goodbye,
J.M.[23]

Between Christmas and New Year's Matkin had time to write again, this time to his brother Fred. The wonderment of China concludes.

Dear Fred,

I received a letter from you Nov. 3d at Manilla, & another here last week, & was very glad to hear you were getting on comfortably. You tell me you have not heard from me for several months, I wrote you either from the Fiji Islands, or Port Albany, Cape York, I am not sure which place.

I wrote to Willie last mail, & I hope you will write to him occasionally; the news of him at Manilla was a great surprise to me & is a sad affair altogether, but we can't alter it now. I think you improve in your writing every letter, & I should like to read some of your essays; you did not say on what subject—I suppose on "Chopping Wood", or "The best way to sell damaged Remnants" &c, &c. You Drapers are a regular moral lot according to your account, & would almost do for missionaries. I see you addressed me Singapore, India, & a good many put Singapore, China; but it is in neither country. Singapore is a small island in the Straits of Malacca—opposite Sumatra & belongs to Great Britain. We have taken in all our Provisions, & Coal, & have had a thorough refit, & we expect to leave here Jan'ry 4th for the Phillipine Islands, again calling at New Guinea &c, on the way up to Japan. So we shall soon be in the broiling weather again; for my part I would sooner go down to Kerguelen's Land again, than into such unbearably hot latitudes. Our ship was inspected on Monday by Admiral Sir Chas Shadwell, & we also had the Russian Admiral on board. . . .

I told Mother to send my last letter on to you, about Hong Kong, & also my Malayan letter to Father. Our address until May is Yokohama, Japan, & a telegram is to appear in the English papers early in April announcing our future address after Japan. We expect it will be Vancouver's Island, but are not sure; you will probably see it as you read the Papers.

Our X.mas passed away very pleasantly, but the weather was so very warm that it seemed less like an English X.mas than our two previous ones—in the Bay of Biscay & off Prince Edward Island. China is a good country for provisions, & a cheap one—Turkey & Geese for 3/. Our X.mas dinner comprised Geese, Turkeys, Ham, a fine large Plum pudding—made by our joint efforts & a great success, & a fine Plum Cake made by our Parsee Baker, "Nowrojee Dowrogee". I am in the Chief Petty Officers Mess, we number 7 & 2 boys to cook &c; for all these extras we have to pay; & we do so out of the Provisions we save, such as biscuits, salt horse &c, which Government allows so much a pound for. We generally save about £3 a month, & when we are in harbour live like fighting cocks—puddings, potatoes & all sorts. We often go in for Curried Cats & Dogs here, as they are great delicacies in the "Celestial Empire".

¶There was some fighting here on X.mas day between the Russians, Prussians, Yankees & English, but it is thought nothing of out here, & when our ship goes out the foreign sailors will all cheer & their Bands will play as if they were sorry to part with us. When there are no foreign sailors to fight with, our men fight with each other—"Challenger's" against the "Iron Duke's" & perhaps the next night they will invite our men on board their ship to see a "Play" or a "Nigger Party". They only quarrel [when] in their cups,[24] & to relieve the monotony. I suppose you went home at X.mas, & would be the only one there of the "children." I was at St. John's Cathedral last Sunday: it was nicely decorated with evergreens, & reminded me of Oakham Church.

¶Our new Captain's name is Thomson; he brought with him a Piano & 2 fiddles, & also his favorite tom cat. The men take the cat on deck at night to fight with our tom cat we brought with us from the Cape; & 'Thomson's' tom generally gets licked. The Troopship "Adventure" arrived here the day before X.mas, from Sydney, & brought 3 of our old Deserters back again. They had been imprisoned for 3 months at Sydney, & said they had come back for their X.mas dinner.

¶There were 2 serious shipping disasters here last week—the wreck of the Steam ship "Mogul" outside the island, & loss of 18 lives; & the total destruction by fire, some little distance north, of the American mail boat "Japan", & loss of probably several hundred lives. The "Japan" was a ship over 4000 tons, & runs between here & San Francisco. You will read accounts of this in the Papers. The "Alaska", sister ship to the "Japan", was wrecked here during the late terrible typhoon.

¶Our men who have been to Canton describe it as the dirtiest & most over crowded city in the world. There are nearly 2 millions of people within its walls, & the streets are so narrow that only 4 people can walk abreast; there are no vehicles in the city; the heavy work is done by porters who carry enormous loads across bamboo poles on their shoulders—& also by the river & numerous canals. The walls are high & very wide, & you are not allowed outside the gates after 7 PM. The river is more crowded than the streets, thousands & thousands of people live on the water entirely; there are all sorts of shops & bazaars on the water, & even Theatres & Singing houses. [There] are lots of European merchants at Canton, & also at Shanghai, 700 miles north. Yokohama, Japan is 1,600 miles northeast & steamers run every week. Pekin is about 1000 miles north, & inland, but you can reach it by the river Tien tsin. It was taken by the British soon after Canton, & its capitulation concluded the long China war. The Chinese then, according to Treaty,—threw open 5 Ports for European commerce, & since that time they have voluntarily opened several others. Still, in social progress & enterprises they are a long way behind the Japanese.

¶The Bamboo is the most useful tree in this part of the world, & is put to an incredible number of uses. I was watching some large Junks the other day, unloading fat pigs from Canton, & it was done without any bustle or noise. A sort of wicker cradle was slung across 2 bamboo poles & carried on the shoulders of 2 men; 2 pigs were then placed in the cradle on their backs, & carried away to be killed.

The Barbers here often ply their trade by the road sides, & at the street corners are professional ear-cleaners, whose stock in trade consists of a 3 legged stool & a wooden skewer. The latter instrument is thrust into the patient's ear, & screwed round several items so as to extract any cockroaches &c. The charge for both ears is about a half-penny, & very picturesque these groups look at the street corners. I have had supper once or twice at a Chinese Restaurant in "European Style"—as they call it. There were about 8 dishes consisting of Ham & eggs, Steak & onions, grilled Fowl, Mutton Chops, curried cats &c, tea or coffee, tarts, oranges, bananas, & cigars. The whole affair 25 cents or one shilling, & you may make a very good supper if you are not too particular. The fish, eggs, & fowl are always good & wholesome.

¶Some of the English sign-boards over the Chinese shops are very funny. One over the door of a stone-masons reads as follows: "Sun Wing Hing will erect a most handsome monument over the grave of any person". This shop is on the way to the cemetery & the sign is both appropriate and impartial.

¶When I was on shore the other day I looked into a Chinese Joss House, or place of worship. The building was similar to other buildings but on the roof facing the road were a whole lot of ugly porcelain gods. Inside, facing the door, was a sort of gilt Altar covered with carved wood & ivory gods & all sorts of queer animals &

reptiles. Several lights were burning, & the whole affair reminded me of the "Transformation Scene" at a X.mas Pantomime. There was one worshipper kneeling before the altar & a Priest officiating. In front of the worshipper was his offering consisting of a large dish of Roast Pork. The priest kept pouring some liquid out of a tea-pot & handing it to the worshipper, who prostrated himself & placed it on the altar. Neither of them spoke, but the Priest seemed to keep a hungry eye on the Roast Pork, & looked rather impatient as there was another worshipper at the door with an offering of Bananas & oranges. The Priest wore no particular dress & appeared to take it very coolly. These offerings &c are the Priest's perquisites & would account for his being such a stout party. The Priests sell small sticks of scented wood—like cinnamon sticks—to the people for them to burn in their houses and junks to their own particular gods. The Chinese worship the moon, & have special services when it is at the full.

¶I had a walk the other afternoon to the top of Victoria Peak. There is a fine zig zag road leading to the very top, & it is so cleverly made that a horse can go up. You can ride up & down again in a sedan chair for a Dollar. The road was made by the Chinese & must have taken a long time & cost a lot of money; these sort of jobs the Chinese can do better & cheaper than any other people, & the Great Wall of China & the "Grand Canal" will testify to the stability of their work.

On the summit of the Peak is a signal staff & a telegraph leading into the town, so that as soon as a vessel is sighted and made out, it is known in the town long before she arrives. A Gun is fired from the summit on the arrival of the English mail. There are several houses up there, & a cottage for the Governor during the sick & hot season. There is also a Barracks for sick soldiers, & there is always a fine breeze blowing up there. The view from the summit is wonderfully striking. You can see the wild, mountainous looking coast of China extending north & south for miles & miles, & away inland, the peaks of mountains over 8000 feet high. Looking out to sea you can count scores of lofty sterile looking islands belonging to the "Ladrone" or "Thieves" group. From the mouth of the great Canton river you will see scores of junks standing across for Hong Kong loaded with provisions & merchandise. Right under your feet is the city, & bay with its crowds of shipping, & opposite is the barren desolate looking peninsula of Kowloon— which belongs to England. Altogether it is one of the most wonderful panoramas you can conceive. Altho' the coast of China has such a forbidding appearance, there is no country in the world so fertile, or so well cultivated. Victoria Peak is 1,875 feet above the sea, but I would sooner go up there 4 times than ascend that Banda Volcano of Goomong-api once. All the roads of Hong Kong are in beautiful order, for there are no vehicles of any sort to break them up.

¶About one mile from the town of Victoria is a gully between 2 ranges of hills: here vegetation flourishes & the whole aspect of Nature is changed. No one—to look at the general barren appearance of Hong Kong—would imagine there was such a pretty spot on the island. Out this way is the cricket ground, & a small but well kept Race-course. The horses that run are chiefly military officers' horses. This gully is called very appropriately "Happy Valley". Here also are all the Foreigners' Cemeteries; the Chinese cemetery is on the opposite side of the island, & looks very bleak & bare, the stones all lying down & no vegetation. The first & largest is the Mohammedan cemetery which is surrounded by a wall, & also looks very bare; the stones there are all lying down. There are many Mohammedans in Hong Kong, the Sikh Policemen & all the Lascar soldiers are Mohammedans, & there are lots of Imaums or Priests knocking about in long robes & silk skull caps. On the other side of the Mohammedan wall is the Catholic cemetery, also surrounded by a wall, & not quite so bare.

Just beyond the Catholic cemetery is the Protestant—the best kept of the lot & the prettiest. The Chinese look after all these burying grounds & are good gardeners. In the Protestant cemetery are buried scores of brave soldiers & sailors who were killed during the long China war, & there are scores who have died from the effects of the climate. Many stones there are to seamen who have been killed by Chinese Pirates on this coast. There are a great many American men of war's men buried there & lots of merchant seamen of all nations & flags. Some of the monuments erected by the different regiments & ships are very large & handsome. The "Calcutta" flag-ship during the China war has one to the memory of over a hundred of her officers & crew. It is a common thing when a seaman volunteers for this China station, to hear some old sailor sing out "There goes another Candidate for the Happy Valley". The favorite walk of the European community on a Sunday afternoon is out this way—when I say walk, I mean ride in a sedan chair, as walking is not the correct thing in Hong Kong. Merchants, hotel keepers, clerks, soldiers, & their wives always use these chairs, & dress & live very extravagantly, & generally above their means. I have not seen anything in the papers yet about the new Arctic expedition.

I hear that the English mail has left Singapore, so we shall probably wait for her; the distance from there [to] here is 1450 miles, & they generally come in 5 days if the Monsoons are not blowing too strongly against them. There is a fine line of French mail boats running here, from Marseilles, via Suez & Bombay; they also call at the French possession of Saigon, Cochin China. The first class fare from here to England in the Mail boat is £98. The principal export to this country is Opium, which comes from British India; there is an immense profit on it, & fortunately it won't grow in China. The weather here is very warm & dry, but liable to sudden change in winter. It is very cold up at Japan now, & the north of China, but it will be getting warm by the time we arrive there. I don't think we are going to Petropauloptsky at all now, being so much behind time we shall probably go direct from Japan to Vancouver's, so we shall get no more cold weather until we are coming thro' the Magellan Straits, next January—[18]76, & that is the summer season down there. We went outside the harbour last Tuesday to Adjust Compasses, & have a little Dredging, taking with us about 40 Ladies & gentlemen. In the afternoon there was dancing on the upper deck & I think the visitors enjoyed that, & the luncheon most. We had a fine view of the Ladrone Islands & the China coast. One of the British Regiments here is going to the Fiji Islands now that group belongs to us. I expect the Fijians will decrease rapidly during the next 10 years. We shall not be able to send or receive any letters for 3 months after this. After we leave Japan we may consider ourselves Homeward bound, for we shall only be about 12 months after that. I thought about having a Picture of the ship amongst the ice done in oil here, but I think I must wait until we reach Japan.

I am writing home this mail, but not to Charlie, so you must send this on for him to read. Mother will send hers on to Willie. I have also written to J.T.S. Remember me to Walter Freeman, & to all the Stamford Friends; I wrote to J. Seaton some time ago.

With kind love to Uncle, Aunt & Cousin Lucy & to yourself,

From Your Aff^te Brother,

Joseph Matkin

P.S. You did not say whether Uncle Tom was better after his accident some time ago. A month's money was paid us yesterday & the Chinese traders—with that sublime pecuniary instinct—got scent of it & swarmed alongside with curiosities &c &c for sale.

P.S. January 1st, 1875

A French Ironclad called the "Montcalm" came in last evening, & also a Prussian Frigate. The French Admiral saluted the British Flag with 21 guns, & the British Admiral with 13 guns. This was returned by the Flagship "Iron Duke" & the Fort ashore. The French Admiral then saluted the Russian Admiral & the Prussian saluted the French. The Russian Admiral then saluted the French, & the French the Prussian. The whole affair occupied half an hour. The Flag of the nation being saluted is hoisted at the Fore of the ship that is firing.

January 3d

One of the Russian Frigates left yesterday, "homeward bound", she saluted everybody before leaving, & the 2 Russian ships cheered each other. An Austrian Frigate is just anchoring & is saluting all the Admirals. We went outside again yesterday to Adjust Compasses, but do not leave finally until the mail arrives, probably on the 6th.

January 5th, 9 PM

The mail has arrived & brought me a letter from J.T.S. & one from Mr. Sleath & Harry, also a Mercury. We leave at daylight for Manilla.

Goodbye,

JM[25]

After refitting and replenishment, *Challenger* returned southward through the Philippines to New Guinea. There is a hiatus in Matkin's communication of nearly two months—a letter to his mother from Manila has not survived. By the time he wrote again, the ship had arrived in New Guinea. Some of the missing days are accounted for at the beginning of his next letter to Swann. Thereafter, the concurrent letter to his mother continues, which (following some of Matkin's strongest language about the various ethnic groups he has encountered) takes us to Humboldt Bay on the northeast coast of New Guinea, then east to the Admiralty Islands, and finally north across the equator again through the Caroline and Mariana islands.

On this leg of the voyage, terrestrial explorations seem to have superseded oceanographic. A planned visit of some days at Humboldt Bay had to be abandoned because the natives were unfriendly. The Admiralty Islanders were quite hospitable, however; *Challenger*'s landing there was the first by a scientific party. Sounding and trawling continued, and on 23 March the incredible depth of 4,550 fathoms was reached in what is now the Mariana Trench—nearly twice the depth of previous soundings.

H.M.S. "Challenger"
Humboldt Bay, New Guinea
February 24th, 1875

Dear Tom,

I had not time to write you from Manilla, as we only had a few minutes notice that there was an opportunity for sending letters.

We left Hong Kong on the 6th of January for there, and had pretty good weather across the China Sea, during the voyage we sounded at an average depth of 2000 Faths. On Sunday the 10th we sighted the large and mountainous island of Luzon, at about 100 miles north of its capital—Manilla. The wind falling we got up steam, and when about 60 miles from the town, we sighted a small dismasted vessel, and went out of our course to examine her, she proved to be a small Spanish coasting schooner, and from her appearance looked as if she had been plundered by Pirates. We towed her in to Manilla, and being of no value, she was given back to her owner, who lived there and had been out looking for her for some time. Her name was the "S. Mayana" & proved to have been robbed by Chinese and Malay Pirates, and her crew murdered.

We anchored for the second time in Manilla Bay on the 11th: the place looked as pretty and sleepy as ever, and on going on shore one evening found the same stagnant peculiarity which all Spanish—and most Catholic countries seem to exhibit,—soldiers, priests, ladies in carriages, bands playing, church bells ringing, natives hard at work, and white folks doing nothing but smoke, perspire, and drink coffee.   *   *   *

Our ship brought the news—per Telegram, received at Hong Kong the day we left, of the elevation of Alphonso to the kingdom of Spain.[26] The government at Manilla refused to credit this information, or to allow of its being published in the papers until they had received two telegrams of their own confirming the same, which will probably reach them during the present year.

¶We found a Spanish Frigate, a Prussian Corvette, and a Russian Frigate, lying in Manilla bay, and the British gun boat "Elk" which sailed the next day for Singapore, and took a small mail for us to be forwarded to England. The Prussians are well represented in their Navy in this part of the world; lots of their Officers, as well as Spanish & Russians visited our ship during the stay, and our Officers gave the usual dinner party before we left. Our new Captain is a regular ladies' man, and a frightful old growler, & spinner of incredible yarns, commencing—"When I was a midshipman 20 years ago." He is also fond of his cigars, & good living.[27]   *   *   *

Dear Mother,

I hope you received my Manilla letter dated 12th Janry all right. The vessel we picked up off Luzon was given back to her owners at Manilla. My share of her consists of the ink with which I am writing this letter, which was found on board the vessel by one of our men.

¶We only stayed 3 days at Manilla & sailed on the 14th for Zebu [Cebu], another of the Philippine Islands 300 miles south. We steamed thro' some very intricate straits, with small islands sprinkled therein as thick as mushrooms. We passed several lofty islands, & the weather got warmer every day, the Thermometer averaging 90° in the shade. On the 17th we passed hundreds of sea-serpents[28] floating on the surface of the sea; they were about 2 feet long on the average, & beautifully marked with black & yellow stripes; they were very active in the water & it was not an easy matter to catch them. On the 18th we sighted Zebu & Matan [Mactan]—an adjoining island. We steamed thro' the Philippine straits which separates the 2 islands, the scenery on either side being beautiful, & anchored off the town of Zebu the same day. The strait is only half a mile wide near the town, where it opens out into a capacious bay of great extent. Zebu is 100 miles in length & averages 40 in breadth; it is of volcanic origin & tolerably high, while the opposite island is of coral formation & scarcely above the water's edge, but very fertile.

¶Magellan discov'd the Phillipine Islands in 1521 & on the island of Matan met the same fate from the hands of the natives as our own Captain Cook at Owhyei [Hawaii] 250 years later. On the spot where he was murdered he was buried, & there is a large monument over his grave. Our Photographer went to Matan to take a photograph of his tomb, & I have obtained one copy. In the great square of Manilla there is also a monument to Magellan.

¶Zebu is a great tobacco growing isl'd, most of the tobacco being forwarded to Manilla to be made into cigars. There was a fine British merchant ship lying at Zebu, from Sydney with coal—en route for China to take in a cargo for England. It seems strange that Sydney should be sending coal to the Philippine Islands, when it lies almost on the surface of the earth in several islands. The Spaniards are too lazy to work the mines, & the islands are as far behind the rest of the world as they were 200 years ago. The trade is chiefly in the hands of British & American merchants. In all Spanish colonies & in most Catholic countries (except France, where the Church is subordinate to civil authority) you find this peculiar stagnant quality.

I was in conversation with a Spaniard at Zebu, belonging to Cadiz, who had been in Australia & America. He said the Philippine Islands was the oldest of all European colonies, & the most backward; & this he attributed to the inherent laziness of the natives. If he had said "to the strange deterioration of the Spaniards since the days of Magellan, Cortes, & Pizarro" he would have been nearer the mark. The natives are undoubtedly lazy, but the Spaniards do very little but smoke, drink coffee & fan themselves. The only hard workers in the Philippine Islands are the Chinese, & they are very numerous. But a Chinaman does not work for the public good, or any body's good but his own. He studies no man's opinion, he wants no man's approbation; & he cares not what you think of him so long as you pay him & let him alone. In such countries as Australia & America where they know how to utilize these people they are invaluable, if not too numerous.

¶There was also a French barque at Zebu, but she sailed the day before us, for London, with cigars & tobacco. There is a British, American, & German Consul at Zebu for it is a place of considerable trade. We obtained fresh buffalo[29] & vegetables there from a German Contractor, Fruit was very plentiful, & cigars could be bought by the hundreds for old clothes, knives, or razors—these articles being luxuries, & unobtainable by the natives of Zebu. I got 1100 for some knives & fancy articles; if it were not for the Custom House I should bring them home.

¶The natives of Zebu are much handsomer than those at the other Philippine Islands we have been to; they are more like Polynesians than Papuans. The women are very fond of dress without being at all overburdened with it. The sea bottom round Zebu is covered with gauze looking creatures called "Venus' flower baskets" which are spun by a species of water-beetle, & look like columns of frozen lace.[30]

I went on shore at Zebu one evening & had a very pleasant walk into the interior. The town was nothing to look at: there was the usual quantity of Catholic Churches in front of which fireworks were sent off every evening for the amusement & instruction of admiring crowds of natives, while the bells of all the churches dis-

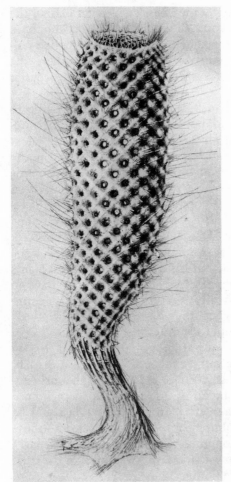

Venus' flower basket sponge, *Euplectella aspergillum*, drawing by J. J. Wild. Reproduced with the permission of the Hydrographer of the Navy. Photo courtesy of the Natural History Museum, London.

coursed discordant music. We took on 75 tons of coal on the 20th, & on the 23d a Spanish steamer arrived from Manilla with official news of the political changes in Spain, & appointing the following day as a general holiday & jubilee on account of the said changes. So on the 24th flags were streaming from all the houses, the church bells rang louder than ever, & fireworks were going off to a frightful extent. The natives were very well satisfied with their new King & would apparently be ready for any further political changes at a very short notice.

¶We left Zebu on the 24th for Mindanao, the second sized of the Philippine Islands, & almost as large as England. On the 25th we passed the large island of Bohol, & on the 26th touched at the small volcanic island of Camiguin & landed a party thereon. The island was about a mile long: one end was flat, fertile, and inhabited; the other a black, sterile, burning cinder rising 1500 feet above the sea, & smoking night & day. It was far more active than any volcano we have seen, & dense volumes of smoke ascended from caves level with the water's edge without affecting the temperature of the sea thereabouts.

¶We left again the same evening & sighted Mindanao the next day. For 2 days we sailed parallel with its western side, & on the 29th anchored for the second time off its capital—Zamboanga. We brought Dispatches to the Governor there from Zebu, concerning the new King, & on the following day the 6 Spanish vessels lying there were decorated with flags, & the town itself dressed out to celebrate the auspicious event. The same evening we steamed across to some adjacent small islands 10 miles south, & the next day landed an Exploring party thereon, while the ship steamed away to make some Magnetic observations. We returned to Zamboanga the same night.

¶On the 31st, Sunday, I went on shore & had a walk into the country. I found that the Rice crop had been gathered in since we were there before, & the fields under water, it being the rainy season. Cockfighting is the great Sunday amusement at Zamboanga; the fighting takes place in the centre of the principal street, most of the naval & military officers looking on & betting on their pet bird. The birds belong to the natives & to the Chinese, & the owners back the bird for so much. Cock fighting is about the only amusement I ever saw the Chinese engage in, but as they are great gamblers this [is] not to be wondered at so much. I was on shore again before we left & examined the old Fort, over 300 years old. I think I described the place in a previous letter. A party of our Officers, Naturalists, & seamen went on a 4 days Excursion into the interior of Mindanao, but as they only penetrated about 6 miles, they can't know very much more about its capabilities & resources than myself. They shot some monkeys, & other queer animals, & brought a tiny monkey back whose mother they had shot, but in about 10 days it pined away & died. Fire-flies were very numerous at Mindanao.

¶On the 3d Febry we got up anchor & steamed to Port Isabel on the same island, for coal. In getting up the anchor we hauled up two other anchors & lengths of cable attached, which, from their corroded appearance must have been there many years. Port Isabel was a native convict settlement, & the convicts carried our coal on board; they were rigidly overlooked by Spanish soldiers. We took in 100 tons, making 240 on board to take us all the way to Japan, via New Guinea, Admiralty Islands &c., against the N.E. Trade Wind. We returned to Zamboanga on the 4th, & found a Spanish Frigate flying an Admiral's flag at anchor.

¶On the 5th we took in some breeding goats & pigs for landing on uninhabited islands, & left Zamboanga for New Guinea. On the 6th one of the pigs foolishly committed suicide by jumping out of his cabin window into the sea. On the 8th we sounded off the S.E. point of Mindanao at 2,025 fms, & on the 9th lost sight of the Philippine Islands altogether. On the 10th we sighted the Meanchis Islands,[31] & trawled off Tulor, the largest of the group. The depth was only 500 fms, & the Trawling was very successful. Tulor was 200 feet above the sea level on its highest point, & looked very fertile. A canoe flying the Dutch flag came out from the island & brought several beautiful birds, which were bought by our Naturalists. The natives were almost black, & seemed of pure Papuan origin. On the 12th we sounded at 2,550 fms but got nothing, the bottom being of pumice stone.

¶On the 13th we were getting close to the Equator, & were out of the Trade Wind region into that of Equatorial Calm and constant precipitation. This region extends for about 300 miles on either side of the Line, & is one of the most uncomfortable & depressing portions of the Earth's surface. Evaporation, owing to the great heat of the sun, is going on night & day to an incredible extent, & the vapour ascending into the cloud region is met by an upper current of cold air from the Pole, which condenses it again, & causes the constant rains that are here met with. Every few hours the rain comes in torrents; I never saw it rain to such a degree elsewhere; & mariners have described the surface of the sea in this region immediately after a heavy downfall of rain, as being almost fresh enough to drink, for a short time. Sometimes it rains on one side of the ship & not on the other, & you can see the rain squall coming for miles before it reaches you. But it is not only the rain that makes it so uncomfortable, it is the damp oppressive nature of the atmosphere which seems to take all the life & energy out of a man. The atmosphere is so hazy that you can scarcely see 10 miles; the sun does not seem so hot as the damp air night & day, & the inside of the ship is like an oven, especially if we are under steam. Your clothes & boots are covered with blue mould & require brushing every day. Men subject to Rheumatics suffer terribly, & can tell when the rain is coming long before it comes. Strong squalls of wind come with the rain, & these assist ships across this region in a few days, but we are not like other ships. We are in this region now & shall be for the next month, as we steer east when we leave New Guinea.

¶On the 17th the temperature of the sea at the surface & the air in the shade was exactly the same—86°. The sea tempᵣ seldom reaches higher than this, & if you are steering for the Equator, it varies as much as 5° a day sometimes. We were 7 days sailing 200 miles, & crossed the Equator in Long. 138° E on Sunday 21st at 9 AM. I might say we flopped across it, sailing about half a knot per hour, & it rained the whole day. The same night we got up steam as there was no sign of any wind, & proceeded for New Guinea, distant 100 miles south. On the 22nd we sounded at 2,020 fms, & passed enormous masses of dead mangrove trees & brush wood &c, on the surface of the water, & Dolphins & Sharks became very numerous.

¶Yesterday at 12 o'clock we sighted New Guinea out of the fog about 12 miles away. The land was high & covered with a dense vegetation, & as the weather cleared a little, several mountain peaks over 6000 feet high appeared above the fog. On the opposite side of the island are ranges whose peaks exceed 16000 feet in altitude, & what there may be inland no man knows. At 6 PM we steamed into Humboldt bay, & dropped anchor about 3 miles from the shore. We soon saw some lights dancing about on the beach, & at 8 PM 2 canoes came off to the ship, but for a long time they would not come alongside. A blue light was lit, & tho' it startled the natives at first, they soon acquired confidence & came alongside close enough to receive several little presents, & bundles of cigars. They were entirely naked, & tattoed, & painted in the most horrible manner, & altogether were the fiercest & most warlike looking race we have yet seen. They were armed with spears, bows & arrows, & stone hatchets &c; they could all say "Segar, Segar", and smoked away with great delight. At 9 PM they returned to their village to fetch others, & all night long we could hear them jabbering, & gathering round the ship in great numbers.

¶New Guinea or Papua was discov'd by Torres in the middle of the 16th century, & the strait which separates it from Australia is named after him. The island is 1200 miles long & averages 250 broad, & is more than 3 times as large as England. It is the headquarters of the Papuan or Austral-Negro race which on this island alone is supposed to number several millions. Of its interior nothing definite is known; one or two Travellers have succeeded in penetrating for a short distance into the interior from the opposite side of the island, but the determined hostility of the natives prevented any connected exploration. The natives being broken up into different tribes hostile to each other proved the greatest drawback. One German Traveller estimated the mountains in the interior as exceeding 20,000 feet in altitude, & of course where there are such mountains, there must be rivers of considerable size. The western side of the island being contiguous to the Dutch & Spanish possessions, has been visited by their ships at intervals, for the last 200 years, & it is certain that the Chinese were acquainted the island much earlier. In exchange for iron & fancy goods they obtained from the Papuans ambergris, tortoise-shell, small pearls, & birds of Paradise. No permanent settlement has ever been made on the island, & even its coasts have not been properly surveyed. The French navigator Bougainville surveyed Humboldt bay & other parts of this side of the island in 1769, & the Russians have also visited it during the present century. One or two British men of war have touched here within the last 20 years,—one, the Basilisk has just gone home after making a survey of the coast opposite Australia, with a view to making a settlement in that direction from Queensland.

¶A year or two ago a report was spread in Australia that Gold had been discov'd in New Guinea, so a party of diggers & Adventurers started from Sydney as the Pioneers of the new El Dorado, & some of them never came back. The natives killed some & the climate killed more, & the expedition returned to Sydney without seeing a sign of the precious metal. The place where they landed is over 1000 miles southwest of Humboldt bay. Those missionaries we saw at Cape York make occasional visits to the southern portions of the islands, but as yet their progress has been small & confined to that small space. The work of Evangelizing the Papuans will be one of difficulty & danger, & will necessarily take a long time. The natives in the interior are supposed to be a very different race to those near the coast, where they are frequently mixed with the Malay family.

¶By daylight this morning about 30 canoes containing over 150 naked warriors, were alongside the ship. They were a dark brown colour with long wooly hair, & were tall & well built men. Most of them were tattoed, & painted with red & blue ochre, had large bones thro' the nose & ears, bone ornaments round their arms & legs, & necklaces of human teeth. The Chiefs were the most painted, & had a row of cock's feathers round the crown of the head. Many had spear & arrow wounds in various parts of their bodies which had evidently been received in battle. Their canoes were something similar to the Fijians, & they brought off Cocoa nuts, bananas, & native war implements for trading purposes. Cigars & old iron hoops were most sought after; they did not care for Beads or old clothes or fancy articles, & had no idea of pipes or tobacco. I should think they have been visited occasionally by vessels from the Philippine Islands, trading with cigars. The iron hoops they straighten, sharpen, & make into hatchets & knives. Their own knives & axes are made of that hard green stone of which brooches are sometimes made in England.[32] The row they made alongside was deafening, & the trading in Papuan war implements has been going on all day very extensively. No women or children have been alongside during the day, but we can see them down on the beach. Their huts are neat looking conical structures over 30 feet high, & built of bamboo & grass thatch. They all chew betel nut to a frightful extent, & have no teeth, or voices owing to this habit. Still, they were a finer looking race than the Arroo Islanders, & far superior to the Aboriginals of Australia.

¶At 7 AM we got up anchor & steamed in closer to the settlement; the Band playing all the time. The Music startled them at first but they soon ceased to notice it; they yelled with delight when the anchor came up, but they evidently could not make out what it was that moved the ship. All the boats were got out at 9 AM, & steam got up in the steam pinnace preparatory to a week's stay surveying, exploring &c. The boats crews were all armed & at 10 oclock the Captain left the ship in his gig, for the native settlement, taking with him several little articles as presents. Just as he approached the village a crowd of naked women rushed down to the beach, armed with spears, bows & arrows &c, & frightened our brave Captain on board his ship again in a very short time, after collaring his presents & everything else they could lay hands on. The women were much uglier than the men, & quite able to take care of themselves. Our Captain is rather a Lady's man too, fond of dress, good living, & drink, & a great spinner of incredible yarns commencing "When I was a midshipman 20 years ago, I was a sad scamp &c". Another boat tried to land at another village & was scared in exactly the same manner. They evidently knew the use of fire arms & it was the sight of those alone that prevented them from proceeding to active measures. This afternoon the steam pinnace, with a party of Naturalists, Officers, &c, effected a landing on an uninhabited part of the island, & shot a few birds, & gathered a few botanical specimens; but they saw no Birds of Paradise, & the natives seemed to have none either. Probably these birds are not found on this side the island.

I should think, judging from the determined resistance met with by the landing parties, & the persistent refusal of the natives to come on board to have their portraits taken, that kidnapping has been carried on here at some previous period by vessels from Australia. The boats were all got in again at 4 PM, & we are now steaming out of Humboldt bay for the Admiralty Islands—400 miles further East. So our stay here has been very short, & we shall add but little to the scanty knowledge concerning New Guinea.

March 10th, at the Admiralty Islands, Lat. 2°30′ S

We lost sight of the mountains of New Guinea on the 26th, & passed Tiger & Matty islands. We have made the usual slow passage across, & have had wet miserable weather. On the 28th we passed Bertrand Island[33] & sounded at 1,025 fms. We sighted this group on the 2nd Mch, & anchored on the 3d in our present position. In the year 1648 Tasman, the Dutch navigator visited these islands, but they were probably visited by the Spaniards at an earlier date.[34] Bougainville visited them in 1767. They are a considerable cluster situated to the W.N.W. of New Ireland; only one is over 100 sq miles in extent: this is tolerably elevated & of volcanic origin;[35] the remainder are small level islands of coral formation. They are all fertile, & many are inhabited, but the total population does not exceed 4000.

¶We are anchored in a bay between the large island & several smaller ones surrounded by extensive coral reefs, over which the sea rushes & roars with tremendous force, especially at night. Our officers have surveyed this bay & called it Nares bay. Several of the native canoes full of naked savages, came out to meet us, & have been alongside all the week trading with their native war implements, cocoa-nuts, bananas, yams, &c, for iron hoops &c. They had no idea of Tobacco or cigars, but iron hoops & hatchets were prized more than anything; & there is scarcely an iron hoop in the ship. They are great Thieves, & are not at all afraid of coming on board to have their portraits taken. They are something like the Papuans, but not so fierce looking, & are evidently mixed with the South Sea Islanders. There is a great diversity of feature & colour amongst them too. They are more used to the sea than the Papuans, & swim & dive like fishes. They chew betel nut, & their weapons are something like the Papuans, but made of different material. They fight, one island & tribe against another, & there is no sign of any missionary ever having been here. They wear Mother of Pearl Shells round the neck as a Charm. The islands are out of the track of shipping & are seldom visited. The natives are also tattoed, & painted & wear bones thro' the nose & ears. The disease called Elephantiasis is very prevalent here.

¶The other morning I bought a cocoa nut with the young tree growing out from it, & on opening it was never so surprised with anything like it before. The young tree holds the nut suspended by long white fibres which envelope the inner shell. It is the province of these fibres to supply the young tree with nourishment until it no longer requires it. On opening the inner nut I found the whole of the milk absorbed, & the nut quite dry, but full of a sweet pithy substance. The Cocoa Palm flourishes best in a sandy soil, & bears fruit in about 3 years. Shooting Parties have been away every day to the large island which is not inhabited, as there are no cocoa palms on it, & have shot enough Pigeons (of a Parrot species) to last all hands 2 days. The natives act as guides, they lie down when the

Village on Wild Island, Admiralty Islands, 1875; from a drawing by J. J. Wild (Natural History Museum, London, Photo No. 484). Photo courtesy of the Natural History Museum, London.

gun is going to be fired, & when the bird falls they yell with delight. The Admiralty Islanders are Polygamists; one of the chiefs offered our Captain 2 of his best looking wives for an American Axe. Yesterday 2 couple of the breeding goats were landed on the large island; there to increase & multiply; there were pigs there already.

¶I was onshore yesterday for a few hours; the isl^d is about 2 miles long, & ½ broad, & contains about 600 inhabitants.[36] They live in large huts, & chiefly on cocoa-nuts, fish, yams, & pigs. The women are the ugliest human beings I ever saw, & are horribly painted and tattoed; the men are handsome compared to them. They [the women?] wore a bunch of Palm leaves, & when about 60 of them danced a war dance to the music of tom toms, it was slightly suggestive of a jubilee in the "Infernal Regions". {In a physical point of view they are superior to the aboriginal women of Australia.[37]} We saw 2 life size, carved wooden gods—male & female—at the entrance to one large hut, but no particular reverence seemed to be paid to these Deities. Last night we gave the natives a specimen of what we could do, by firing off several shell & Rockets at 9 PM, & which frightened them very much. We are just leaving for the Caroline & Ladrone islands further North.

April 8th, Latitude 30° N, Longitude 137° E

We lost sight of the Admiralty Islands on the 10th, & sounded on the 11th at 1100 fms; the Equator was crossed on the 12th under sail, weather wet & miserable. On the 13th the depth was 2,650 fms, & on the 14th & 15th we were under sail. On the 16th the depth was 2,500 fms, & on the 17th we kept under sail. On the 18th we had to steam as there was no wind; & the weather was roasting. On the 19th we sounded at 2,325 fms, & Trawled successfully. On the 20th we picked up the N.E. Trade Wind, & tho' it was dead against us we managed to sail 5 knots per hour lying close to it; & we were soon out of that Rainy region into beautiful weather. On the 21st we sighted one of the Caroline Islands a long way off, & sounded at 1,850 fms. On the 22nd the wind was so far ahead that we were driven to leeward of the Ladrone Islands, & we were to have called at Guam for fresh provisions. As it was we bore right away for Japan, sailing 6 knots per hour.

¶The 23d was a great day on board, at 6 AM we sounded at the enormous depth of 4,600 fms, but as there was a doubt about it, the line was hauled up again. It was sent down again more heavily weighted, & 2 patent Thermometers were attached. The depth was decided to be 4,550 fms (5 1/6 miles)—the greatest reliable depth ever obtained.[38] One of the Therm'trs burst owing to the tremendous pressure on it; the other registered the temp'tr on the bottom as 35° (or 3° above freezing point for fresh water), & the temp'tr at the surface was 80-1/4°. The position of this great depth is Lat: 11°16' N, Long: 143°15' E, about midway between the Caroline & Ladrone Islands.[39]

On the 24th we sailed about 100 miles, & sounded again on the 25th at 2,300 fms; on the 27th the depth was 2,525 fms, 29th 2,450 fms—April 1st, 2,500 fms., & yesterday 2,500 fms.

On the 1st April we Trawled & carried away the trawl. On the 23d we must have dropped our sounding lead in some deep ocean valley, almost as much below the surface of the sea, as the Peak of Mount Everest is above it. On Good Friday we sighted the first vessel we had seen for 6 weeks. We signalled her—the "John Nicholson", from Newcastle (Sydney) to Manilla with coal—45 days out. We have looked for several reefs & one or two islands that were in our course, according to the Spanish Charts, but we could not find them. There are no English charts of that portion of the Pacific. On the 6th there was a partial Eclipse of the sun visible here between 3 & 5 PM. We are 350 miles from Japan to day, & the weather is foggy, cold & wet. We are wearing our warm clothing again. The Therm[ter] which stood at 80° last week, is now at 54°, & the surface temperature of the sea which was 85° at the Equator is now 60°. We are hoping to get in by Sunday, & are hungry for fresh provisions, letters, & newspapers. Our latest papers from England are November 23d '74. A copy of this has gone to Barrowden, Swadlincote & to Mr. Daddo; you must forward it to Charlie & the other boys, as I have not time to write them this mail. Let Walter Thornton have it also, to read. I shall write to him from Yokohama next month. I will now conclude, hoping all are well, & with best love,

From Your Affec'te Son
Joseph Matkin[40]

Having passed to the west of the Mariana Islands, *Challenger* continued north to Japan, arriving in Tokyo Bay ("Gulf of Yedo") on 11 April. For the crew this period of more than a month between the Admiralty Islands and Japan was the most boring, dreary segment of the voyage.

After so many uncomfortable days at sea and expectations of fresh exotica in Japan, Matkin was greeted by the tragic news, received through his aunt and cousin rather than from the immediate family, that his father had died just before Christmas—nearly four months earlier. In writing to his mother, he understandably had little say about the voyage. Concerns about the family and his own future suddenly loom large. As we would expect, the tone of this letter, and the one to brother Will written the same day, is quite unlike any of the others; by Victorian mourning standards, however, it is moderate, controlled.

> H.M. Ship "Challenger"
> Gulf of Yedo
> April 12th, [18]75

My Dear Mother,

I have heard of our great loss, but it seems hard to believe that Father has been buried nearly 4 months. I have not heard from you yet, or from any of the boys, & I am anxious to know all the particulars. If there had been but a word of message from him, it would have seemed less hard to bear.

I hope you will not fret too much, but look forward to the time when we shall all meet again, in a happier world! It is our first great trouble, & I am thankful to think that Father lived long enough to see us all provided for, & properly educated. Few children in our station of life have had so much spent on their education & start in life. I often think of Father's pleasure in seeing us off to Billesdon every half year; & coming to meet us on our return. I know by his manner that when he said 'good bye' to me as I came away, he seemed to think it was for the last time.

We only anchored here last night & the latest mail came on board some hours before the old ones. The only letter I had by the first mail was from Aunt Lizzie which said "J.T.S. wrote & told you of your poor Father's death"!

This was dated Feb^ry 4th, & J.T.S.'s letter of 20th Decr did not reach me until 9 PM. It is not 24 hours since I learned it all, but it seems weeks. Poor Father would not get my last letter to him, from Hong Kong in Novr last. I have his last letter to me at Cape Yorke, & he still complained of his chest; I see that he posted me a Mercury on the 14th Decr. It would be a sad X.mas for you all this year. I am writing to poor Willie; he would feel it so much, not being able to come to the funeral. I shall write to Fred next mail; you must send this on to him; of course if Charlie is at home now I shall not need to write specially to him. I hope the baby is well; it is pleasant to think that Father lived to see one of his grandchildren.

By the time you get this we shall be "Homeward bound" & you will be able to count the *months* instead of years as they fly past. I have fully determined to leave the Navy when I get back; there will be nothing to hinder me; I shall have a little money & what is better a <u>good character</u>! & I hope we shall all be settled down in England, & have many happy years yet in the old home!

I have a few pounds in the ship's bank & if you want money just at present you have only to say so. I shall let it remain in the bank otherwise. You remember me breaking poor Father's favorite cane just before I came away! I had bought a nice Malacca cane at Hong Kong thinking to replace it.

I hope you will have a nice grave-stone over the grave; you must tell me in what part of the cemetery it is. If you have a picture of the cemetery I should like one. I had a letter from Uncle Joe dated 30th November: there were 3 from J.T.S. & one from Walter Thornton Dec 14th. I have had several papers, but I have not read them yet for I have been busy & confused.

I was glad to hear that Mr. Pykett was at the funeral for he & Mr. Thomson were Father's most intimate friends. Aunt tells me that Miss Wildman had more family trouble; it seems all bad news this mail & I was in such good spirits when we sighted Japan last Sunday morning (yesterday).

I hope her sister is better again, & that she keeps well herself. I could scarcely fancy the old home without Miss Wildman. From what Aunt said I suppose Charlie is at home for good. I am so glad to think that the business is going on as usual, & I think the arrangement the best that could be made. Poor Father had provided for everything, so far as he was able; it has been terribly uphill work since we boys grew up, & in everything that conduces toward an upright manly Christian life, he left us a bright example, & a good name, & we'll keep his memory green.

It is a pity that Charlie could not have lived at home, but perhaps it is better as it is, & you will be able to assist them in family arrangements &c, & will be less lonely with Charlie at home. I hope you will like Harriet & that Charlie will have a comfortable home. The mail leaves this evening, & another is due from England next week. I am anxious for a letter from you to know that you are well, & getting over it a little. If anything were to happen to you, there would be no such thing as 'Home' for me or for Willie & Fred.

And now I will conclude this sad letter, hoping to write a little more cheerfully next mail. With kindest love to you, & all the boys, & Miss W.

> I remain,
> Your Affectionate Son,
> Joseph Matkin[41]

Dear Willie,

The sad news of our dear Father's death only reached me at 5 PM last night, & I can scarcely realize it all yet; to think that he has been buried nearly 4 months is very hard! but it was not less so for you, being in England, & not being able to go to his funeral.

I have not heard from Mother yet, but I hope to next week. I keep fancying that she is ill & fretting for some one to comfort her. I hope you will write as often as you can, & write hopefully and cheerfully. I do not know all the particulars &c yet of the funeral, & the letter from J.T.S. written Decr 19th to inform me of the sad event did not reach me for some hours after one from Aunt Lizzie dated Febry 4th, assuming that I already knew of it.

Those are the only two letters I have received since Father died; but I have had several papers.

It would be but a sad X.mas for you all last year, & you would feel it most being alone!

Poor Father would not get my last letter from Hong Kong written in November last. He wrote to me at Cape Yorke, & then complained of his pain, but wrote cheerfully.

I see he posted me a Mercury on the 14th Decr. I have long thought that it was "Heart disease" that troubled him, but some times people live a long time & suffer from that complaint. I believe that poor father thought when he said 'good bye' to me, as I came away, that it was for the last time.

I was glad to hear that Uncle Joe & Mr. Pykett were at the funeral: Mr. Pykett & Mr. Morrison were Father's most intimate friends. I hear too that Miss Wildman has had more troubles & long journeys to sick beds, so it is all bad news this mail. Of course you & I are so situated that we can get no sympathy from those around us, & we can most appreciate the comforts of home.

By the time you get this we shall be 'homeward bound', & by this time next year shall be near England. I hope we shall see each other next year, &, as I tell Mother we may all have many happy days together yet. Father has given us all a good education & start in life, & we owe everything to him: he has also left us a good name, & a good example. I hear that Charlie is to come home & do the printing &c. & I think it a very good arrangement, but it is a pity he was married so soon in one sense,—he will scarcely be able to live at home. Still, his being at Oakham will make it less lonely for Mother, & I hope they will all get on together. Of course Miss Wildman is staying; I don't know what we should do without her. I tell Mother I hope there will be a nice stone over Father's grave: we can at least visit that if we could not see him buried!

I see by the papers that you are having a very cold winter at home this year, & you must feel it more up at Edinburgh. I am expecting to hear from you as to how you are getting on &c. I know you have a deal to put up with. Your regiment does not go abroad, I think, so that is one good thing. I have fully made up my mind to leave the Service when we get back: my 5 years are up in August this year. I hope to get a situation as a clerk or some thing of that sort; & I shall have a little money as a stand by.

If you have your likeness taken this summer be sure to send me one; & tell me what leave you are allowed &c., & whether you are obliged to wear your regimentals when on leave. We find it very cold here, just coming from the Tropics. Mother will send you my long letter written during the last voyage; & I hope to write to you more cheerfully next time.

The mail leaves to night: I don't think you need send this home as I have just written. With best love, & hoping this is our last great trouble.

From Your Affec^{te} Brother,
Joseph Matkin

P.S. We shall be here for some weeks yet refitting &c.; & shall then go south to Nagasaki &c &c. thro' the inland seas. We shall leave Japan finally about the end of June, either for Vancouver's Island, or Honolulu, Sandwich Islands.

At any rate we shall go to Honolulu from Vancouver's so our address will be either this one or the other. You will probably see by the papers. After August it will be Honolulu, & after October Valparaiso. Write as often as you can, at least once to each place.

I wonder how poor Fred is getting on.[42]

Two weeks later Matkin resumed in earnest his normal expository tone in a letter to his cousin full of Japanese history, geography, and foreign relations. *Challenger* had moved from Yokohama to Yokosuka for dry-docking.

Challenger's visit found Japan in the midst of the modernization following the Meiji restoration of 1868. Matkin had been reading extensively and seemed anxious to transmit home his new knowledge. The

Japanese people were the most likable of all he had encountered on the voyage; they and their history fascinated him. Curiously, there is no mention of his father's death nor of Swann's letter conveying the first news thereof. Perhaps an intervening dispatch from JM was lost or not copied into the letterbook, although the opening lines of the letter below indicate no discontinuity.

H.M.S. "Challenger"
Yokoska
April 30th, [18]75

Dear Tom,

We arrived here at 6 o'clock, on the night of the 12th and received our letters and papers soon after, and all were most eager to read the news. The railway accident was a most terrible affair,[43] and I think the English companies might learn a great deal from the Japanese, altho' the railway between Yokohama and Yedo is on the American system. The marine disasters of the "La Plata", and "Cospatrick" were pitiful events, as regards the latter I hope more survivors may turn up, and that we may learn more particulars as to the cause of the fire, and the means taken to overcome it. How the newspapers gloat over the details and horrors of the event![44]

¶Since my arrival here, I have been "up to town"; for Yedo is a town, and excepting London and Paris, there are no European cities larger[45]; however, I am not going to give a description of its wonders in this letter, the next one must be reserved for that. We are now in the Japanese Imperial Dock as Yokos[u]ka,[46] a town and bay in the Gulf of Yedo, 18 miles east of Yokohama, for the purpose of re-fitting, &c.

¶This letter is a sort of introduction to the descriptive one which is to follow, containing an account of European progress, &c. in Japan, up to 1856,[47] when the country was opened for commerce and intercourse with all nations. It is gathered from various books. In the year 1278, Marco Polo the great traveller of Venetia was at Pekin soon after Kublai Khan had over-run China; on his return to Europe in 1295, he pointed to the eastern margin of the Yellow Sea, & said there was a great island there named "Xipangu" or "Cipango", peopled by a highly civilized and wealthy race, who had bravely rolled back the tide of Tartar conquest in the days of Kublai Khan. His countrymen smiled with incredulity at this wondrous tale, and for several subsequent centuries Europe knew but very little more about Japan. It was to search for the mythical land that Columbus sailed from Spain in 1492, and discovered the great

American continent, which, he for a long time thought was the same fabulous Xipangu. Before Europeans found their way round the Cape of Good Hope, China was allowed to send 15 Junks annually to and from Nagas[a]ki,[48] a port at the extreme south of the Japanese Empire, for trading purposes.

The Japanese group was inhabited long before the commencement of authentic records; whether peopled by refugees from China, the Corea peninsula, or from the great deserts of Scythia is not certain. The ancient history of Japan says that 650 years BC a hero known as the divine warrior, firmly established a dynasty which has flourished to the present day, in a line of 122 male and female monarchs. Of their reigns far better records exist than the oldest European Empires can boast; the early monarchs combined in their persons the double office of high priest, and generalissimo. Chinese historians assert that Japan was voluntarily tributary to the Celestial Empire, but this may be doubted. Kublai Khan after he had subjected China, sent envoys to the Emperor of Japan demanding his submission, which was refused, and when the great Tartar Conqueror invaded the country with a Mongol Chinese Army, he was boldly met by that of the Japanese, and totally routed and driven back. The Mikado, or Emperor, was assisted at this juncture by a "Tigoon",[49] or Assistant Emperor, who took command of the armies while the Mikado performed his spiritual part, of praying for their success. The office of Tigoon was then made hereditary, and since then there has always been a "Spiritual" and a "Temporal" Emperor in Japan, until 1868, when the present Mikado defeated the reigning family of usurpers, and ascended the throne as sole Emperor. After the Mongol hordes were driven back, the Japanese retaliated by ravaging the coast of China, and for several subsequent centuries, their barques were the terror of the China Seas.

¶The Portuguese were the first to re-discover Xipangu, when Fernando Mindes Pinto accompanied a Japanese barque from Ningpo, China, and reached Kiu-Siu, Japan. His report on reaching Europe in 1542 was hardly believed, but from then to 1579, and after the Spaniards were firmly established on the Phillipine Islands members of Spanish and Portuguese missionaries—under the great Francis Xavier, who was afterwards martyred in China, visited Japan, and converted one million of the natives.[50]

¶In the year 1579 Taiko Tama[51] was the Temporal Emperor, who one day asked a Spaniard, how it was that the King of Spain had managed to possess himself of half the world?—The reply was—

"He first sends priests to win the people, his soldiers then are sent to join the native Christians, and the conquest is easy." Taiko Tama was so much struck and alarmed that he swore "no priest should be left alive in his dominions." In 1587 the first edict for the banishment of the Catholic missionaries was issued, the native Christians were sent away to fight in the Corea, and were decimated. In 1596 the edict was reversed against X.ns. All missionaries were ordered to quit the country, twenty-three disobeyed, and in 1597 suffered death at Nagas[a]ki,[52] and were duly canonized in 1627 by Pope Urban, the 8th. Taiko Tama's warrant has been preserved, which reads as follows—"I have condemned these prisoners to death, for having come from the Phillipine Islands to Japan, under the pretended title of Ambassadors, and for having persisted in my land without my permission, and preached the Christian religion, against my decree, I order, and wish that they be crucified in my city of Nagaski."—A few Portuguese were allowed to remain in Japan, but they were obliged to disown their religion, & conform to the native habits and laws. In 1767 the Abba Sedotti[53] learnt the Japanese language from refugees at Manilla, and sailed from thence with several missionaries, to make another attempt at converting the Japanese. He was landed on the coast, and from that hour has never been heard of. Christianity never flourished in Japan after this.

In the year 1600 the Dutch made their appearance there, notwithstanding the hostility of the Spaniards, and the bulls of the Pope, and were allowed to establish a Factory at Firando [Hirado], and from that time until within the last 30 years, they have remained master of the field, as regards trade with Japan. The pilot on the Dutch ships on their first perilous voyage was "William Adams" a native of Gillingham, Kent. They were two years on their passage out, the fleet being separated and dispersed near the Straits of Magellan. The "Erasmus" Adam's ship was the only one that reached Japan. The Portuguese misrepresented the character of the new arrival, and endeavored to have the crew treated as thieves and robbers saying "that if justice was executed upon them it would terrify the rest of their countrymen from coming there any more.["] Adams, and one of the Dutch sailors was brought before the Emperor, in whose presence Adams spoke up manfully —"I showed him," says he, "the name of our country, and that our land had sought out the East Indies for a long time," and after explaining the purely mercantile purpose of the voyage, the King asked him whether our country had wars? "I answered him, Yea, with the Portuguese, and Spaniards, being at peace with all other nations." This was the best reply he could have given, & probably saved the lives of the whole party.

¶They were treated kindly by the emperor, but they and their stout barque were never to leave Japan. The "Erasmus" was ordered to the city of Yedo, the Temporal Capital of Japan. Adam's merits were so appreciated that he obtained great influence, and when in 1649[54] the next Dutch ships visited Japan, considerable privileges were conceded them, thro' "the good offices of Will Adams." By those ships he sent letters to England; and had been 15 years in the country when he learnt that a vessel bearing the red cross of Old England had reached Firando. She was the "Clive" of London, belonging to the East India Company, (then in its infancy) and commanded by Captain John Saris, furnished with a letter from King James the 1st, and suitable presents for the emperor. The "Clive" left the Thames on April 11th, 1611, & reached Firando on the 11th of June, 1613, having spent 2 years trading on the way. Adams was sent from Yedo to Firando, to greet his long expected countrymen: thirteen weary years the old man had looked forward hopefully, and at last his prayer was heard. He, with Captain Saris, and 10 Englishmen, started for Yedo, bearing the royal letter, and presents, and the influence of Adams soon obtained from the Emperor a favorable treaty, granting to England the most important privileges ever conceded by Japan to a foreign country. In 1613, an English, as well as a Dutch factory, was established at Firando. The English from political reasons very soon withdrew, & so avoided the trouble that overtook the rest of the European residents in Japan. Will Adams was employed at Yedo in building ships for the Emperor, and was advanced to the rank of Imperial Tutor, being acquainted with geometry, and mathematics. He acted as interpreter to the East India Company, and possessed an estate about 2 miles from where we are lying in dock. He left about £800 in his will, to be divided between his English, and his Japanese family. There are people in this neighbourhood who claim to be his descendants. About 3 years ago, his tomb, and that of his Japanese wife, was discovered by an Englishman living at Yokohama: and all Europeans visiting Japan endeavor to see the grave of the first Englishman who reached the country. It is on the summit of a hill on his own estate, being buried there at his own request, and the Japanese Mrs Adams lies beside him. I went yesterday to visit the tomb, there is a Bud[d]hist Temple near it, the Priest of which says that Adams used to worship there. The hill commands a splendid

Tomb of Will Adams, Yokosuka, Japan; from W. J. J. Spry,
The Cruise of her Majesty's Ship "Challenger" (London:
Sampson Low, Marston, Searle, and Rivington, 1876),
p. 276. Photo courtesy of Scripps Institution of Oceanography.

view of the Bay of Yedo, and of the surrounding country: the scenery is magnificent, and is considered the finest in this part of
Niphon. There is a religious feast observed at Yedo in June, to the
honor of Will Adams, the "Man of Kent".

¶After his death the English factory at Firando was abandoned,
and in the following year the persecution against the X.ns again
began, & ended in the withdrawal of the English, the expulsion of
the Spaniards, & imprisonment of the Dutch to the island of
Decima, Nagaski, where they were not allowed to worship according to their religion, or even to observe the Sabbath. Under these
conditions the Dutch factory continued at Decima, until the
recent treaties with European Powers. Nagaski is at the extreme
south of the Japanese Empire. Firando is a little further north. The
Empire of Japan—or Niphon, as the natives call it, consists of four

large islands, and about 3000 small ones—many scores of which we passed coming into Yokohama. The large island of Niphon, or Japan Proper, an island over 800 miles in length—Sikok—Kiu Siu, & Yesso.[55]

From the shores of Japan to those of China is 450 miles in a direct line. The area of Japan is about 160,000 sq. miles (one fifth larger than Great Britain and Ireland together) and the population is estimated at 30 millions. It was formerly governed by some 200 Daimios, or native Princes, who were tributary to the Emperor, but very turbulent and independent; but in 1866[56] this feudal system was abolished, and the present Mikado is sole Emperor (Temporal & Spiritual) and he frequently sees the Ambassadors at Yedo, or Tokio, as the natives call it; which means Eastern Capital. It has only been the metropolis since 1600, before which time it was a paltry fishing village. The ancient capital of Japan was Kamakura, an insignificant little town about 8 miles from Yokoska, and which I could see from Adams's Tomb yesterday. In the 16th century according to the Jesuit Priests, it contained 200,000 houses.

¶Japan is fertile, and beautifully cultivated: it is rich in minerals, including Coal; possesses plenty of magnificent harbours, and its seas abound with fish. Agriculture is the principal native pursuit, rice is largely cultivated and exported. The chief manufactures, &c. are those of silk and cotton, lacquering or Japanning, and porcelain. The internal trade is very extensive and rigid regulations are in force for protecting and encouraging home industry. Foreign trade was far from being popular until the last few years. In 1853 the American Fleet under Commodore Perry, forced its way up the Gulf of Yedo, and after proceeding to active measures forced a Treaty in 1854, between Japan, and the United States. The Ports of Osaka, Kanagawa, Yokohama, Nagaski, & Hakodadi were thrown open for foreign trade in 1859,[57] when all the great European Powers arranged separate Treaties with Japan. The 2nd Treaty between Great Britain was concluded by Lord Elgin in 1859, soon after the fall of Canton. There was a "Spiritual", and a "Temporal" Emperor of Japan at that time. The former lived at Micao,[58] a city 350 miles further south, and according to Japanese etiquette the "Spiritual" Emperor never leaves his Palace.

Our Gov't in 1859 presented him with a fine steam Yacht, to travel round his Palace and grounds. Osaca [Osaka] is the seaport of Micao: Micao has nearly a million inhabitants. The Japanese are of a yellowish complexion, with thick black hair; are not much like the Chinese in appearance and quite different to them in temperament, being a cheerful, contented, jolly people. I consider them the most wonderful and interesting race we have yet met with, their curiosities, and manufactures in wood, and bronze work, are infinitely more artistic, and pretty than any thing the Chinese ever produced; but must tell you more about them in my next letter. The weather here is rather warmer than England in the middle of the day, and certainly colder in the evenings and early mornings. Volcanoes, hot springs, sulphur springs, &c. are very plentiful in Japan, and earthquakes are thought nothing of. The atmosphere is wonderfully clear here, and the Peak of Fusi-hama [Fujiyama], 60 miles inland, looks scarcely more than 6 miles away, at sunrise, & sunset.

¶Fusi-hama, or the "Matchless Mountain" of Japan, is the prettiest mountain in the world; it is 11,500 feet high, & stands in the middle of an immense plain, so that we can see it two-thirds of the way from the summit; it is covered in snow more than half way down. In August, there is scarcely any snow on it, and then you can easily ascend it. Lady Parkes, the British Ambassador's wife has been to the top. It can be seen 100 miles out at sea, and was sighted a few weeks ago by the Captain of the "Sylvia", English man of war, from the summit of another mountain 140 miles further north. It is quite a baby among mountains, having made its appearance according to reliable Japanese authority, one evening about 500 years ago, its birth was accompanied by a tremendous earthquake, and for 400 years, Fusi-hama was a volcano of the most active kind, but at present it is only a moderate smoker.

¶The British Ambassador here, Sir Harry Parkes,[59] lives at Yedo; he has been out here many years, and has done more towards opening the country for trade than any other European. The Gulf of Yedo, is one large harbour, 40 miles long, and 12 broad, containing scores of beautiful little coves, and bays, where ships can lie securely.

About 30 miles down the Gulf, on the southern shore, stands its Seaport Yokohama, where ships lie that could not get up to Yedo, which great city is on the opposite side. It is 12 miles from Yedo by water, & 18 by rail, & contains over 300,000 inhabitants, including about 3000 Europeans & Americans, & about the same number of Chinese, who are considered foreigners in this country. There are not as many vessels lying in Yokohama bay, as at Hong Kong, but still a large number; the British, French, & American Flags being most numerous. There are 8 good sized Japanese men of war lying there, 6 here at Yokoska, & others elsewhere. The Flag Ship was built at Aberdeen, & is nearly as large as the "Audacious". Most of the Japanese men of war were built in England or America, as their dock yard has not got in proper swing yet, it is very compact, & situated in an impregnable & secure position, has been made by French Engineers, & the leading hands are Frenchmen. The ships are manned entirely by Japanese sailors, trained at the Naval College at Yedo, by English Men of War's men, who receive from £300 to £500 a year. The Officers are partly English and American, as well as natives. The scenery round here is magnificent, something like the Isle of Wight, on a larger, and grander scale. Will continue by next mail.

Yours faithfully,
Joe Matkin[60]

The following week Matkin continued his travelogue of Japan in a letter to his mother. Some pages are missing, but the narrative continues in the concurrent letter to his cousin. Religious practices and the new Japanese railway dominate his attention.

Curiously, after three years of lengthy letters, in Japan Matkin had grown conscious of their length and here apologizes. But they have become a matter of honor; moreover, there is a plan to combine them all under one cover eventually, we are told.

H.M. Ship "Challenger"
Yokohama
May 7th, [18]75

My dear Mother,

The mail has not arrived yet but it is expected to day & I trust there will be a letter from you. We have been so busy lately getting ready for sea, & leave here on the 10th for our cruise in Japanese waters, so I will get this letter off before we leave. Of course you would receive my letter to Fred by last mail, & this is intended to describe Japan as I have found it thus far.

I know you all like to know what these countries are like; & as I don't like to send you a slovenly or imperfect account, the letters are rather long. Of course I compile the letter from my log-book, & I hope to send you the 2nd volume from the Sandwich Islands, & bye & bye, when I get settled at home I shall write the whole in our book for myself, & you can keep the original.

We do not have such a crowd of junks round us here as at Hong Kong, & there is no regular floating population here at all, the sailors & fishermen all live on shore, the same as in other countries.

The Japanese junks are built stronger & rigged differently to the Chinese; they have one large square sail & they are also propelled by sculls. You see no women or children in these junks, working like horses, neither do you see them working in the fields, or doing any sort of rough work. Their duties are confined to spinning, cooking, nursing & smoking. They all smoke in this country, but their tobacco is very mild, & the pipe holds only a few grains. The married women blacken their teeth & shave off their eye-brows.

¶In appearance the Japanese are a small clumsy looking people, with heavy regular features, & a look about the eyes & mouth as if they had been suddenly wakened from a sound sleep, still they have a healthier & fresher look than the Chinese. They appear to be of Chinese Tartar extraction, & yet there is something of the Malay about them. Their dress is a long dark woolen robe reaching to the knees, with arm holes, & open in the front but fastened round the waist. If it is cold they wear several of these garments one over another. The women dress the same as the men, only the dress is open round the neck, & there is a sort of panier behind. They wear no shoes or stockings in the country unless it is dirty weather when they wear wooden clogs. In the town the people wear neat straw plaited sandals & socks shaped like a mitten with a thumb hole for the big toe, which holds the sandal on by a sort of loop. The men in the town wear tight fitting pantaloons, exactly like the harlequin at a X.mas pantomime, but in the country they wear only the long robe, & the men working in the fields have only a strip of linen round the waist. Neither sex wear any head covering; in cold weather they wear a scarf round the head, & in wet or hot weather they have umbrellas ingeniously made of bamboo & prepared paper. The poorer classes wear no under-clothing whatever; the men generally have the crown of the head shaved but wear no pig-tail, & the women dress their hair something like the Chinese women, & if they have any jewelry, arrange it in the hair; but they are not fond of cheap trinkets, & their dress is generally of some quiet colour. They paint a good deal however, but some are very handsome without paint & have fine complexions. In Yedo & Yokohama many of the better class of men wear European dress, & you see many in a sort of half & half rig, like the native priests of Tongatabu. But they seldom look well in our dress, & their clothes have a 2nd hand Jewish appearance. None of the women wear European costume.

¶Bread & meat here are rather dearer than at Hong Kong, but the meat is better than any we have had since leaving home, & the prices are rather cheaper than in England. The vegetables are the finest in the world, even better than in China. You can buy carrots & turnips 3 feet long & the same thickness all the way down. Green peas &c are just coming in, but fruit is scarcely ready yet; wild flowers are very plentiful, I have just gathered lots of violets in the country. We have Provision boats & Curiosity boats along-side every day, & you may see some wonderful things in the way of cabinets, work-boxes, tea caddies, & knick knacks of all sorts— scarcely as cheap as in China, but far prettier & more original. If we were only allowed the space, we might bring wonders home from the different countries, but as it is we have to stow everything in our chests. The Japanese practise the same system of extortion as the Chinese, but they are always good-tempered & civil. They are very clever & ingenious in imitating anything, & they already make loco-motive engines, & marine engines of good size in Japan; you see lots of American sewing machines in use. Of course labor is not so plentiful in Japan as it is in China, & there are not so many helplessly poor people. I have never seen a professional beggar yet.

¶A few years ago the Japanese seemed as if they were going to adopt European habits & dress entirely; & English & American newspapers began to speculate as to how long it would be before they learnt the English language & altered their whole national system &c. But latterly a reaction appears to have set in, & they are fast falling back into their old habits & ways. The Japanese are a more united people than the Chinese, & a better governed people, but they are frightfully taxed, & the burden all falls on the poor classes. However there is a great reform going on, & there are already native newspapers printed at Yedo which are commencing to agitate, & to "Want to know, you know", & to ask why these things should be. I learnt this from translations in the English papers. The Japanese wash & get up linen beautifully & are very quick & neat in their ways.[61] * * *

The shores of the Gulf of Yedo look more like England than those of any other country we have visited; there are lots of batter-ies on the commanding points, and some immense earth forts in front of the city of Yedo, built on small & conveniently situated islands. The town of Yokohama stands in a hollow, and we cannot see much of it from the ship, more of Kanagavu [Kanagawa] is visi-ble on the opposite shore. The Japanese sailors are dressed like our men, and their officers similar to ours; their national flag is white, with a red ball in the centre. They do not wear pig-tails but most have the crown of the head shaved, have no sign of any beard or whisker, and a long beard is greatly admired in this country, especially by the fair sex. Lots of leave has been given to the crew since we arrived:—more than they have money enough to take advantage of.

¶A few days after we arrived 2 clerks belonging to the French Bank here absconded with 37,000 Dollars, and bought a small vessel, & hired a crew to take them to Manilla; as soon as their absence was discovered, a steamer was sent after them, as it was known the wind had been unfair for their getting out of the Gulf of Yedo. The Steamer took with her a Sub. Lieut. and 10 armed men from H.M.S. "Thalia", which is stationed here, and came up with the schooner the same night—becalmed. Before the vessel was boarded the two clerks went down to the cabin and shot themselves,—one through the head, and the other through the heart, after which they shortly died, and the money was recovered. One was an Italian, the other an Englishman: they had been several years in the bank, and held a very considerable position in the European Society of Yokohama, but gambling and betting proved too much for their integrity. The Europeans here, like those at Hong Kong live very extravagantly, and try to cut a dash. They must have their Horse racing, Club-houses &c, the same as in England, only about 10 times dearer, and they are little gods in their tin pot way, but on returning to their native country many of them find that they are rather ordinary personages after all, & not unfrequently return again to the place of their former glory.

On the 19th of April, I went on shore for the first time, and returned again on the evening of the 20th. There are three Daily English papers printed in Yokohama, and one weekly, also a "Japan Punch" for the select few; which is a feeble imitation of its English prototype; for unless you know everybody of consequence among the Europeans, you don't see exactly where the joke lies, and soon get bewildered.

There is an English, a French, an American, a Dutch, and Japanese Imperial Post Office, also an English, and a French landing quay, and an English and French quarter of the town—called English and French Hataba. There is also a Chinese quarter, where the Celestials may be seen scraping money together, and minding their own business, for they are under the same regulations as other foreigners. The English community, and the foreign Consuls, &c, live in the most healthy part of the town, on some heights called the Bluffs, from where a fine view can be obtained of the town, and the Bay of Yedo. Their houses are built almost entirely of wood, for earthquakes are not rarities in the country; but most of them are very pretty erections, and have nice flower gardens attached. Japan is a fine country for wild flowers, and I have gathered lots of violets in the fields. There are about twice as many hotels in Yokohama as there ought to be,—English, French, & American,—very large buildings too, and considering that all Europeans of any standing have private houses, it is a mystery to me how they are supported. There are only three Churches, an English, Roman Catholic, & a Presbyterian, which is an American. In this respect Yokohama is scantily supplied compared to Lisbon, Bahia, and Manilla.

¶The first thing that strikes one in Japan, after visiting China, is the beautiful roads, and clean streets: there are none of those fearful smells that you find in Hong Kong or Manilla. The streets are wide and straight, and kept clean by the people, each householder sweeping and watering his frontage. Proper labourers mend the roads, and there is an efficient Police force, but they appear to have very little to do; there is no drunkenness in this country, and every one can read and write the Japanese language, so one of the girls at the Tea-gardens told me—who spoke English, and said they went to school until they were over 13.

¶The Japanese houses are built almost wholly of wood and plaster, generally 2 storys in height, with carved gable ends and roof. The fronts and backs of them can be opened and closed at pleasure for ventilation, &c. and the interior is kept beautifully clean. The rooms are partitioned off with wooden partitions, sometimes only paper, and the windows are made of thin but strong paper, which will keep out wet, and yet admit light. The floor of the house is raised about two feet, and covered with beautiful white rush mats, on which cushions are spread at night for sleeping. The Japanese pillow is simply a small wooden trestle which they cover over with a roll of cloth or paper; the pillow itself is exactly like those used by the natives of the South Sea Islands. There is a Porch inside the door before you mount on to the raised floor, in which they leave their shoes or sandals, wearing either their socks, or going barefooted in-doors. They expect all visitors to take off their boots unless they are very clean indeed, though they admire European boots & shoes more than any other article of dress. Are the most hospitable people I ever met with.

¶Paper enters largely into the use and domestic wants & arrangements of the Japanese, and they are unrivalled in their manner of utilizing straw and paper to the best advantage. You can buy a Mackintosh cape made of prepared paper, and capable of keeping out wet, as well as another, for about 2/, and you could not distinguish it from an ordinary one. Over these capes the fishermen spread thick straw over-coats to keep out the cold in winter; and by spreading a bamboo frame from the head to the stern of the boat, they make it water tight & warm by fixing movable sheets of straw. They have a charcoal fire in all the boats, & when hauling in their nets they beat a wooden rattle to charm the fish. The fires in the houses are all charcoal; the fireplace is a movable one, and generally stands in the middle of the room. The charcoal comes from the great interior forests where many thousands are employed in its preparation. Of course there are no chimneys to the Japanese houses. The horses, which appear to be very scarce in Japan, are shod with plaited straw shoes. Oxen and sheep are far from plentiful as there is no demand for them, pigs are met with everywhere, and poultry & game there is a good supply of. You seldom see a Butcher's Shop, but fishmongers are numerous, as the natives live chiefly on fish, rice, and eggs. There is a great number of Ravens, may be seen on half the housetops in Yedo.

¶Except European private carriages, there are scarcely any horse conveyances, the passenger traffic is performed by neat little cars called in Japanese "Jinrikishaws" (our sailors thought "Jenny Rickshaws" best). They are drawn by one man who stands between the shafts, and trots along very much faster than a London Cab horse. These cars are cushioned, have hoods for wet weather, lanterns at night, and are numbered like Cabs. They are an American introduction, and have appeared within the last 20 years, and are far more comfortable & proceed faster than the Hong Kong sedan chairs. Their number in Yokohama and Yedo alone, exceeds 200,000, and the men who draw them are mostly from the country, so that the agricultural pursuits have suffered in consequence. You can ride about five miles for a 1/: money here is principally Mexican, & American silver dollars, & Japanese silver, copper, & paper money which is very difficult to understand, and is being withdrawn for a more convenient system of coinage. The Europeans here, as at Hong Kong, invariably ride if they have many yards to go. Soon after I landed I went to the Railway Station, a large stone building with every convenience for passengers, and directions printed in English, French, and Japanese. The railway is 18 miles long, and is I believe the only one open in Japan, but they are now talking of making one from Yedo to the northern part of the island.

The present one was opened about 12 years ago, is a double line, and winds round the shores of the Gulf of Yedo all the way. It calls at 4 towns on the journey, and is greatly patronized by the Japanese; there are no tunnels, but lots of bridges: the Station masters and Engine drivers are mostly Englishmen, but the clerks, and all the other officials are natives; these and the Police all dress in European clothes. The fare to Yedo, 1st Class is 5/, 2nd Class 2/6, and 3rd 1/3. No European travels by the latter, and are only used by natives, but many of them occupy 2nd, and even 1st. The carriages are constructed on the American principle, the entrance being at the ends, and not at the side as in England, and as they are of glass the Guard can see every Carriage interior as far as the engine: and there is a platform running the whole length of the train, so that if he is wanted in any particular carriage, he can walk outside the cars, & not through the doors of each. The seats are all round, and the centre is open for traffic, without disturbing the rest of the passengers. As both sexes smoke, there are spittoons let in the floor of the carriages. The sides as well as the fronts are all glass, so that the carriage is entirely open in fine weather, and there are Venetian blinds to keep out the sun & dust. The trains run every half hour from 7 AM to 10 PM and takes an hour to accomplish the journey to Yedo. You have a fine view of the gulf on one side, and the surrounding country on the other, which is another advantage you do not have in English carriages, unless they are empty, and then you only catch a glance as you shoot past. The country is richly cultivated, and there are houses, and market gardens all the way: rice is the principal cultivation, and the fields are under water at present. The men working in them are entirely naked except a strip of linen round the waist.

The religion of Japan is Buddhism & Confucianism chiefly, the latter, with Chinese letters, writing, &c, was introduced AD 285. Buddhism found its way through China, & Corea, from India, about the year 552 AD. In travelling in Japan the tourist will continually meet with Saints, Shrines, Buddhist Temples, wayside idols, images, and tablets of various kinds, the outward evidence and superstition of the Japanese people. In a Buddhist Temple are, drums, bells, candles, images, books, and a variety of Altar ornaments. There are many "off-shoots" & "schools", &c, off the original doctrine. They worship Buddha, and a host of minor deities, and pray to a vast number of saints; they believe in the transmigration of souls, have a high moral code, and are influenced in their actions by the fears & hopes of the future. You will often come across a shrine with candles burning in front of it, at the corner of a street, and each household appears to have one, and favorite deity the same as in China. There are Buddhist Temples all over the country, generally built on an eminence surrounded by trees, or else in the midst of a garden. There are Temples in all the towns, and of course Priests to look after them, who are generally rich, and are unfavourably looked on by the Government on that account. The largest and finest of the Temples at Yedo was burnt down a few years ago, by Gov^t completely it is thought, as it was getting enormously rich. Besides Temple Gardens, there are large Tea Gardens, on the outskirts of the city, where there are nice walks, and all sorts of amusements for the people, and scores of Tea houses for refreshment. Every day they are thronged with holiday makers, and the Temples with worshippers, and on special Saint's days (which occur about as often as in Catholic countries) half the population is to be found there.

¶I took a Jinrikishaw from the station to the largest existing Temple, about five miles from thence, but still in the city. The building stood in a garden, at the bottom of an avenue, and was an enormous structure of carved wood, gilded, & lacquered beautifully, and would be a great wonder in Europe. In shape it was like no other building I have seen, and seemed to be all projecting gables, something after the style of a Swiss Cottage on a large and elaborate scale. It was about 100 ft. in height, and covered 3000 sq. yds. of ground. The wood carvings about it represented all sorts of griffins, dragons, fabulous serpents, & birds, &c, carved & gilded very beautifully. There were several gilded idols to be seen inside, and Priests capering about as bad as your Ritualists at home: there were steps leading up to the Temple, on which the people knelt and prayed to their particular gods, and put their offerings in a large receptacle for the purpose. The interior could only be seen from a distance, as it is fenced round with lattice work, and only opened on special occasions. I could see therein all sorts of pictured allegories, representing the effects of different crimes, and their punishment, &c.

¶The Temple grounds are fenced in, and at the entrance you may see from 500 to 600 Jinrikishaws, waiting for the return of their passengers. About three fourths of the people in the gardens are women and children in holiday dress. Occasionally you may see the Father & Mother with all their family drive up to the entrance, looking as if they had come to enjoy themselves, and meant to do it. Every one looks cheerful, and contented, and always ready to laugh, even against themselves. In front of the Temple was a large carved and gilded Dove-Cote, and hundreds of the most beautiful Pigeons I ever saw flying round the roof of the Temple, and amongst the people as tame as Chickens. The visitors bring all kinds of food for them, corn & rice principally, but am not able to say whether they are held peculiarly sacred, or are only tolerated as pets. The people are not at all bowed down to their religion, & if they see a foreigner laughing at their devotions, they join in it quite heartily. In one place there were figures of wax, and bronze images, representing all sorts of torture, and horrible punishments, executions, monster demons, &c. and every body seemed to laugh at these. One was an enormous Dragon (something like S. George's) attacking and trampling under foot a whole Japanese army; another a minotaur attacking a Fleet in the bay; one represented a Tiger about to devour a woman for doing something wrong; and a still more remarkable one of an enormous bird flying away with a Queen on its back. The metal work and gilding of the creatures was beautifully done, but the wax figures were not equal to Madame Tussauds. There were also live monkeys climbing up poles, archery tents presided over by pretty young ladies, children's toy shops (greatly patronized) and lots of Tea shops.

¶From the Temple I went to some large Tea Gardens, which were thronged with holiday-makers, and here the same arrangements &c, were to be found; there were penny peep-shows, the pictures of which were beautiful, representing splendid scenery, and Temples inland; the Mikado's Palace, and himself in his state robes: the Tykoon (office now obsolete) or "Spiritual Emperor", in his Palace at Miaco: Japanese armies besieging castles, and all sorts of native scenes & personages. The Tea garden was situated on an eminence from which a great part of the city could be seen, but not the whole, by a long way. You must remember that there is no smoke hanging over Yedo, the fires being all charcoal, so that the atmosphere is very clear: there are no factories or large buildings in the way anywhere.

¶Yedo covers a larger space of ground, in proportion to the inhabitants than any other city. The streets are on the average, wider than those of London (excepting the larger thoroughfares) and there is not more than one family occupying a single house. The city covers over 70 sq. miles of ground, and the population is about a million and a quarter. Some of the streets are as much thronged as the great London streets, but as there is no carriage traffic, the passengers move along much quicker in such wide thoroughfares;

those of Yedo would be very suitable for Tramways, and they are talking of adopting them. These gardens are so pleasant and rural that you could fancy yourself in the country; but the busy hum to be heard all round, and the dense crowds moving under your feet are features that indicate convincingly to the traveller that he is in one of the largest cities in the world. The houses of the richer Japanese in the suburbs: most of them are finely carved, and have nice gardens attached.

¶The girls at the Tea-houses are very sedulous in soliciting customers, & very kind and attentive in their manners. The tea is very nice, and almost colourless, being newly gathered: it is drunk without milk or sugar, and in the tiniest China cups imaginable. They bring you a Teapot to yourself, with fancy cakes, eggs, and fruit, if it is in season. If you choose they will play the Guitar, and sing to you, but their music is not inspiriting, and I fancy they charge extra for it. The cost of a good tea is about 9d, but they will get more if they can, appearing to think it an excellent joke if able to impose upon you.

¶I forgot to mention that there was [a] large bronze god in the Temple gardens, which seemed to be credited with extraordinary healing powers; as Mothers would bring their scrofulous children, and after rubbing their affected parts, stroke the corresponding part of the deity, but without any visible result. The old gentleman's brass face was partly smoothed down with constant rubbing, giving the features a most benevolent appearance.

¶There are several salt water canals at Yedo, spanned by good bridges, and the bay facing the city is crowded with fishing smacks; the fish-mongers living in the adjacent streets, and are very numerous, they are very cleanly, & you can buy almost any kind of fish. Enormous porpoises weighing half a ton, great turtles over 300 lbs. in weight, and all sorts of shell fish, including good oysters. Near the Railway Station was the Japanese Naval Training College, and in this neighbourhood are a few European hotels, and here live those Europeans who are allowed to reside at Yedo, their number is very small, I did not see ten from the time I left the Yedo Station until I returned again at night, reaching Yokohama at 6 PM after having spent a most wonderful day. There is a Telegraph from Yokohama to Yedo, and a cable to Europe and America via Shanghai and Singapore, and across the Atlantic another via Pekin, and across Siberia to St. Petersburgh, but the Pacific has none yet. I stayed at the Temperance Hotel at Yokohama all night, and spent the following day looking round that place, returning on board at night; have not been on shore again yet, but shall do so before we leave.

We have re-fitted and are now provisioning to last us to Vancouver's, have taken in our coal, and are almost ready for sea again; we were a week down at Yokoska, and I was on shore several times, and visited lots of little villages. In that vicinity there are plenty of Temples, particularly one at Dai Butsu of Buddha,[62] in which there is one of the largest images in the world, except perhaps the famous Colossus of Rhodes. It represents the god Buddha, and is cast in bronze, it stands over 44 feet high, and will hold 60 people inside. There is also the ancient Capital of Japan, Kanagawa,[63] & in that neighbourhood have been fought some of the most terrible battles recorded in Japanese history: to which people the town and district is classic ground. Not less interesting to us is the solitary grave of Will Adams, the one Englishman whose name and memory stands out from the obscurity of 250 years, as the representative of our nation in Japan; and who, when such men as Francis Xavier, and the numerous Spaniards and Portuguese who were contemporary with him are utterly forgotten, still has descendents, who are proud to claim him as their progenitor. From his tomb you can see the beautiful island of [blank]. Yokoska is a good sized town and the people very cleanly. There are eight bath houses, where both sexes bathe promiscuously; their sense of propriety being sacrificed to that of cleanliness. The natives could speak French; at Yokohama they spoke English; and down at Nagasaki they speak Dutch as well as Japanese, of course!

May 9th

There is no mail in yet, and we are to leave to-morrow afternoon for Kobi [Kobe]—300 miles south, shall probably be back here by the 10th of June. I was on shore again yesterday, and had a beautiful walk out into the country, to Mississippi Bay, with numerous potations of Tea on the way, there being houses for its sale everywhere: the natives pay about 1/2 for a cup. I went through a whole lot of Tea fields; the tree is something like a sloe bush, and the leaves are similar. The San Francisco Mail boats are the fastest and most regular steamers that come to Japan: they run every fortnight now, and have purchased three of the White Star Line, the "Oceanic", the "Belgic", and "Gaelic", the former will be here next month. They run from San Francisco to this port in 18 & 19 days, and sometimes bring mails from England in 35 days. The P & O Company's Steamers take 50 days from Southampton. My old ship the "Audacious" is expected here every day. Another Japanese Man of war came in yesterday—she was formerly H.M.S. "Malacca".

I have read of some of Dr. Kenealy's[64] Vagaries in Parliament, to day, from Telegrams in the English Newspapers. Latest information concerning "Challenger". Leave Yokohama, June 16th, arrive at Vancouver's, July 29th; leave there August 10th, arrive at Honolulu Sep 8th; leave there Sep 18th arrive at Tahiti, Oct 19th; leave there Oct 29th, arrive at Valparaiso, Dec. 11th; leave there Jan. 4th/76, arrive at Falkland Is[lds], Magellan Straits, Feb. 4th; leave there Feb. 12th, arrive at Monte Video Feb. 24th; leave there March 7th, arrive at Tristan d'Acunha, March 26th; leave there the 28th, arrive at Ascension I[sld], April 12th; leave there April 20th, arrive at S. Vincent, Cape Verdes, May 10th, leave there the 17th and arrive at Portsmouth, June 16th, 1876.[65]

May 10th
Mail just leaving. Devil fish are very plentiful here, and are said to [be] eaten by the natives.

Yours faithfully,
Joe Matkin.[66]

From Yokohama, *Challenger* sailed south and west to Japan's Inland Sea, stopping at Kobe and returning to Yokohama on 6 June. The report to his cousin, the last letter from Asia, is full of enthusiasm for the Japanese people and their scenic islands. Displaying more ethnocentrism than was usual for him, Matkin judged their future progress as dependent upon Christianization and the adoption of the English language.

H.M.S. "Challenger"
Kobi
June 1st, 1875

Dear Tom,

We left Yokohama on May 12th and arrived here on the 15th, after a rather stormy passage round. The coast of Japan is very rough, and not unlike that of Barbary: we had to anchor on the 14th in Osaka Bay, a pleasant anchorage off the mainland of Niphon. The bay contained 13 good sized villages and the cultivation all round them was wonderful. The fishermen in Japan tie their nose up at night to prevent ague, fever, &c.

We entered the "Inland Sea" of Japan at 1 PM on the 15th, and were at once in smooth water. This sea or strait is over 300 miles in length, and separates Kiu Siu from Sikok[67] at its south eastern extremity, and Sikok from Niphon[68] in the north. It varies in breadth from a quarter of [a] mile, to 20 miles, and contains therein some 2000 islands of all sizes, between which there are hundreds of different channels where small vessels can push through, and of course plenty of good bays and harbours. The navigation for a ship like ours is very intricate, and large ships always anchor every night, & generally take a Japanese Pilot. It is only within the last 15 years that large vessels have adopted this passage, and it took several years properly surveying by the British, French, & American vessels. It cuts off about 100 miles of the voyage from Yokohama to Nagasaki and Shanghai, and enables the numerous Junks to run from Nagasaki, &c, to Kobi, in comparatively smooth water.

¶Kobi is just inside the northern entrance of the Inland Sea, and is 330 miles from Yokohama, & 400 north of Nagasaki. It is situated in an enormous gulf or bay in the main island of Niphon, called the Bay of Osaca, which is about as large as Hobson's Bay, Melbourne, and contains the ports of Hiogo, Kobi, and Osaca, besides scores of fishing villages. Osaca stands at the extremity of the bay, Kobi is 20 miles to the south, and Hiogo is close to it, the two towns running into each other, but Kobi was substituted for the latter as an open Port, on account of its better anchorage, the same as Yokohama for Kanagawa. The two towns look something like Wellington & Cape Town, as to position, being situated at the foot of an amphitheatre of hills, but they look more like English than those places, and the population together is about 200,000. Osaca contains about 300,000, and with Hiogo was one of the Treaty Ports opened in 1859: but the anchorage is bad, and ships have to lie a long way from the town, so in 1867, Kobi was opened, as it afforded better facilities for shipping. There are a good many Europeans here, but it is not such a place for trade as Yokohama.

¶The early Ports opened by the Japanese for Foreign Trade were the worst they could choose, and they did all in their power to disgust foreigners from coming. The Pilots were encouraged to run the ships aground, and an extraordinary system of espionage was practised towards them. Officials were set to watch them wherever they went, and the shopkeepers were not allowed to take foreign money, or pay away the native coin. If a foreigner purchased anything, he, and the shopkeeper went with the goods to a Govern^mt Official, who priced them and received the money, paying the poor tradesman again in native money, and cheating him frightfully, until they dreaded to see a foreign customer enter their shops. Officials boarded ships long before they came into port, and had to find out why they were coming, and how long they meant to stay, &c. If it were a man of war they had to report all particulars concerning the Armament, number of men carried, and what was her object in coming, &c. They had to write all this down, and another set of officials kept a check on them, and if there was the slightest sign of anything suspicious in their reports, or behaviour, they were handed over to the executioner. The Captain of an American man of war, writes in his book[69] an account of his entry into the Port of Nagasaki, about 20 years ago. Long before he was near the place a two-sworded Official was seen coming out in a small boat, and making signals with his face to be taken on board, but the "Stars and Stripes" sailed on and took no notice of his frantic gestures, until after the ship had left him behind, when he was seen to kill himself by falling on his own swords, rather than survive his disgrace, which by the law of Japan, would have included his family, as well as himself. The American Captain remarks "that the most awful sign of despair he ever saw depicted on a human face, was on that of this poor baffled man, as the ship shot past him."

¶Soon after the Treaty with the Americans in 1853, there was a terrible Typhoon along the Japanese coast, accompanied by an Earthquake which destroyed an immense amount of shipping and property, as well as lives. A deputation of Japanese Officials waited upon the European residents to inform them that such sort of occurrences were very common in Japan, and that a repetition might be shortly expected, so that if they would avoid the catastrophe, they had better leave the country as soon as possible, and never return. The deputation was as civilly received, and was about as successful as that of the three Quakers who went to St. Petersburgh to advocate Peace, just before the Crimean War. Since that time a real change has come over the Japanese people, and they are encouraging Foreign Trade more and more every year. They are undoubtedly of an inquisitive and commercial turn of mind, and only want a few missionaries to make them a great people. There are now a few in Japan and China, but it will be many years before Christianity has established a footing in either country. I think the English language will be taught in the schools of Japan for the future, as well as the native for the latter is not sufficiently copious to allow any one to express himself on any subject unconnected with Japan, without introducing a host of new words. It would not be possible to translate the Bible, without first arrang-

ing a grammar, and making many new words: so it is highly proba-
ble that our language will eventually be spoken. The new Japanese
silver and copper money is very pretty, and similar to ours at Hong
Kong. I am a great believer in the future of this country, but until
Christianity is taught I dont think the people are prepared for such
radical changes, as the American and English newspapers presage.
These writers seem to forget that the people are influenced and
controlled by the Priests, more than by the ruling Powers.

In 1867 just before the present Mikado ascended the throne there
was a general rising in this neighbourhood against Europeans in
general, which was doubtless incited by the Priests. A boat's crew
belonging to the French Man of War "Duplex" was murdered, (cut
in pieces) & their bodies sent off to the ship. This occurred at
Osaca a port 20 miles down the bay, but the men were buried at
Kobi. The French Admiral demanded from the Japanese authorities
500 Dollars for the friends of each of the men, and a certain number
of heads as satisfaction for the outrage. The heads were forthcoming in any quantity, collected promiscuously from the different
prisons, and the money was also eventually paid. The British
Ambassador Sir Harry Parkes, was at Osaca at the time, and had to
fly for his life, some of his mounted body guard being wounded by
the excited Japanese people. Sir Harry has had several narrow
escapes of his life in Japan, and China, but he is a very courageous
man, and not easily intimidated.

Just before the China Wars he was carried naked round Japan in
an iron cage: he was Secretary at that time to the British Ambassador, and was then Mr. Parkes.

¶On May 17th I went on shore here & looked over Kobi and
Hiogo, which are something like Yedo on a smaller scale. There are
Consuls of all nations here, lots of Europeans and Chinese, a few
European Hotels, a Catholic and a Protestant Church, and there
are some fine walks here. Lots of Curiosity shops abound in Kobi,
and the goods were quite different in design to those at Yokohama;
the town will soon be lighted with gas; there are not so many
Jinrikshaws here, but there is a Race course for the fashionable
Europeans, and a Daily paper printed. In the vicinity of Kobi there
are two beautiful Waterfalls situated in a deep gully between the
highest hills. There are scores of Tea houses up there which command beautiful views of the Waterfalls, the Bay and City of Osaca,
and several islands at the entrance of the "Inland Sea". Every fine
day these islands show out most distinctly, and the Tea houses
are crammed with people chattering, smoking, Tea-drinking, and

studying nature. I have been up there several times, and always
enjoy it. The weather is much warmer now, and the people wear
brighter garments, and all use pretty painted Parasols to shelter
themselves from the sun, so that a crowd of Japanese holiday makers forms a very picturesque scene. The waterfalls are each about
150 feet in sheer descent, and are the prettiest of the kind I ever
saw. There are several Gardens and Temples near Kobi—all of
which I have visited. There is a Temple of the Sun at the foot of the
hills, crowned by an enormous gilded globe, representing the sun.
There is also a Temple of the Moon built on the summit of a steep
hill, over 2000 feet high, however it was erected up there I can't
imagine, as most of the materials would have to be brought up
from below by manual carriage, for the road is narrow and steep,
and there are 250 stone steps leading up to the Temple after you
have almost reached the summit. The building itself is of carved
wood, but the interior is magnificent. We had to take off our boots
before entering, but were allowed to touch and examine all the
Altar ornaments, which, with the Altar itself were of bronze and
gilt work beautifully carved, and there were massive gilded candle-
sticks weighing 400 or 500 lbs; any amount of candles and tapers
burning, as well as beautiful smelling wood which scented the
whole place. A large steel mirror stood on the Altar which was supposed to represent the moon, and any amount of food offerings in
the shape of rice, fish, &c. were placed on shelves near the Altar,
which affair looked something between a Royal Throne and a large
State bed, and near it was the finest sounding bell I ever heard, it
was a sort of large metal vase on another hollow metal vessel, with
a large silk cushion between the two, and it was beaten with a
wooden mallet. A journey up to this Temple is considered a sort of
salutary pilgrimage, and you see lots of old women toiling up this
road every day most perseveringly. The Japanese have large burying
grounds outside the town: their monuments are solid stone pillars,
standing on a flat base, and another flat stone fixed on the top; the
column bears the inscription in deeply engraved characters. The
Chinese grave stones are generally shaped like a horse-shoe, &
occupy more space than the Japanese. The European Cemetery
contains the graves and monument of the murdered French seamen, and also those of the crew of an American Admiral's Barge,
who, with himself were drowned at Osaca in crossing the Bar,
about the same time as the French massacre. There is an upright
tablet for each of the 10 seamen, and a large monument in the cen-

tre to Admiral Lee, whose body, a few days after he was interred was taken up again, and sent over to America in a man of war for interment.

¶I have been on shore several times here: one afternoon I visited a large Tea-roasting shed belonging to a company of Chinese merchants. The sheds contained some 300 small coppers in rows; and I thought at first the women were washing clothes, for they were half naked. Each copper had a charcoal fire burning underneath it: the inside of the Copper was black leaded, and I fancy that has something to do with colouring the leaf. Each copper contained about 4 lbs of green Tea leaves, which the women shook up with both hands continually. Chinese overlookers superintended the firing process, after which it was weighed and packed for export to San Francisco. Most of the Japanese Tea is consumed in America, but it is not so good as the best China Tea. The Tea shrubs are like small gooseberry bushes, and bear about three crops of leaves a year, the leaves are like those of the sloe. Our ship's Band used to land sometimes of an Evening, and play in front of the British Consul's house causing a great gathering of natives to hear the music.

The Japanese have a native spirituous drink, called Saki, which tastes like strong cider with a dash of spirit in it, and will produce intoxication. In some of the Japanese Temple grounds there was a white horse, which had a carved stable, and a man to look after it, & seemed to be greatly venerated: the people fed it with small plates of beans, which were sold by the man who looked after it. Every Japanese house contains a shrine and a particular deity, shut up in a sort of cupboard, in which offerings of fish, rice, &c, are placed, and changed daily. The family pray in front of the shrine, whenever they feel disposed, but I think the women do the praying for their husbands, who smoke approvingly in the background.

¶On the 20th of May, I went on shore with two ship-mates, and took the train to Osaca, which railway has only been opened 12 months, and is still a great source of amazement and admiration to the people hereabouts. It is only a single line with sidings at each station for the Trains to pass each other. The same regulations and prices are in force, as on the Yokohama and Yedo line: the engine drivers are all Englishmen, and the rest of the officials Japanese. The carriages are English (built at Birmingham) and are not half so comfortable as the American built cars. The railway is 22 miles long, winding round the shore of the Bay, and calling at 4 stations. The train occupies over an hour running the journey, and passes through a beautifully cultivated and very populous country; there are villages everywhere, and I should think this part of Niphon is more populated than any part of England.

¶Wheat and barley are almost ready for cutting, and also rice, but they get two crops of that every year. The rice fields are under water for several weeks after being sown, but at present they look like an ordinary barley field, the ear is exactly like that of barley, only shorter and twice as thick. When ripe, all the corn, rice, &c, is pulled up by the roots, and threshed almost directly. The grinding is accomplished by water mills, on quite a different system to ours in England. The force of the falling water raises a large upright beam which falls into a small pit and crushes the rice; this is a very rude and primitive contrivance, and I have no doubt it will very soon be superseded by steam crushing mills. The country people seem to grow a sufficient quantity of Tea for their own consumption, and dry it in the sun; the firing process described previously is only effected on that which is intended for export, and is done, I am told, to prevent the Tea from getting sea-sick. If it were not served in this manner it would lose its virtue entirely before it reached England.

Osaca proved to be a very large and dull sort of place. I have been there twice. The streets are long and rather narrow compared to those at Yedo, some of them were five miles in length, & there were lots of large wooden bridges over the numerous creeks. There are scarcely any Europeans there, and the people stared at us as if [we] were great curiosities. There were scores of public bath houses conducted on the same principal as at Yokoska. Few vessels go to Osaca now, as they are unable to get close up to the town. About 40 miles inland from it is the old "Spiritual" Capital of Miaco, which contains nearly a million inhabitants, and beyond there, some 50 miles, is another large city called Kiola.[70] The new railway which is being made from Osaca to Yokohama, a distance of 320 miles, is nearly finished as far as Kioto. The European and the native name of the Japanese towns is frequently different.

¶There was a large Japanese Citadel at Osaca, a really strong fortress surrounded by very thick high walls, encompassed by a deep wide moat, and flanked at each angle by a large pagoda shaped tower. It covered some 200 acres of ground and was guarded by Japanese soldiers armed with long swords. I saw no sign of any cannon on the walls, and the soldiers seemed to have no rifles. I and my friend entered one large Temple at Osaca, which was crowded with worshippers, but we had to come away quickly or should have disturbed them with laughter. About 30 yellow-robed skull-shaved priests, were singing the most doleful, monotonous slow music I ever heard in my life, commencing with a single voice pitched very high, & ending in a succession of deep howls and groans, in which the congregation joined with loud fervency. It takes a good deal to make me laugh in a foreign country, for I generally wear my features screwed up into an unconcerned indifferent sort of focus, but on this occasion had to run for it although I stood it better than my friend.

¶Every fishmonger's shop in this country contains stewed "Devil-fish" which are considered great delicacies in Japan, and are very numerous in its seas. During the last few days there have been many processions in the streets of Kobi, for May is a sort of festival month. Two enormous carved wooden cars have been dragged about the streets by hundreds of young children of both sexes, in fancy costumes of cotton, silk and paper. At night each child carried a red paper lamp, and the cars were lit up. The carving on them was magnificent, and there were some splendid embroidered silk curtains to them. The cars are kept at the Temple, & are dragged by a long rope. There were also processions of young girls in fancy silk dresses, straw hats, flowers in their hair, and painted face, and were accompanied by a native band of fiddles, gongs, and tom-toms, and in each street they went through a queer sort of posture dance, moving like marionettes. There were other processions of women dressed as men, & men as women, but they were of a noisier character. I dont know what these affairs all mean but the streets are crowded with people, and everybody seems highly delighted and interested, except the Chinese community who appear to take no notice. In walking along a street in Japan, you will see in the porch of each house, all the clogs and sandals belonging to the inmates, as the people never enter in their shoes. Occasionally you may see a few pairs of flash English shoes among the group, and you can conclude they belong to some British Sailors taking tea up stairs.

On the Queen's birthday our ship was illuminated, and the British Consul gave a Ball; this afternoon our Officers gave a Luncheon & Dance party, & the old Japanese Governor, and his native interpreter came. They wore the usual—black dress clothes, & white chokers, sitting upon the bridge in arm-chairs the whole time, trying to look as if they enjoyed it. On the 25th May we left Kobi for a four days cruise in the Inland Sea, returning again on the 29th. We did not go to Nagasaki after all; it being too far, but went about 120 miles thro' it, and came back, having anchored every night, and landed parties of Officers at the settlements. According to the Treaty no one is supposed to land except at the Treaty Ports, but these orders are not so strictly carried out now. Every where we anchored boat loads of native men & women, with children, came off to the ship, & seemed vastly pleased & astonished with what they saw.

Of the scenery in the "Inland Seas", I can truly say, that it is the finest I have ever seen. There are nearly 2000 islands sprinkled therein of all sizes & shapes, and the main islands of Niphon and Sikok, which are tolerably high lands, are generally hidden from view by the smaller ones. The narrowest part we passed through was about 4 miles broad, but lower down it is less than a mile. The number of Junks and fishing vessels to be seen in the Inland Sea is something enormous, and exceeds anything I have seen off the coasts of England or China. The men in them are almost naked, and of a copper colour, to see them sculling their boats along, and hear them singing their wild discordant choruses, is to be irresistibly reminded of the South Sea Islanders, & the Malay fishermen: but there the similitude ends; the marvellous and unrivalled cultivation to be seen in every direction around, affords abundant testimony to the superior industry and civilization of the Japanese people.

¶Some of the islands are very lofty, and at every point rounded, the scenery seemed to change like a grand moving Panorama. At one moment we seemed to be passing one of the Tropical islands of the South Pacific, and before the ship had fairly got past it, some miniature Kerguelen's Land would start up—a mass of sterility & desolation! In the morning one could imagine that he was steaming through Cook's Straits to Wellington, and in the evening he would find himself at anchor in the counterpart of the Bay of Naples—scenery more varied and diverse in character it is impossible to conceive. The number of villages past during a run of 50 miles, is extraordinary, and the cultivation is something wonderful. Islands rising quite abruptly to a height of 1500 feet, were cultivated to the very top, & the produce would support 50 such little villages as were situated at their foot, whilst—others similar in size, shape, & altitude, and scarcely a stone's throw away—were pictures of solitude & sterility!

¶We leave Kobi to-morrow for Yokohama. We have on board now an amphibious animal called a Salamander, purchased from the shore; I believe there are some in the Crystal Palace Aquarium:

they are something between an Alligator & a Sea Serpent, and would astonish most people.

Yokohama, June 16th
We arrived here on the 6th after a very rough passage round, and received our mail.

The other evening our ship's company were invited by the Temperance Society of Yokohama, to an Entertainment of Songs, Recitations, &c, with the addition of refreshments, in the shape of Sponge Cakes and Tea, Strawberries and Cream, which were provided by the Ladies of Yokohama. The other British and American Ship's Company's were invited, but the affair was got up specially for the "Challengers" Crew, and our Band was in attendance.

Group of Japanese visitors aboard *Challenger* at Yokohama (Natural History Museum, London, Photo No. 504). Photo courtesy of the Natural History Museum, London.

Yesterday we went out to sea for three hours with a large party to shew them how the Dredging and Sounding operations were carried on. There was a grand Luncheon & Dancing on the Quarter Deck. Besides a large party of ladies and gentlemen resident in Yokohama, there were among the company the French & the Japanese Admirals, with other Naval Officers, the British Ambassador, Sir Harry Parkes, and Lady Parkes, & several Japanese State Officials, including the Prime Minister of Japan.

The latter party were dressed in European Costume, with Wellington Boots outside the Trousers. Sir Harry Parkes is a fine looking man, and so is the Japanese Admiral.

We have taken in seven more hands to fill up our complement, they came from the "Audacious" which ship came in on Saturday from Shanghai, and is now in the Imperial Dock at Yokoska.

The distance to Honolulu by the route we are to go will be 4,500 miles, & they have allowed 45 days for the journey, averaging 100 miles per day, including sounding operations, &c.

The harvest is all over in this country now, and the weather is very hot, & peaches, &c are nearly ripe.

Our photographer took a picture of the Company on board yesterday.

> Believe Me,
> Sincerely Yours,
> J. Matkin[71]

After a final ten days in Yokohama, *Challenger* began its Pacific crossing on 16 June 1875. The expedition was still one year and 118 oceanographic stations away from landfall in Britain. The Asian segment, from Hong Kong to Japan, had begun with troubling tales from home for Matkin and had ended with the even graver news of his father's death. Yet his Asian experiences were probably the most striking and memorable of the voyage. And fortunately, the sad news was now behind him. The fabled beauties of Polynesia and adventures in Chile and Argentina lay ahead.

Polynesian Islandfalls
JULY 1875–NOVEMBER 1875

Midsummer of 1875 found *Challenger* sailing, and sounding almost daily, on a track due east from Japan to a point in the north-central Pacific whence the ship would turn south toward Hawaii. The passage was uneventful, even monotonous: Lieutenant Spry reported that "very little of interest occurred from day to day, and the results of the trawling and additions to the natural history collection were very scanty."[1]

Just before the turn south and two weeks before reaching Honolulu, Matkin began another round of letters. Hawaii would be agreeable, especially after "4500 miles of watery solitude."[2] Honolulu, in fact, seemed a place worth settling down in for awhile. The lack of any mail from home, however, was disheartening.

> H.M. Ship "Challenger"
> North Pacific
> July 13th, [18]75

My dear Mother,

We expect to reach Honolulu by the 25th of this month, so I will commence my letter as usual. We left Yokohama at 3 PM on the 16th June, with the homeward bound Pennant flying, & the band playing "Home Sweet Home" as we steamed past the other men of war. I can't say that I felt particularly lively as we lost sight of mount Fusihama & the coast of Japan. I was thinking how altered everthing seemed since we sighted Japan on the 11th April previous, & were so pleased to see the letters come on board {from which I first heard of my paternal loss at home, casting a gloom over everything now[3]}. Any letters that reached Japan after we left, will be forwarded to Honolulu, & we shall get them before leaving the Sandwich Ilds.

On the day before we left Yokohama, the United States ship "Kearsage" arrived; this was the vessel that sank the "Alabama", in action off the coast of France.[4] One man deserted the day before we left, & one man died the day after we left, from the effects of poison taken by himself whilst in a state of intoxication. He was a Scotchman—James Macdonald—and his rating was Sick Berth

Steward. He had been many years in the Navy, & joined the "Challenger" at Sheerness. He was an intemperate man when he could obtain liquor, & this was the man who was robbed whilst drunk at St Thomas's, West Indies, & wished to be brought on board in a newspaper & analyzed to see what kind of drink they had sold him. When at sea he was all right; & a very steady, saving, industrious man; he earned over £30 by a sewing machine which he bought at Melbourne. In harbour he would squander away an incredible amount of money, & was never quite sober, but at sea he was more careful over a penny than the generality of Scotchmen; & to hear him holding forth to the sick seamen about the evils & lamentable effects of drunkeness & immorality, you would think him a paragon of excellence & virtue. He was also a dabbler in scientific pursuits & of an enquiring disposition, & he was a great admirer of the poet Burns. I could not help thinking, as they lowered his body into the sea, of the Epitaph that Burns wrote for himself, & how applicable it was to the deceased.

> "Is there a man whose judgment clear
> Can others teach the course to steer,
> Yet runs himself life's mad career
> Wild as the wave?
> Here pause—and through the starting tear
> Survey this grave.
> The poor inhabitant below
> Was quick to lear and wise to know
> And keenly felt the friendship glow
> And softer flame:
> But thoughtless follies laid him low,
> And stain'd his name!"

¶He was buried on the 18th, & directly after the funeral we took soundings at the enormous depth of 3,900 fathoms, 4-½ miles, the second deepest sounding we have ever obtained.[5] We have come nearly 3,500 miles since leaving Japan, as we have not steered direct for the Sandwich Islands. We have been steering due East nearly, & tomorrow the course will be altered to the South, so we shall soon be in hot weather. We have 950 miles more to run, & are at present under steam with calm pleasant weather. We have taken a fine line of soundings, having sounded about every 200 miles. From Japan the depth has been 17th June 1875 fms, 18th 3,900 fms, 19th 3,675 fms, 21st 2,900 fms, 23rd 2,300 fms, 24th 2,575 fms, 26th 2,800 fms, 28th 2,900 fms, 30th 2,775. July 2nd 2,050 fms, 3d 2,500 fms, 5th 2,950 fms, 7th 3000 fms, 9th 3,050 fms, 10th 3000 fms, 12th

2,675 fms. The dredging and Trawling has been very unlucky, for the bottom of this part of the Pacific is very rough; being of volcanic formation. Several Trawls and some 10,000 fathoms of thick line have been carried away, the value of which is nearly £500. A few good specimens have been obtained, & lots of great boulders of pumice stone which contain the fossil remains of animal life. Serial temperatures are also taken, & the strength of the current (if any) is ascertained. We have had good winds, but damp, wet, & unpleasant weather; this we shall leave behind shortly & pick up the pleasant North East Trade Winds. {Our Lat. to day is 38° N. Long 150° W.[6]} We are rather ahead of our time for reaching Honolulu, so far, & the Captain says he will drive the ship home by the first week in May 1875.[7]

We have only seen one vessel on the way, the American mail boat "Mohawk" from San Francisco for Yokohama, on the 1st July. {We are now on the Pacific Admiral's Station, which extends to the 180° of long. and includes the coast of America from Alaska to Cape Horn.[8]}

Last week with us contained 8 days, & we had two Sundays as the Admiralty allow us no pay for that day. One of the Sundays, your Sunday—was kept as a holiday. Ships that sail round the world & steer east all the way, gain 12 hours when they reach Long. 180° E or the opposite meridian to Greenwich; they then take an extra day in that week which leaves them 12 hours behind Greenwich, & that they also gain again on the voyage back if they still keep to the East.

¶I expect to send you the second volume of my Journal from Valparaiso next X.mas. I thought this volume would have lasted me home, but China & Japan occupied nearly 100 Pages. My friend Miss Midji Maru said as I paid her the night before leaving Japan, for my 55th cup of tea, that "she very sorry when Challenger go." Miss Maru & her married sister kept a Tea-house on the highest hill near Yokohama, which commanded a fine view of the town & bay. She could speak very tolerable English, & was on that account a martyr to every intelligent traveller—like myself. I calculate that I asked her several thousand questions as to statistics & customs &c., of Japan. I think I pumped her pretty dry, & I was able to tell her more about the productions, population, position of the cities &c of Japan than she herself knew.

The Japanese are very ignorant about everything of that sort, & scarcely know the size & extent of their own country. Miss Maru was quite in a fog about everything ten miles out of Yokohama, &

concerning the history of Japan she could only relate her & her Papa's experience. If I asked her about her religious views & opinions she would only laugh & point to the English Church, & to a large neighbouring Buddhist Temple, & say "that is your church, & this is mine", after which she would change the subject to the prices of lady's clothing in England. Of course I was particularly well up on that subject: I think I told her that a complete costume for a respectable English lady would cost something under £2000 including a set of teeth & some back hair. Miss Maru was very pretty & very genteel: her manner of bowing and receiving payment for refreshment, I have never seen equalled. I have her portrait, & so have about 60 others in the ship, I find. Her ideas on the position of the Sandwich Islands were very vague, & to give her an idea of their whereabouts, I could only jerk my finger out toward the Pacific & say, it's a month's sail out there.

I see by the papers that the Americans have managed to possess themselves of one of the Sandwich Islands as a naval depot for their fleet, & have persuaded the King to refuse this privilege to any other Power. This is another smart piece of Yankee diplomacy, & I dare say that they will succeed before long in what they have been long scheming for—the annexation of the whole group.[9]

July 24th

We expect to reach Honolulu on Tuesday 27th as we are only 300 miles away. We have picked up the N. East Trades, & the weather is much hotter, but pleasant. The depth since the 17th has been about the same, but we have lost 1500 fms more line. We shall probably sight land on Monday evening for most of the islands are lofty—being of volcanic origin. The Pacific Islands generally lie in clusters & groups. The Sandwich group numbers 13, 8 of which are inhabited. Their total area is about 6000 square miles,[10] & the population 60,000. The group was discovᵈ by Captain Cook in 1778, & named after the 1st Lord of the Admiralty—Lord Sandwich. Cook estimated the population at 400,000, so that in all probability these people will have entirely disappeared before the lapse of another century—like the natives of Tasmania, who also numbered several thousand when Cook visited there. The Kanakas—as they are called—were described by Cook in his Journal as the finest, the most civilized & the best disposed of all the South Sea Islanders. I believe Captain Cook treated them fairly, & was a good humane man; but the conduct of several of his officers & crew was cruel & disgraceful, & in a squabble at Owhyee in 1779 caused by the impolicy of some of his crew, Captain Cook fell a victim of the sudden resentment of the natives. Owhyee or Hawaii, is some 200 miles S.E. of Oahu & the other islands, & was not discovered until 1779, when Cook returned from the Arctic to the Sandwich Islands again. It is the largest island, its area being over 4000 sq miles & the population 27000, & it contains the stupendous volcanoes of Mauna Loa & Kea—each over 13000 feet high & still active. The Kanakas belong to the same family as the Maories & Tongans but are unmixed with the black Papuan &c. The present King is Kalakaua, elected in 1874. The islands are just inside the Tropics, & very fertile, producing sugar, coffee, cotton, arrowroot, rice, cocoa, bread fruits & numerous West Indian fruits. They are well situated for trade, & the seat of considerable commerce, chiefly with America. Honolulu on the island of Oahu, is the capital of the group.

July 27th at Honolulu

We sighted Land at 3 AM this morning, & by 5 AM the islands of Molokai and Oahu were in sight. We dredged between the two islands all the morning, but got nothing, so proceeded onward under all sail & steam. The islands were lofty & very irregularly shaped, & they looked about as large as Fayal, Azores. At 2 PM we passed Diamond Head at the South Eastern extremity of the island, & at 3 PM we anchored outside the reef in front of Honolulu, & made a signal for a Pilot to take us in. A native Pilot soon came out, & took the ship thro' the shallow channel between the great barrier of reefs which extend all round the island.

¶We anchored by 4 PM within a stone's throw of the town, & the Mail was at once sent on board. There were no letters for me as usual, but as another mail is expected in tonight from San Franciso for Sydney, I live in hopes. There are a few whaling vessels here, & the American flag ship "Pensacola." The usual Contractors, Washermen &c., came on board at once. Fresh Meat, Vegetable & Bread are each 2½ per lb & Provisions in general are pretty cheap. Butter made on the island is only 1/3 per lb & this is a great rarity in the Tropics. The Meat is quite as good as in Australia, but Fruit is not so cheap as it should be, & washing is 5/- per dozen.

¶About half the native population of Honolulu came down to the wharf to see the ship moored, & after seating themselves as comfortably as possible, gave themselves a half holiday apparently, & took a good long view of us for about two hours—then went home to supper. From my short experience I should say that the Kanakas are about the most cheerful, good natured, indolent & impulsive race of people under the sun. One old gentleman came on board & introduced himself as Mr. John Williams,[11] shook hands, & said he had the pleasure of seeing me before at Calcutta. I told him I had never been there, so he shook hands again & said it must have been my brother, & told me to come & see him at any time. The Kanakas, or Hawaiians, are in appearance the very counterpart of the Maoris of New Zealand, & much darker than the Friendly Islanders, being as brown as Malays. Still they have a pleasanter look than the Maoris, & are much superior to the Fijians or Admiralty Islanders. They are a fine made people, & dress in European costume. I have not seen any women yet except at a distance. Of course the Hawaiians are X.ns, converted by American missionaries chiefly, & they are undoubtedly the most civilized of the Pacific races, but, unfortunately, it is this civilizing process that is killing them so fast. They are a very sensitive people, & one of our seamen who has been here before assure me that they can lie down & die whenever they feel disposed. So unless they intermarry with some other race, say the Chinese, or unless they introduce the Mormons into their country,[12] they will soon belong only to the past. There are lots of Chinese at these islands, & a very hardworking useful lot of people they are: I am sure it is a good job they are so numerous. I see by the Papers that the Measles has carried off about one third of the Fijian natives since we were there, so the English & Chinese will soon have it all to themselves in that group.

¶The island of Oahu has a very pleasant rural appearance, but the scenery is certainly not Tropical as seen from the ship; if it were not for a few cocoa palms dotted here & there, one could fancy the island to be one of the Azores. It is not to be compared to Tongatabu or the Fiji Islands, & the town of Honolulu has a very straggling, unimposing appearance. {The weather is much hotter here; Thermometer 85° in the shade, and over 120° in the sun.[13]} The ship is already full of flies, & the ravenous but playful Mosquitos are only waiting for us to turn in before commencing their onslaught. I shall tell you more about Honolulu after I have been on shore. The Mail for Sydney is in sight, I am told, & will be in by II PM, so in the morning I hope to get my long expected letters.

July 28th

The mail is in & there was not a single letter or newspaper for me from anyone; now I shall have to wait until we reach Valparaiso

Honolulu and the Valley of Nuuanu, 1875; from T. H. Tizard, H. N. Moseley, J. Y. Buchanan, and John Murray, *Narrative of the Cruise of H.M.S. Challenger with a General Account of the Scientific Results of the Expedition*, 2 vols. in 3 (London: HMSO, 1885), vol. 1, pt. 2, p. 761. Photo courtesy of Scripps Institution of Oceanography.

in November. This mail left England July 1st, & as the mails only leave San Francisco once a month, we shall get no more at this place. I need not say how disappointed I am, for it is nearly a twelve-month since I heard from you or any of the boys, or from Miss Wildman. I suppose you have taken to writing once a year now, & then all of you together; this is a very kind & impartial arrangement. I scarcely know yet the particulars of poor Father's death, & none of our ship's mails have been lost anywhere. If you have any bad news to tell, it is better to tell it at once than to leave people to "think & fear". I suppose I shall get a heavy mail at Valparaiso, but we shant be there for three months yet.

I see by the papers that the Arctic expedition has sailed. I hope they will be successful, & I should very much have liked to have gone. The London News contained a good portrait of Captain Nares & some pictures from this ship of New Guinea & the Admiralty Islands, which I hope you saw. It was in the middle of June that these pictures appeared. We are now taking in 240 tons of coal, & as soon as it is in the crew will be granted 48 hour's leave. Our back wages were paid today in American Dollars value 4/2 each. We had the American admiral on board this afternoon, a rough farmer looking old gent. H.M. Sloop "Petrel" arrived to day from Panama. We have 2 officers & 3 seamen from this ship gone up the Arctic, all in the "Alert"; they left us at Hong Kong. We heard last night of the death of Lady John Franklin[14] per telegram received at San Francisco. The death only occurred in London on the evening of the 18th, & the San Francisco Papers for the 19th contained accounts of it all. This was very quick work. The distance from San Francisco to Honolulu is 2,300 miles, & takes steamers 8 days.

August 4th

I have been on shore here twice, & went horse riding both times. Honolulu is quite a civilized place—more like England than any place we have been to since leaving Wellington. There are some fine shops of all kinds, & a very considerable white population, principally American settlers, & merchants, Consuls of all nations, missionaries of all sects, & invalids from America & Australia. There are several churches—Hawaiian or native, Methodist—Wesleyan—American & Roman Catholic. The native church[15] is the largest, & is just like an English dissenting church, was nicely seated, & contained a good organ. The King's seat is under a crimson velvet throne {canopy[16]}, & in the centre was a large spittoon containing several ends of cigars, & other pews contained the same so I suppose the Hawaiians smoke in church. It was a week day when I looked in so of course I can't say that they do smoke during Service, but they certainly smoke in the church. There are only four hotels allowed in Honolulu, & they have to pay a very heavy tax to the Government, which makes the drink very dear, & promotes temperance.

¶There are lots of Chinese keeping shops, & they nearly all have native wives, & generally very pretty children. They work hard & are looked down on by the natives on that account. Of course the natives do work, but in a desultory sort of manner, & they want very high wages. About 40 of them under an American overseer helped to coal this ship: their pay was 5/- per day, & the overseer offered them another 4/- to work until 10 PM one evening. The natives laughed but refused the job, & told the Yankee to do it himself.

¶The natives can nearly all read & write the Hawaiian language & most of them can speak English. Their books &c are printed in the Hawaiian language & they only talk English to strangers. They live in nicely furnished houses & are altogether quite as civilized & as comfortable as the working classes in Europe. Their living costs them very little, & they work when they like, & on a Sunday they dress in the latest fashions, & strictly observe the Sabbath. They are religious & straight forward, without being at all gloomy or strait laced. The town is laid out exactly like an English town, with good horse roads, for there is a large carriage traffic. On week days the men wear a shirt & trousers, & go bare-footed; the women wear a long coloured loose night gown, reaching to the ankles, & generally no shoes or stockings, & their hair is worn either in two long plaits or hanging loose down the back. They wear natural flowers in their hair, & also in their little straw hats. This is their week day dress; on a Sunday they dress rather more expensively than our English lady bound to a Flower Show, & they are very fond of high lace up French boots. The real native women are not so handsome as the Tongan women, but the half-castes are very handsome, & many are married to Europeans. There is an upper, a middle, & a lower class, even among the T~~ongans~~ Kanakas.

¶The principal amusements are bathing & horse riding; all the natives can ride, & they can all swim; you may see bathers bathing & throwing their children into the sea before they can walk, to teach them to swim. I believe there are more horses on the small island of Oahu than in the county of Yorkshire. They breed so fast that every native who wishes can have one. Every decent white person keeps a horse & carriage, & there are horses grazing all over the island. The women ride the same as the men astride of the animal, & to see them on a Saturday afternoon in their white muslin dresses galloping along the streets by scores, with their long hair streaming behind, & a baby in front is to see a pretty sight. Saturday afternoon is a general holiday with the natives, & they dress up & trim their hair & their horses with flowers before going out. They generally use the clumsy Mexican saddle, & the saddle & bridle costs as much as the horse. You may buy a horse for a pound but you can't hire one under 2/- even for an hour. The last horse I rode was a Donkey at St Michael, Azores, but I have been riding twice here, & shall go again to morrow. Honolulu has been thronged this week with sailors on horse back, & I am sure it's a great mercy no one has been killed.

¶I must not forget to remark on the number of pigs, dogs, & cats there are on this island—especially cats. You see cats in all the houses, in all the gardens, on all the walls & roofs, & following people about the streets, & Pigs are to be met with everywhere. I saw a party of natives cooking a pig, & they asked me to partake, but I wasn't hungry. They cooked him whole, just as he was killed, in a pit full of red hot stones, & as soon as he was sufficiently cooked they poured cold water on him. Cocoa nuts are rather scarce here, but bananas, oranges, mangoes, guavas, melons, & limes are plentiful, but they are brought from the large island of Hawaii.

¶The natives are very fond of elaborate funerals. I met one on Saturday afternoon, with plumes & hearse in grand style, & the natives in the deepest mourning, but chatting loudly & affably as they marched along. The cemeteries are pretty & the natives are fond of grave stones. The finest building in the place is the King's Palace built in 1872 by Kamehameha the 5th. It is something like a gentleman's Hall in England with a square church tower in the centre.

¶The State officials &c here are nearly all Americans; there is a fine native Cavalry regiment, one of Infantry, & a Military Band. The population of Honolulu is 70,000; of whom about one third are foreigners. You can't hire a horse, or buy a glass of beer, or do any sort of bartering with the natives on Sunday; it is the Sabbath with men & beasts, & they considered it a great sin some time ago, when an English, a French, & an American man of war all sailed from the island on a Sunday morning. They don't mind a vessel coming in on a Sunday, but they can't approve of one sailing. Our men who were on shore last Sunday could buy nothing, nor get any drink, & they say that they walked about the town with their pockets full of money like a lot of codgers. {For such a large town I never saw quieter or more orderly streets: the shops are nearly all shut after 7 PM & the streets are as quiet as on Sunday Evenings in England, and there is evening service in nearly all the Churches.[17]}

Most of the houses are built of wood, nicely painted & with flower gardens attached. This island is not a fertile island, for the soil is nothing but lava cinders, & everything has to be cultivated. I consider these natives far superior to those of the Philippine Islands in civilization & social progress, &c. altho' they were cannibals less than 100 years ago. The first Missionaries arrived here in 1820 in the brig "Thaddeus" from Boston, United States. The white ladies here, of course, ride horses the same as in England, but where everybody rides astride, a side saddle looks ridiculous. Lots of the native women are over 6 feet high, & generally as tall as the man. As compared with a Chinese or a Japanese town there seems a great scarcity of children in Honolulu, & the girls seem twice as numerous as the boys. There are newspapers printed here, & a monthly Magazine {but they are after the American style, and the tone is not equal to our own local papers[18]}.

¶Yesterday we had several of the native Ministers & State Officials on board, & today we have had King Kalak[a]ua[19] himself on board with his son & several officers. Kalakua is a stout man, about 40 years old, almost as black as a negro, & in figure very much like a small Edition of "Sir Roger".[20] He was dressed in an old civilian suit of clothes & an old straw hat, but his son & his officers were well dressed. All our Officers were in full dress, & the Band played the "Sandwich" Anthem, & he was received just as we should receive Queen Victoria. He stayed to Luncheon & had his portrait taken afterward. . . .

I should not mind living here for a few years; I should keep a horse, & a litter of cats, dogs & pigs, the same as other X.ns here. The founder of the Hawaiian Kingdom was Kamehameha the 1st who conquered and ruled the islds for 20 years. He was born 25 years before Cook discovered the islands, & was a very remarkable man. His memory is still regarded by his countrymen, & he is

King Kalakaua and *Challenger* officers, Honolulu, 1875 (Natural History Museum, London, Photo No. 516). Photo courtesy of the Natural History Museum, London.

often called the Napoleon of the Pacific. Before his time the islands were subdivided into petty kingdoms. He was succeeded by four other Kamehamehas in succession, then Luna[l]ilo, who died last year & left a widow who ought to be Queen, but the Americans succeeded in getting Kalakua elected to the throne. King Kamehameha the 2nd with his favorite wife visited England in 1824 in a British merchant ship, & was very kindly received by King George the 4th. He and his Queen died in England of the "measles", within 8 days of each other & their bodies were brought back to Honolulu by Captain Lord Byron, cousin of the poet, in H.M. Ship "Blonde" & buried here. Kamehameha the 4th also visited England in 1849 when he was heir apparent.

¶We expect to leave here on the 10th for Hawaii, Tahiti, Society Islands, Marquesas Islands &c, before reaching Valparaiso. I must now conclude, with very best love to you all, hoping you'll write more regularly.

From Your Affectionate Son,
Joseph Matkin

P.S. I enclose list of future addresses.

I am writing to J.T.S., to Swadlincote, to Walter Thornton & to Mr. Daddo by this mail, & shall write to the boys as soon as I hear from them. Send this to Stamford & to Willie. We have just received from the "Petrel" an animal called a "Terrapin" or Land Tortoise, brought from the Galapagos Islands. The Petrel has several which are going to the Zoological Garden, London; these animals are nearly extinct {as the whale ships kill them for food[21]} & are only to be found on two of the Galapagos Islands. Ours is dead, & weighs 300 lbs., but they have been known to weigh 1000 lbs & to live 100 years. I have been to the English Cathedral here, but it is a very poor one for a Bishop's Church, & ritualistic forms & postures are greatly in vogue. I also went to the American Free Church, the service in which is something between the Wesleyan & the Dutch Reformed Church at the Cape of Good Hope.[22]

The concurrent letter to his cousin concluded with remarks about church ritual rather more pointed than he apparently felt appropriate for his mother's reading.

Aug. 10. . . .

On Sunday I went to the English Church, a very poor one for a Bishop's, & the service was burlesqued by that mixture of millinery, & tom foolery called "Ritualism". In the Even, to the

American Free Church, which was quite a healthy change, in an opposite direction from the "posture" & "imposture" practiced at the Cathedral.

The American Illustrated Papers I have seen here, are full of copies from the "Graphic" & "London News", and the local papers contain the novels of the best English Authors, printed in Chapters, & without the Author's permission. I mentioned this to an American, & he said, "Of course they do the same in England with ours, or they can do if they like."

Yours Sincerely,
J. Matkin[23]

Despite the absence of mail from home, Matkin began another round of correspondence two weeks later when *Challenger* arrived in Hilo Bay, on the island of Hawaii. These letters cover the ship's passage from Hawaii south to Tahiti and east to Juan Fernandez Island and finally the arrival at Valparaiso. They do not contain quite the abundance of detail typical of his earlier letters; he makes no mention, for example, of the bushels of manganese nodules dredged up en route to Tahiti, or of the hundreds of sharks' teeth and cetacean ear bones brought up soon after departing for Valparaiso. Perhaps Matkin's enthusiasm for writing was flagging a bit in response to the lack of letters from home. That problem would be happily remedied, however, upon arrival in Chile.

H.M. Ship "Challenger"
Hilo, Hawaii
Aug. 19th, [18]75

My dear Mother,

On arrival here I received a Mercury & a Cornish paper from Miss W[ildman] brought from Vancouver's Island by the Flag ship "Repulse" & altho' a letter would have been most welcome the Papers were better than nothing. We left Honolulu on the 11th Aug. for this island, distant 200 miles & arrived on the 14th, having had to steam against a head wind. Before leaving we received from H.M.S. "Petrel" two of the Galapagos Turtle I told you about us seeing last. One weighs nearly 300 lb & the other—a small one— about 50 lb. The large one can walk away with a man on its back. They are very ugly creatures—the head being like a Serpent's & the hind feet like an Elephant's, the fore feet are shaped like the flippers of a seal, & there are five toes on each foot. They live entirely on fruit & green stuff,[24] but do not masticate their food, they swallow it exactly like a Python would a monkey. If possible we shall bring them home alive. The men polish their shells with bees wax & oil, when they are cleaning their guns on a morning.

¶We sounded on the passage across about 12 miles from the island of Mauii at a depth of 2,075 fms. It was a lofty island, but inferior in altitude to this island of Hawaii, which is the highest in the Pacific. On the night of the 12th the blazing volcano of Mauna Loa was visible, but we were 90 miles away. We were steaming parallel with the island of Hawaii all day on the 13th, for the town of Hilo is on the Eastern side of the island & the prettiest of all the ports of the Hawaiian islands. The island is about 75 miles in length & 55 in breadth, & seen from the sea has a beautifully verdant appearance owing to the large amount of rain it receives thro' the attraction of its high mountains. In the manner that it sloped upward from the sea it reminded me very much of Madeira, but the great rents & upheavals everywhere in the soil were more indicative of its a volcanic formation origin.

¶On this morning of the 14th we anchored in the bay of Hilo about a mile from the town & close to the American flag-ship "Pensacola." The natives that came out to us with a Pilot were wearing the "Repulse's" hat ribbons on their straw hats & they told us that the "Repulse" had gone round to Kalakekakua [Kealakekua] bay—where Captain Cook was murdered. They abused the Americans frightfully, said they never had any money to spend & that the "Repulsers" were rolling in Dollars, & that Admiral Cochrane had behaved like a father to them, & had thrown his ship open expressly for them & supplied boats to bring them off, as well as refreshment.

The "Repulse" is homeward bound so I suppose the Admiral wished to leave behind a favorable impression. Admiral Cochrane is the grandson of the famous but ill used Admiral, the Earl of Dundonald.[25]

¶The Americans thro' their great political influence here, succeeded in getting Kalakua elected to the throne against the wishes of the natives, & there were terrible riots in Honolulu at the time. The natives have settled down now to their new King, but when, after his return from America, he tried to pass a law ceding to them —the U. States, the Pearl harbour on the island of Oahu, as a naval depot, they fiercely refused to part with a foot of their soil & the discontent & distrust of foreigners is generally gone due to their new rallying cry "Hawaii for the Hawaiians." America gave us the light, said a native priest in a sermon at that time, but now that we have the light we should be left to use it for ourselves.

There is no doubt that the Americans have had all the work of civilizing & educating the Hawaiians, & they have done their work well, but this does not establish their right to the islands if the natives would prefer British protection. The ancient Chiefs of these islands are all dead & foreigners seem to think there will be no more Kings after Kalakua, but the remnant of the people will, at his death, decide what form of government they would prefer. When that time comes there will probably be news about the matter. Since the last election, the Americans do not feel safe without one of their ships of war cruising round the islands.

¶The town of Hilo is much prettier than Honolulu, the houses are generally white & surrounded by trees & gardens; it is a small town of about 3000 inhabitants including a couple of hundred white people. There are American, English, & Roman Catholic missionaries & Churches, a native Governor, & plenty of Chinese. The latter work chiefly on the sugar plantations which are very numerous on this island, & at the back of this town are mills where the cane is crushed. Sugar growing & cattle & sheep farming are the principal pursuits here, but the sugar planters are losing because the native laws do not permit them to make Rum, & this article is the most profitable item in a sugar planter's income. There will be plenty of Rum made there bye & bye I know: the island is as large as Mauritius & that island produces half the sugar & rum consumed in Europe.

The Cattle farmers are on the slopes of the mountains, where grass is plentiful & the climate suitable; some of the farmers have 20,000 sheep & thousands of cattle, the main profit of them being the wool, tallow, & hides; for meat is so abundant & cheap that it hardly pays for bringing in from the station. We paid 2d per lb for beautiful beef but the regular inhabitants pay less. Fruit was plentiful, especially pine-apples & mangoes—a delicious Indian fruit.

¶We had one good view of the mountains the day we arrived, they are 25 miles apart, Mauna Kea—the extinct volcano was covered with a small patch of snow, but the active one of Mauna Loa was free from snow. They are each over 13,700 feet above the sea & slope upwards so gradually that a mule can climb to their summits. They are not conical shaped like Teneriffe or Fusihama but merely rise from a chain of mountains which is 10,000 feet high for several miles near the culminating point. There is an active crater

near the summit of Mauna Loa, & an immense burning lake of fire
in its side, about two thirds of the distance from its base. This
wonderful lake of fire is what all travellers coming hither try to go
see. Nearly all our officers have been, but as it is a two day's journey & you require horses & native guides, none of the ship's company have been able to go, & I was disappointed.[26] But to compensate for this, last Sunday night the upper crater of Mauna Loa burst
out brilliantly, more than it has done for the last 6 years, & we had
a splendid view of the strange subterranean fire. We have not seen
it since, & the people say that you may be here for months without
seeing it burning. It is the upper crater that the people fear the
most, for when it bursts out the burning lava rushes down the sides
of the mountain toward Hilo. There have been about six eruptions
during the present century, the last & worst being in 1868. A good
many of the foreign settlers left Hilo then, for Honolulu thinking
the town was going to be buried like the "Cities of the Plain":
those who could not leave sent men out on horseback toward
the mountain to see what course the stream of fire was taking. It
came direct for Hilo, burning & destroying everything in its
course, but, about 10 miles from Hilo, it entered the Earth &
flowed underground until it reached the end of the town, when
it burst out of the ground & ran into the sea, forming a long
peninsula of land which is now covered with cocoa-palms & vegetation.

¶Horses are as plentiful here as at Honolulu, & every native
seems to possess one. I was on shore yesterday afternoon with two
ship-mates, & we went riding, but there are no good roads here like
those at Honolulu. The whole surface of the island is lava cinders
&c, overlaid with a thin soil which is very fertile on account of the
abundant rain. Water is plentiful & there are numerous creeks &
rivers in which groups of natives may be seen bathing all day long.
We bathed & rode, & ate fruit & sugar-cane & joked with the
natives. They are about the most hospitable ever we have seen, &
so cheerful & pleasant. The young ladies decorated our horses & us
with flowers & brought us fruit to eat, while they had a gallop
themselves.

We saw lots of children surf-bathing, or riding in to the beach on
the top of the great rolling waves which break all along the shores
of the island: they balance themselves on these waves & shoot in
to the beach at the rate of 40 miles an hour.

We are just steaming away from Hilo, & shall soon have seen the last of the Hawaiians & their pleasant land. We are bound for Tahiti—distant— 2,200 miles South.

September 15th, South Pacific, Lat. 10° S

We are some days behind time again & are still 300 miles from Tahiti owing to the light winds & the numerous soundings we have had to take.

We have sounded about every other day, the average depth being about 2,650 fms: the Dredging and Trawling have also been successful.

We had the usual unpleasant & hot weather in the rainy latitudes near the Equator, but caught the S.E. Trade Winds 300 miles North of the Line & still have them—tho very light. The Equator was crossed at midnight on the 5th, in longitude 151° W.

We have had another sad death on board since leaving Hilo, & this time it was one of the Scientific gentlemen—a fine healthy looking young man too. Dr. Von Suhm joined the ship at Sheerness in '72 as a Naturalist under Professor Thompson, & was very clever in his profession—quite an enthusiast in hunting up Natural History specimens at the different islands we have visited. About a week ago he was laid up with a severe cold which produced Erysipelas[27] in the head & soon caused delirium. On Sunday he was so bad that we could not have Church, & the bells could not be struck; he was worse on Monday, & in his wanderings gabbled away in German, which Mr. Buchanan, Chemist on board, could understand & answer. He talked of the battles he had helped to gain during the Franco-Prussian war, in which he served as a common soldier, & was wounded. On Monday morning he got out of bed & said he must pay a visit to his father & mother; & he died at 3 PM the next day. He was buried at 9 AM yesterday morning, the officers & ship's company being in full uniform. We sounded soon after at the depth of 2650 fms. Professor Thompson helped to carry him to the gangway & was very much cut up about it, as were all the Scientific Staff.

On the day before he died we got ready for anchoring at a small group of islands called the Carolines,[28] so that he might be kept more quiet, but we passed over the exact spot where they should have been—according to the Chart—without seeing anything of them, so we proceeded onward for Tahiti. Several of the small groups in the Pacific are positioned incorrectly, we found the Admiralty Islands a long way out.

Dr. Von Suhm had a brother at Manilla who used to be constantly with him during our two visits there. He was a native of German Holstein & about 26 years of age. All his books & effects will be sent home from Valparaiso to his father, but letters will be sent to his friends from Tahiti.

Sunday, October 3rd, at Tahiti

On the 16th Sept. we brought up from the bottom of 2,350 fms the head of a harpoon & several whale's teeth; so we might have dropped our dredge on the skeleton of some hunted whale. On the 17th we witnessed the most beautiful sunset I ever saw in my life & at 4 AM the next morning we sighted Tahiti & Moree [Moorea], another of the Society Islands—at the Distance of about 60 miles. The isl^{ds} were lofty, the highest peak of Tahiti, called the "Diadem," being over 8000 feet above the sea level. The elevated portions of both these islands is very irregular—a mass of pinnacles, ridges & gullies thrown up by volcanic agency, in a wonderfully curious & striking manner. We sounded between the two isl^{ds} at 1525 fms, then steamed in towards Tahiti & anchored within 200 yds of Papiete [Papeete], the Capital town of the island.

¶Tahiti or Otaheite was discovered in 1767 by Captain Wallis R.N. & in 1769 Captain Cook visited it to observe the "Transit of Venus". He visited it four times & discovered the surrounding seven islands, which he called the "Society Islands" after the Royal Society which was the instigation of his coming hither, as it is also of the voyage of the "Challenger". The appearance of Captain Wallis at this island in 1767 did not create among the natives so much surprise as you would imagine, for it was but the realization of a prophecy of one of their sages. This personage had foretold that at some distant period a canoe should visit their shores without outriggers. I think I told you that a Polynesian canoe was balanced on the water by an overhanging wooden framework like a hurdle, so of course a vessel without this appendage was a great curiosity, & when it did appear in the shape of H.M. Ship Dolphin, the Tahiti islanders soon made friends & traded with its crew.

¶The natives of Tahiti & the adjoining islands have always had the name of being the handsomest of the Pacific races; they are also the laziest, the most cheerful, & the most immoral. They are lazy because there is nothing to induce them to work: their island produces spontaneously everything that they require for sustenance, it is naturally the most fertile island in the Pacific or elsewhere. The Tahitians are wonderfully like the Hawaiians, but are

whiter, taller, & better featured. They are just as kind in disposition, of a similar temperament, & they speak nearly the same language, but they are even lazier than the Hawaiians, fonder of gay
clothes & music; not so well educated & far more immoral. Their
island is frequently called the "Gem of the Pacific", which it
undoubtedly is, & it is also called the "Corinth of the South Seas"
on account of the universal immorality, & the inherent lightness
& gaiety of its people.

¶This island & people have caused more desertions & punishments in the British Navy than all the rest of the islands of the
Pacific together. Captain Cook lost several men here for some
weeks, & only recovered them by a stratagem. It was this island &
the pleasant memories it excited among the harassed and op-
pressed crew of the "Bounty", that caused that ship's company to
mutiny near Tongatabu, in 1789, set their wretch of a Captain with
19 of his officers & men adrift in an open boat, 3,600 miles from any
European settlement & return with their vessel to Tahiti again.
Here many of them settled down & married native women, while
the remainder, 9 in number, taking with them some Tahitian men
& women, sailed away to the south east, & found a home on Pitcairn Island—1000 miles away,—where they remain undiscovered
for 20 years. Meantime their Captain—Bligh had, after suffering
incredible hardships succeeded in reaching the Dutch island of
Timor, by way of Torres Straits; from whence he proceeded to England, & gave his own version of the Mutiny. The "Pandora" frigate was sent in search of the mutineers by the Government, & discovered the portion who had remained at Tahiti. They were torn
from their wives & children, & put in irons on board the "Pandora" for passage to England. This "Pandora" was totally wrecked
on the great coral reef near Cape York, but the whole of her crew
reached Batavia in the ship's boats, & eventually arrived in England, where the mutineers were tried & several of them hanged.

¶The Pitcairn mutineers included the ringleader—Lieutenant
Fletcher Christian: they destroyed the "Bounty" & settled down
on their fertile but almost inaccessible little island. They lived
very comfortably together for about 3 years, when a great quarrel
occurred about the women, five of the mutineers, including Christian were killed by the native men, & these latter from motives of
self-defence, were also killed by the remaining four seamen. Soon
after this two of the seamen succeeded in manufacturing a native
spirit & were continually intoxicated: one of them was killed by a
fall from a cliff, & the other was killed by his comrades for the sake

of quiet, leaving but two of the original mutineers—a midshipman & an able seaman. A great change then took place in the moral character of these two men: they discovered an old Bible & a Prayer book, & this led them to alter their mode of life. They observed the Sabbath, taught their wives & children, & began to lead honest Christian lives. At this juncture Young, the midshipman died, leaving John Adams, a rough uneducated English seaman the sole protector of the colony of women & children. How he did his work all the world now knows: he succeeded in establishing such a community as had been the dream of poets & the aspiration of philosophers; & he died in 1829 at the age of 69, Chief Magistrate & Pastor of Pitcairn island, in the presence of his children & the flock he had guided so well. His youth had been marked by terrible & unusual events, but his end was peaceful & noble, & I think there are few lives in History, so strange & instructive as that of 'John Adams', the Mutineer. In 1831 the Pitcairn islanders having increased to 87, many of them removed & came to Tahiti where they were kindly received by the present Queen Pomare. But the manners & morality of their Tahitian cousins proved quite repugnant to the Pitcairn people & they returned again to their island. In 1856 the colony having increased to 137 was removed with the general consent to Norfolk island near New Zealand—the convict establishment there having been previously removed. There the majority of them are now but a few returned again to Pitcairn island, & we shall probably visit them on our way to Valparaiso.

¶In 1842 the French Government took forcible possession of the Society & Marquesas Islands, & established their seat of government at Tahiti. This transaction was nearly the cause of a war with Great Britain but the storm blew over, & the two powers bound themselves by Treaty to respect the neutrality of the Hawaiian Islands. Previous to the French occupation Tahiti had been visited by British & American missionaries from Honolulu who had converted the natives & established schools &c. Even now the majority of the natives are Protestants altho' there are lots of Catholic priests & Sisters of Mercy on the island. In most of the native houses I have visited there were Bibles & other books, all printed by Spottiswood's London in the native language. Queen Pomare was not dethroned by the French for they found her influence very great, & her authority useful, so she receives a Pension from the French Gov^t. & has a palace, Guard &c, but no real authority—that is centred in the Military Govn^r.

¶There is a coral reef in front of Papeete harbour, & a tiny island at the entrance which is covered with Palm trees &c., & further adorned by a small battery of Guns. This island is set apart for the isolation of lepers, for leprosy is as prevalent here as at the Hawaiian Islands, & there is the same fearful decrease of population. The native population of Tahiti is now under 10,000: Captain Cook estimated it at 200,000, in 1770, & he also describes in his Journal a grand Naval review he witnessed here when there were 150 large canoes manned by 7000 Warriors. Tahiti is 50 miles in length, & averages 12 in breadth, & is in Lat 17° S. It is well watered being mountainous, & wonderfully fertile—the general productions being about the same as at the Hawaiian & Fiji groups, but the indigenous fruits & roots are far more plentiful. Sugar & Cotton are produced, but in no great quantity.

¶When we got close in to Papiete, I thought I had never seen such a beautiful place in my life; & now as we are leaving it, I think so still, & the longer we stay the more we seem to like it, & the people—native as well as European. The houses are all hidden by cocoa palms, bread fruit trees, orange trees, & banana trees, & tropical greenery of all kinds, & the great mountains rise just behind the town, their sides being covered with vegetation, & the intervening plains & vallies richly cultivated, & watered by numerous creeks, in which the natives seem to bathe & wash clothes the whole day. Right opposite Papiete, about 20 miles distant is the other mountainous island of Moree, from which the most beautiful oranges I ever tasted in my life are brought & sold in Papiete 50 for a shilling. Behind this island the sun sets every evening, painting the mountains every imaginable colour for about 10 minutes, after which it is at once as dark as it is with you at 9 PM.

¶There were two French war ships lying off the town, one of which sailed the other day for Brest. There is a military Governor, an Admiral & any amount of officials & soldiers stationed here, but of working Frenchmen of the agricultural or manufacturing class there are none. Either they do not care to leave the parent country or they are not encouraged to emigrate, & this is the great cause of French failure in colonization. They failed in Canada, in Louisiana, in Madagascar, & in the East Indies, & they have failed at New Caledonia, at the Marquesas group, & here also. The French possessions in the Pacific must cost the Home Govern^t a deal of money: here at Tahiti there are lots of officials, soldiers, priests, & Sisters of Mercy, all drawing money, & yet there is nothing to show for it. There is scarcely anything exported from the island, & no shipping trade to speak of; whereas if the British or

Americans had them they would soon become the West Indies of the South Pacific. I should think the Fiji Islands are already producing & exporting more cotton & sugar in one year than these islands have done since they were under French protection. Papiete is not to be compared to Honolulu for business or fine shops; what trade there is here is in British & American hands—the richest merchant & shopowner here being an enterprising Scotchman named Mr. [John] Brander, who is married to a native Princess & cousin to the Queen—one of the handsomest women I ever saw. There is no newspaper printed here: the people get their news & letters at uncertain periods from San Francisco in sailing brigs belonging to Brander & Co. However the French people here make no pretence about working: there are a few French Cafes & Restaurants, but no French Shops: they enjoy themselves as much as possible, take things easy & are very hospitable to strangers.

¶I looked over the soldiers Barracks & was very pleased with everything. The soldiers seemed so comfortable & contented & were so civil & obliging I was so sorry I could not speak French:— "I intend to learn bye & bye."[29] Some of the French soldiers came on board our ship & I showed them round: they were very much struck with our curiosities from China & Japan. I also got acquainted here with some very nice English, French, American & Danish people who could all speak English. I and a few of my messmates used to meet them nearly every night at their houses & they entertained us first rate with music &c. They came on board the "Challenger" & saw the ship's Albums containing photographs of all the countries & peoples we have visited. One evening we all went to a grand Ball out in the country, & we did enjoy ourselves. We have felt more at home here than at any place we have been to since leaving Sydney. We have had the French Governor & other grand officials on board, as well as Queen Pomare & the Royal Princes & Princesses to an afternoon dance & luncheon. Queen Pomare is quite an old woman now, & rather reminded me of Grandmother if she were about three shades darker & a good deal stouter. Queen Pomare was received with usual honours,—the yards being manned, & the officers & ship's company in full uniform. As she came over the gangway the Band played "God Save the Queen" & the old lady bobbed at our Captain as much as to say "go along with you." The French Governor & officials were elaborately polite to her, but she seemed more at ease when they went on shore again & left her to see the dancing. She was dressed very plainly & seemed a nice old lady. Her sons & daughters were fine looking persons, spoke English & French & played the piano. The Princesses were dressed in pink muslin generally their dresses being made like those of the natives of all the civilized Pacific Islands. They are what used to be worn in Europe many years ago, & were called "Saques" I think, being shaped like a dressing gown without any waist, they were introduced here by the Missionaries' wives & many European ladies here wear them: they look very nice in hot weather & I should'nt wonder if they were to become fashionable again. Queen Pomare before she left the ship asked to see a sailor's hornpipe danced. She is very much liked by the natives & is rather partial to English people.

¶There is an English Church here where the service is conducted in English in the morning & in French in the evening; also two Roman Catholic Churches. The only schools on the island are managed by the Priests & Sisters of Mercy: the English & American girls here are educated at these schools & speak French as well as English: they also speak the native language. They reminded me very much of the Australian young ladies, but they grow much quicker than they even: at 14 they are as tall & as fully developed as girls of 20 in England & there were several under 16 that were married & had children. When there are a lot of them together & they don't want you to know what they are saying they jabber away in the native language which seems to me to have no backbone to it. The natives here dress more like the Tongans than the Hawaiians, & resemble the former people very much, but they are fonder of pleasure & more addicted to drink—especially the men. I have been on shore here several times & I think the place prettier & pleasanter every time I go. The island seems made for the people & the people seem made for the island. Our crew have had lots of leave here & spent a good deal of money, & an enormous quantity of curiosities have found their way on shore.

There are 5 men now in irons for swimming on shore & remaining all night: one man has deserted altogether, & there are lots of leave breakers to be punished when we get to sea. There are Consuls of all nations living here. Coral is very plentiful & so is the fruit— especially oranges, plantains & bread-fruits. The bread fruit tree is as large as a Walnut tree & covered with fruit which when ripe, is much larger than a cocoa-nut & covered with a tough serrated skin, & eats something between new bread & custard. I like it very much. A British merchant Captain is now in prison here for kidnapping round the islands, & his vessel is close to us. There are lots of Chinese here but they seem the laziest Chinese I ever saw.

One of them told me he should never get back to China again—
"China too muchy work—Tahiti plenty food, plenty wife—no
work." There are no horses here like there are at Honolulu & Hilo,
neither are there any cattle or sheep altho' grass is plentiful on the
hill sides. The cattle & sheep for consumption here are brought
from Honolulu. We are paying 10d per lb for very inferior beef.
Other things are much dearer than at Honolulu, except drink &
fruit. At the French Police Court here the other day a Chinaman,
for breaking into a European house, was sentenced to 10 years penal
servitude in New Caledonia; another Chinaman for deliberately
murdering his fellow-countryman received 6 months imprisonment. All the roads here are shaded with cocoa palms & bread fruit
trees, & it is certainly a lovely island; but there are lots of mosquitos here, tho' no vicious animals.

¶These natives are wonderfully fond of music, & when our Band
plays on shore they dance until they drop & get so excited. The
natives all asked eagerly what the "Challenger" had come for: they
seem to have some vague idea that the British will one day come &
drive the French away. They hate the French & were wonderfully
struck with the Duke of Edinburgh when he visited there in the
Galatea[30] some years ago. At one large native house they showed
me his portrait where he sits surrounded by native girls {leaning
on some native young ladies in a manner that the Duchess would
not like[31]}.

¶At this house I dined with the family in the open air, off roast
pig baked in a hole under hot stone, & roast plantains. They all ate
with their fingers but they accommodated me with a knife & fork,
& refused to take the money I offered them. They had a large portrait of Queen Victoria before marriage, in her state robes. They
asked me if she was a good Queen, & how many children & grandchildren she had. This family consisted of an old grandfather &
grandmother; their married sons & daughters & a host of children:
—about 50 altogether all living in one large 2 roomed house. After
dinner one of the young women asked me if I was a Mormon: &
what I thought of the Mormons? I told her these were never very
numerous in England; & that the Mormon faith was dying out
even in America, the only country where existence of such a community was possible. They seemed rather disappointed at this, &
after a bit one of the men who was dressed in a black shirt & skirt
& silver mounted spectacles of a generally clerical appearance,
brought me a piece of parchment to read, which explained it all. It
read as follows—"This is to certify that Taranei has been duly

elected a member of the Mormon brethren, & is hereby authorized to preach the Gospel of Jesus Christ as it was revealed to his servant Joseph Smith"—Signed by 2 of the original 70 Elders of that Church. He then told me that all the family were Mormons, but this the young women stoutly denied. 'Sig' I believe they lied. They brought me Brigham Young's portrait to look at & to write his name thereon. I wrote Brigham Young—American imposter. I presented the old grandfather with a much prized Sheffield Razor, & came away amused.

¶We received a mail here but there were no letters for me as usual. We are just leaving Tahiti for Valparaiso & in another hour shall have seen the last of the brightest & pleasantest land we have yet seen. I shall post this there, & shall of course get letters. Let Walter Thornton have this to read, then send it on to Fred & Willie. Of course Charlie will see it at home. With very best love to you all, hoping you are all well.

> I remain,
> Your affectionate Son,
> Joseph Matkin

P.S. Novr. 15th, at Juan Fernandez

We arrived here on the 13th, after a smooth passage from Tahiti, sighting on the way the island of Tibuaai [Tibuai], one of the Austral group. Before we had been 3 weeks from Tahiti we were in Lat 40° S, & found it very chilly indeed. This island & settlement reminded me very much of Tristan d'Acunha, & rises quite as abruptly from the ocean to a height of nearly 3000 feet: the hill sides are covered with forest & a scanty herbage upon which flocks of wild goats may be seen feeding. The island is 10 miles long & 5 broad, well watered, but far from fertile: the soil is wretchedly poor, & the island has a very desolate appearance. It was discovered in 1563 by Juan Fernandez, & in 1703 Alexander Selkirk was left here on account of some quarrel with his Captain, & rescued by a Bristol privateer in 1709. From Selkirk's Journal De Foe fetched the material for his interesting & popular romance of Robinson Crusoe.

Anson called here in 1765 & his crew caught several goats, the ears of which were supposed to have been clipped by Selkirk. Soon after Anson's visit the Spaniards from Chili [Chile] settled here & built a fort, the remains of which are still standing. When the Chilians gained their independence at the beginning of the present century they obtained the islands as an appanage of Chile, & established a convict settlement here, which has since been removed to Magellan Straits.

¶At the present time the island is rented from the Chilian Gov^t by Don Fernando Lopez, a Chilian who employs about a dozen Chilian families in Seal fishing & the rearing of cattle. Their chief trade is with the numerous whale ships that call here during the year for fresh provisions. There are lots of fine cattle, horses, & goats here, but there is scarcely any cultivation, & the people seem very lazy. They charge very dear for everything. Meat is 1/0 per lb, & they are very independent. The working men look to me like convicts; the women have an Indian cast of countenance, & the children are very healthy looking. There is no priest of any sort on the island, & they can only communicate with Chili about thrice a year.

The sea round this island abounds with fish, especially fine crayfish, & hundreds have been caught by our men. Our Botanists, Naturalists, Photographer &c have been very busy here {there are some very rare humming birds in the forest here[32]}. We have a young Kid & a young tame Seal which we shall take away with us as pets from Robinson Crusoe's Island, the latter cries just like a child. Juan Fernandez has a wonderfully wild & desolate appearance: I was on shore yesterday for a few hours, & mounted the elevated promontory whence poor Selkirk used to gaze seaward in search of that friendly vessel which was to bear him to his distant home. I could not help thinking of Cowper's poem on the subject. "O Solitude where are the charms which sages have seen in thy face? Better dwell in the midst of alarms, than reign in his horrible place."

At the back of the settlement are some enormous caves in the hill sides, & in one of these Selkirk is supposed to have lived in winter. Many suppose these caves to have been made by buccaneers as a hiding place for their treasures, but I think they are natural or nearly so. The largest is about 50 feet long, 20 broad & 20 high, & the sides & roofs are covered with beautiful green ferns, & stalactites which gives them quite a fairy-like appearance. Hundreds of English & American names are carved in their sides & I added my own to the number. I dined today off the descendent of one of Selkirk's goats, & he ate rather tough. We are just leaving for Valparaiso, distant 360 miles, & the Captain has issued Wine to all hands, as this is the third anniversary of the "Challenger's" commission. The average depth across here from Tahiti is about 2000 fms, & the bottom is tolerably level.

P.P.S. Nov. 23rd at Valparaiso

I received 15 letters on arrival here, some of which had been all round the world, & several papers. I am very pleased with them all, & with the news in general. There were 3 from you, one each from the boys, one from Miss Wildman, 2 from Swadlincote, two from Seaton, London, & 5 from Barrowden. I am writing to Charlie, & shall write to all the others bye & bye. I am so pleased to think that everything is going on as usual at home: you did not say whether poor father received my last letter from Hong Kong or not. I shall send my Journal home by this mail if possible & you must write at once to Monte Video, & say if you received it all right, & remember I want to copy the whole into one book, some day, for my own amusement in years to come. By the time you get this we shall be in the Atlantic again, & you will get letters oftener.[33]

{I have been on shore here already, and hope to describe Valparaiso next mail. The weather is beautiful, & we are getting new potatoes, peas, strawberries, cherries, &c. We can see the snow covered peaks of the Andes rising above the clouds some 80 miles inland. Our cruise is extended until the end of June, so shall not be home before July. Next address Monte Video, & Ascension; you will get letters oftener now. Shall be in the Atlantic by the time this reaches you.[34]}

JM

Across the Pacific at last, and back to continental terra firma. *Challenger's* Pacific crossing had spanned five months, sixty-two oceanographic stations, and more than 12,000 miles. Matkin's experiences ashore in Hawaii and Tahiti had been among his most enjoyable. But as the voyage entered its final six months, he became restless for its conclusion and for the opportunity to leave the Royal Navy and find other work, as we see in the final letters.

7 South American
Denouement
DECEMBER 1875–JUNE 1876

Many of the losses which must be experienced [during a
long voyage] are obvious; such as that of the society of
every old friend, and of the sight of those places with
which every dearest remembrance is so intimately connected. These losses, however, are at the time partly
relieved by the exhaustless delight of anticipating the long
wished-for day of return. If, as poets say, life is a dream, I
am sure in a voyage these are the visions which best serve
to pass away the long night. Other losses, although not at
first felt, tell heavily after a period; these are the want of
room, of seclusion, of rest; the jading feeling of constant
hurry; the privation of small luxuries, the loss of domestic
society, and even of music and the other pleasures of
imagination. —Charles Darwin (1839)[1]

Arrival in Valparaiso broke the gloom and worry that had been induced
by five months without mail from home. We sense a tone of relief in
Matkin's Chilean letters, the first written about two weeks after arrival.
Chile, especially Valparaiso, was much more to his liking than other
Spanish countries he had visited, a sentiment stemming apparently
from his perception that Catholicism was not as strong a force.

H.M.S. "Challenger"
Valparaiso
Dec. 7th, 1875

Dear Tom,
    Since writing your last I have received your October letter, and
expect no more until we reach Sandy Point,[2] which will be about
X.mas day. I also received another letter from home, and think that
all my missing ones have now reached me. I was greatly surprised
to hear of the loss of the "Vanguard",[3] having served in two of her
sister ships, and knowing them to be such fine vessels.
    We are having beautiful weather here now, and fruit of all kinds

296

is in season: Chili is a great country for Cherries and Strawberries, and is not a bad place to live in, altho' there is an earthquake about every other night.

After a longer stay I find Valparaiso much better than I thought, and the people more tolerable also; they are certainly a little more lively than the Spaniards or Portuguese, and not such bigotted Catholics—especially here in Valparaiso where there is a very numerous foreign community. I believe the people up at Santiago, and inland are more "priest-ridden" and prejudiced against Protestants. Religious processions are very common in this city; there was one the other day over half a mile long which completely stopped all ordinary traffic for more than an hour. A short time ago if you met one of these processions you had to take off your hat until it passed; but this is not insisted upon now, and the Protestants have their own places of worship, and enjoy the same privileges as their Catholic neighbours.

¶I believe Chili is the best governed, and the most prosperous of all the States in Sth America, and its history has not been so turbulent and sanguinary as that of the other little republics. In these Spanish settled countries there is always a party fighting against the Government, and a war on hand. Peru and Bolivia have been the worst for this, and I believe the Priests are the instigators and fomenters of all these quarrels.

There were never such upstirs in America or Australia, or even in India where so many different races are peac[e]ably & quietly governed by British laws and impartial policy. I think the Chilians are beginning to see these things, and to keep the priests in their proper position. In a young country like this I did not expect to find so many poor and uneducated people as there are, in which respect it is very different to Australia and America: in Valparaiso there are just as many poor as there are in Lisbon, and they are quite as ignorant; except for the better Classes, schools are rarely to be met with.

Chili was conquered from the Incas of Peru by two lieutenants of Pizarro, the famous conqueror of Peru in 1532.

The Indians have dwindled away by degrees, but as they intermarried with the Spaniards, there are many half-castes. Some pure Indians may be met with at the back of the Andes, and also in Patagonia; but they wont work like the natives of the Philippine Islands, and as there are no Chinese here as yet, the Spaniards have to work themselves; so there are many poor people.

The Chilian Revolution broke out at Santiago in 1814, and after many years fighting, the Spaniards were compelled to acknowledge the independence of Chili. About the same time Peru and Bolivia, and other Spanish States obtained their independence, but they have been fighting amongst each other ever since, & are only now getting settled down. The great influx of foreigners, the railways, telegraphs, and numerous Mail boats constantly running now from England, have done more to establish stability in these countries than any internal influence & improvement; and newspapers have done more than priests for the people.

I have noticed in the shop-windows here Dickens' works translated into Spanish, and also those of many other British Authors. That popular periodical the "British Workman"[4] is also translated and re-published out here, & has a great sale; I believe it is also translated into Russian & Italian.

The two leading figures in obtaining the Independence of Chili, were Admiral Cochrane, Earl of Dundonald,[5] & General O'Higgins, an Irishman,[6]—Cochrane commanded the Chilian Navy, and O'Higgins the Army against the Spaniards. Cochrane's great exploit was the cutting out of a treasure ship[7] from Callas, and at the conclusion of the war the Chilians quarreled with him about some money he had advanced from his private estate, and he left them in disgust. Cochrane was one of the most daring Admirals we ever had: his feat of destroying the French Fleet in Basque Roads,[8] with fire ships, is said to have determined Napoleon against his meditated invasion of England. Cochrane was dismissed [from] the British service on account of some stock-jobbing panic he is said to have caused; but it was proved many years after that he was quite innocent. After fighting for the Chilians he commenced the command of the Greek Fleet against the Turks until the independence of Greece was established & recognised; he then remained unemployed for some thirty years[9] when his innocence was established & acknowledged everywhere but at the Admiralty. The Americans and other nations began to cry out about his ill-treatment, & at length he was re-instated by Queen Victoria, and made an Admiral. He died in 1861, at the age of 92, and was buried in Westminster Abbey with all honours; but it was the Americans, I believe, who, by writing a correct account of his life effectually cleared his good name, and pointed out to his countrymen, & the rest of the world, that the bravest of British Admirals, and the man who had wasted his fortune in obtaining the liberties of others, had been driven by harassing creditors from his country, and died in debt. However bad the Chilians treated him when alive, they are very kind of his memory, and proud of his exploits. Here, in Valparaiso, there are Cochrane Streets, Squares, & Hotels; there is a fine Iron-Clad ship, built in England for the Chilian Gov^t called the "Admiral Cochrane", and in the centre of Cochrane Square, is a fine statue & monument of him in the attitude of declamation; erected last year with splendid ceremony. I think he and O'Higgins have done more for the British name and credit in this country than most people imagine.

The British residents, altho' Protestants, are more liked, and get on better with the Chilians than their Catholic neighbours, the French or Italians.

The richest two merchants in Chili are Englishmen, and the greater part of the busiest portion of this city is owned by an English banker, resident here. In no other city we have been to were there so many foreign residents, a great number of whom are British & Germans; also lots of Italians, French, & Dutchmen; but scarcely any Americans; not being nearly so well represented as in China & Japan. The richest of the merchants are British and Germans, the French & Italians are generally Hotel keepers, some of which are first-class houses. The harbour here is more crowded with shipping than any other we have anchored in; there are some splendid Mail-boats running direct to Liverpool, and a whole fleet of coasting steamers, nearly all of which fly the British Flag, and are manned by English crews. The railway employees, & the mechanics in the Docks are mostly English also.

There are two fine streets in Valparaiso over two miles in length, for the city is built at the foot of a range of hills, and, like Hong Kong is about four times as long as it is broad. In these two streets there are some splendid shops, and tramways extend the whole length of the city. There are three large squares & some public gardens; but altho' the business portion of the city is very tolerable, the remaining larger part is miserable & wretchedly dirty. The houses are mean and built of wood, the streets are narrow, and the people naturally very unclean in their habits and ways. There are some fine Catholic Churches, one, the finest I was ever in, though nothing to be compared with our English Cathedrals; there were some fine oil paintings on the walls, and at the service, the solo singing was very good; there was a large Organ and three small ones. I always notice in these Catholic Churches that the congregation is nearly all ladies.

I have been on shore a good deal on duty, & on pleasure, and like the place much more than I did at first; if I were staying on this station, should certainly learn Spanish. Valparaiso is a much better place than Bahia or Manilla, and more like an English City than any we have visited since leaving Sydney.[10] The Chilians dress very extravagantly, especially the ladies who are tolerably handsome: they have fine dark hair & eyes, good teeth, and very small feet; but a cheerless impassive sort of countenance: they wear no bonnets, and many of them smoke; dancing is their chief amusement. There is a British Consul here, and a resident minister at Santiago. I think it the most expensive place we have been to, a dollar wont go as far as a shilling in England. Wages are very good, especially for mechanics, but there are an awful lot of idlers and loafers about the wharves. The landing wharf is crowded by them all day, and directly a passenger land, they fight over him and his luggage, and demand to take him somewhere or anywhere for half a dollar. There are no Police to check them, for they are the smallest, sleepiest, and most miserable looking lot of policemen and soldiers I ever saw anywhere, and the sailors are not much better.

¶The other evening I went to a Circus, on purpose to have a good look at the Chilians of all Classes, and to see how they took their amusement. The principal feature in the programme was an allegorical representation of the Republic of Chili; it being represented by a painted young female in scanty classical garments who galloped about the ring & kept waving the Chilian Banner in the most insane manner, whilst a small army of Chilian Soldiers, gentlemen and ladies rode round her with drawn swords and kept the rest of the world at a respectful distance. The Band played the Chilian Anthem, and red-fire added to the sublimity of the edifying spectacle, which was re-demanded by the patriotic but excited Chilians, about six times, until the foreign portion of the audience could stand it no longer;—the Frenchmen coughed scornfully, the Germans and Dutch grunted impatiently;—and the British portion winked at each other, and shouted "Encore" after every one else was tired. Directly this was over for the last time a very large Merchant Skipper, weighing some 20 stones, and frightfully intoxicated, made a rush across the ring after the Procession; three Chilian soldiers averaging about four feet in height fisted him and forced him back, but every now and again he would make another dash, and the people would hiss and yell at him until you would have thought he would be glad to get away; but all he did was to make an elaborate bow to each part of the house, and sing out pitifully "Once more, do let us have it once more." He was either an Englishman or an American, and was fined 10 dollars next day.

The population of Valparaiso is over 200,000, and it covers a large area of ground. Santiago is 70 miles inland by rail, and is a very fine laid-out city; most of our Officers have been up.

Sub. Lieut⁵ Lord Campbell & Balfour left the Challenger on Thursday for Santiago, thence across the Andes to Buenos Ayres, and Monte Video, where they will catch the steamer that brings this letter. The journey across the Andes is rather a dangerous one, but more interesting than that by Magellan's Straits. We have a new Sub. Lieut Lloyd, from H.M. Ship "Teredos" out here. The "Repulse" is expected in here shortly with Admiral Cochrane, son of the famous Dundonald, and the Chilians will make a great fuss with him. We have taken in all our coal and provisions to last us to Ascension, and expect to sail on the 9th for Sandy Point.

¶I was on shore on the 2nd (my 22nd birthday) and had a good ramble all by myself. I also posted the second volume of my Journal to Mother, which I think will be more interesting than the last, as it describes a less known part of the world; some day I hope to copy the three volumes into one book. I climbed the hills at the back of the town for a view of the snow covered Andes—the noblest range of mountains we have seen; there are many peaks over 20,000 feet high, and one Aconcagua, is nearly 24,000 feet above the sea level, and the highest mountain in the world that can be seen from the sea-level. I am going on shore again to-morrow, as it is a grand religious festival day, being the anniversary of the burning down of the great Cathedral at Santiago some 10 years ago, when so many ladies were burnt to death. The shops will all be closed, and there will be special services in all the Churches.

¶We have lost a few men here by desertion, and some are also to be invalided, for this voyage is very trying to weak constitutions, but don't think I ever felt better than I do at present, & like the prospect of a cold weather trip better than a repetition of the Tropics. We have taken in a fine Bullock for X.mas, & shall be about in the Same Latitude as on X.mas Day '73 off Prince Edward's Is^ld. The land round Valparaiso is very bleak and sterile, but soon improves as you get clear of the sea.

The Mail leaves to-morrow, so shall write again from Sandy Point, and you will get letters more frequently now, as our long voyages are pretty well over.

Yours Sincerely,
Joe Matkin[11]

Challenger remained in Valparaiso a total of three weeks, departing for the Straits of Magellan on 11 December. Southerly winds first drove the ship westward for a revisit of the Juan Fernandez area. For the final Christmas Day of the expedition the crew was again at sea.

Matkin began his first letters of the new year in the Messier Channel, the inland waterway separating Wellington Island from mainland Chile, in preparation for the ship's arrival at Punta Arenas (Sandy Point), midway through the straits. The following letter to his cousin is complete; interpolations ({ }) are from the concurrent letter to his immediate family, of which only a portion has survived.

> H.M.S. "Challenger"
> Messier Channel, Patagonia
> January 5th, 1876

Dear Tom,

We are gradually getting nearer to Sandy Point, so will begin to get my Patagonian letter ready; after closing your last I received no more from any one, but of course shall get some at Sandy P^{nt}.

We lost six men by desertion at Valparaiso, and sailed from there on the 10th of Dec. but had to return the same day as it was too rough to take the Magnetic Observations required; we made a fresh start on the 11th and had a head-wind which blew us right across to Juan Fernandez, so we sounded and trawled off that island, but did not go in again; the wind still being ahead we were by X.mas day in Lat. 42° S. and Long. 89° W. and the weather began to get very chilly.

¶On X.mas Eve we had a Concert which proved a great success, and it is to be repeated once a fortnight for the remainder of the commission. In addition to the Ship's Band there were 4 Violins, Violincello, Piano, and Harmonium; and of course there were Readings, recitations, dances, & laughable Farces. The Captain played a very nice Solo on the Violincello with Piano accompaniment, and several of the Officers and Scientific gents assisted as well as the Crew.

Great cooking preparations were made the day before X.mas, in the shape of Plum puddings and cakes, and as we had brought a fat Bullock along with us, we did very well in the provision line.

After dinner there was dancing on the Main-deck, and in the Evening Wine was issued to all hands, the Paymaster sent me a bottle of Madeira for my dinner. This X.mas passed as pleasantly as any of the four we have had in the "Challenger", all the more so, I think as it will be the last; but the pleasantness only lasted until tea-time, for the seamen had managed to smuggle a great deal of liquor into the ship before leaving Valparaiso, which was brought out on X.mas night, and soon did its work, more than half of the Crew were helplessly drunk, and the quarreling and fighting was something awful, especially among the Irish portion of the crew, and one poor fellow had his jaw broken in three places. Since leaving England we have lost a great many men by desertion, and have demanded others in their places from any men of war we have fallen across, who have invariably sent us their worst characters— generally Irish { & cockneys[12] }; so of course they are pretty numerous. An old German at Valparaiso, who spoke good English, and had been in Australia and America remarked to me that the worst class of men he had fallen across were the Irish of America, & I think the Irish of London and Liverpool are equally as bad, and much worse than the native. They are nearly all Roman Catholics, and when in liquor are very noisy and eloquent about the wrongs of "Ould Ireland."

On the last day of the old year we sighted the coast of Patagonia, & anchored in a bay off the peninsula of Tres Montes, or three mountains. The coast of Patagonia has a very dismal forbidding appearance even in that latitude; the land being high and covered in interminable forest, without exhibiting a sign of human being or habitation.

The Country is inhabited by wild romantic tribes of Indians, who live by hunting and fishing, some of whom we shall see by and bye. Chili claims sovereignty over the whole of this country, from the Atlantic to the Pacific down to Cape Horn; but there are no white settlers south of Cape Tres Montes, except a Chilian Convict settlement down at Sandy Point. On the 1st January we crossed the Gulf of Penas, and entered this Channel to do some surveying.

Messier Channel is about 150 miles in length, and averages 3 in breadth, so that it is more like a Canal than a strait; on the east side it has the mainland of South America, and on the West—Wellington Island, which is 130 miles in length, and uninhabited. The scenery through the strait is very bold & striking, but presents the same monotonous aspect everywhere, and is very different to the beautiful Inland Sea of Japan. The land on each side runs abruptly from the water to a height of more than 1000 feet, so that very little of the interior of either the mainland or the island can be seen. This Channel is completely sheltered from the wind, and the sea is as calm as the Welland.

The hills are covered in forest, and those over 1000 feet are clothed in perpetual snow, although this is the middle of summer, and we are only in Lat. 49° S. The interior of the country seems a mass of undulating forest clad hills, and snow covered mountains, and by the way it is broken up it is easy to see that volcanic eruptions have been very active here at some distant period.

The land too seems strangely destitute of animal life, you see nothing but wild ducks and geese, and at night everything is as still as death, and seems even more forsaken & uncanny than Kerguelen's Land. The Channel varies in depth from 500 to 50 fathoms in the centre, and there are scores of splendid bays and harbours; but as the Strait has not been properly surveyed as yet, very few vessels come through, but the passage is far preferable to that in the open sea outside Wellington Island.

Sailing vessels could not come through as there is scarcely ever any wind, and the Channel is very winding. On New Year's evening we anchored in a splendidly sheltered bay, and the various exploring, botanizing, and shooting parties landed, but obtained nothing except wild sea-fowls, and some splendid King-fishers. Fish are not plentiful except shell-fish.

We left again on the 2nd & anchored that night in a bay under the lee of the main-land, which was one of the most secluded and romantic looking spots in which we have ever anchored {& was called Gray's Harbour[13]}.

We lay close under the land which seemed to rise right above our mast-head, and was covered in beautiful forest. Early the next morning several exploring, shooting, botanizing, and photographic parties landed, and leave was also given to the crew to wash their clothes in a fresh water creek. Some of these parties managed to set fire to the bush, which spread so rapidly that by 5 PM the whole country was in flames in all directions, and after sunset the scene was grand beyond description, the fire being even more extensive than that we saw at Cape York, North Queensland.

The different parties returned in the evening, bringing in various spoil, but Mr. Wild, the Artist, who generally rambles about by himself, was found to be missing, and his non-appearance up to 7 PM caused great anxiety, as the fire seemed to be spreading over the whole country, and several search parties of Officers and men went out to look for him, taking with them rifles, bugles, &c, and guns were fired at intervals from the ship. All hands were on deck to see the fire, and listen for any tidings of the missing gentleman. One of the parties went round the coast in a boat, and the crew landed at the back of the burning tract of country; here they found the old gentleman, and very pleased he was to see them. It appears that he lighted a fire before going sketching to a distant spot, to serve him as a beacon on his return; but the flames spreading so rapidly, completely frustrated the object in view, and compelled him to make a long detour whereby he lost himself altogether. They brought him alongside at 9 PM and the ship's company gave him three cheers for being so kind as to relieve the general monotony by getting lost.[14]

¶When we left the next day, the fire was burning as fiercely as ever, and the ship was covered with ashes; we proceeded about 60 miles and passed thro' the "Narrows" where the Channel is only 240 yds across, and the navigation is very ticklish; we anchored last night in another well sheltered bay[15] and shot four Seals. In this bay we found a Peruvian steamer[16] which had been abandoned by her crew. She was bound from Callas, I think, to Hamburgh, with Copper ore, hides, leather, and a general Cargo, and in coming through the "Narrows" in Messier Channel, she ran on a rock, and knocked a hole in her bottom.

Her crew managed to bring her into this bay, and run her aground. Soon after a French surveying vessel came along and the Peruvian Crew are supposed to have gone away in her, but this is not certain; anyhow the Frenchmen appropriated all the steamer's boats, sails, and various deck-fittings, as her share of the spoil, and sailed southward. Some 10 days ago, a British steamer came this way, and found her, which of course became her prize; she left her Chief Officer and 3 men in charge, and went on to Valparaiso, whence she will send assistance to repair her, and bring her to that port. With her cargo, I fancy she is worth nearly £10,000, and the crew of the vessel that picked her up only numbers 24. The Chief Officer came on board our ship to dinner last night, & attended the concert which was afterwards given. We exchanged some flour, and preserved meat for beer, butter, & walnuts, part of her cargo. The Patagonians came down to her the night before we arrived, but were soon frightened away by fire-arms. We heard from her that the French surveying vessel was wrecked somewhere lower down, but it may not be true.

We left that bay at 4 AM this morning, and have come about 50 miles, to the end of Messier Channel. We are now anchored between the Madre Archipelago, and the main-land, and shall remain two or three days.

Jan. 12th, in Magellan's Straits

We passed a Chilian man of war on the 6th, and a merchant steamer yesterday bound north. We remained two days in the bay off Madre Island, to do some surveying in the different channels. On the night of the 6th it came on to blow with tremendous force, and steam had to be got up at mid-night as there was great danger of the ship dragging her anchor. The stern of the ship swung within 20 yds of some large rocks, and 40 of a small rocky island; the wind was blowing right on to these rocks, so that it was impossible to get up the anchor and move until the wind lulled; it was about the most dangerous position the "Challenger" has ever found herself in.[17]

We left on the 8th and steamed through Innocent's Channel, into Sarmiento Channel, where we anchored for the night, the next day, being Sunday, we remained at anchor, and on the 10th steamed thro' into Smythe channel, where we anchored; steaming thro' that yesterday, and entered Magellan's straits, anchoring at night in this bay, Port Tamar. A ship entering these Straits from the Pacific, does so at Cape Pillar, some 30 miles west of our entrance from Smythe Channel. We have come all the way from the Gulf of Penas, through a series of channels separated from the Pacific by a continued chain of islands, the largest being Wellington, Hanover, Madre, and Queen Charlotte Islands. The generality of ships keep outside these, for the Channel navigation is very intricate, and must anchor every night. The mountains that rise so abruptly from the sea in this part of the Patagonian coast are a continuation of the great chain of the Andes, which, rising to a little of the north of the Equator, extend southward for 4000 miles, and passes through Colombia, Peru, Bolivia, Chili and Patagonia, disappearing finally in the Fuegan Archipelago. The coast of Pategonia south of the Gulf of Penas, is indented with deep and narrow inlets, which penetrate within the mountain masses in a similar manner to the fiords on the west coast of Norway.

Since we have reached above Lat. 50° S. the scenery has changed again, the land is now destitute of trees, and almost of vegetation of any kind. The islands and mountains present the same bleak aspect,—bare lava and basaltic rock, and except from the numerous cataracts which fall from the hills, you hear no sound anywhere. Of course snow is more abundant, for we are now in Lat. 52° S.

Latterly we have fallen in with a new kind of duck called the "Steamer", which in its manner of locomotion is a perfect masterpiece. It weighs from 20 to 30 lbs and its feathers are so thick that ordinary shot will not penetrate them; they are splended divers but cannot fly. Their wings are double jointed and they move them backwards and forwards like a beam-engine in the old steam boats. Our Steam Pinnace chased one the other day, but as she could only steam 10 knots an hour, and the duck 17—the latter won the race. It generally takes about a dozen close shots to bring one down; they are not good eating, but our men had a taste of them.

¶We had a heavy thunder storm the other night with frequent lightning, and snow and hail all at the same time. We passed through some very narrow places yesterday, and are now anchored in a bay of King William IVth Land, and Desolation Island,—part of the Fuegan Archipelago forms the southern side of the Strait, which here is about five miles wide. This morning our anchor broke clean in two by dropping on to a hard rock when let go, and the ship drifted within 50 yds of the rocky shore before the other anchor could be let go, or steam got ready. We are now anchored in a bay which is completely surrounded by mountains, and must at some time, I should imagine, have formed the crater of an extinct volcano, like X.mas harbour, Kerguelen's Land. In this latitude it rains & snows every few hours, and a dense fog continually hangs over the land; which weather we shall have until we reach the Falkland Islands next month.

Ferdinand Magellan the discoverer of these straits was a Portuguese by birth, but sailed in the service of Spain. In 1519, the Molucca Islands being claimed by the Portuguese and Spaniards also, Magellan thought he could reach them by sailing round the southern part of America, and started from Spain with five ships and succeeded in reaching these straits, they are 360 miles in length, and he lost two vessels on the passage through. He was 110 days crossing the Pacific, when he discovered the large group afterwards called the Philippine Islands; at one of which—Matan—he was murdered by the natives: (we saw his tomb and monument there.) Only one of his fleet the "Vittoria" succeeded in reaching Spain, by the way of Cape of Good Hope, and this vessel was hauled on shore and long preserved as the first ship that circumnavigated the globe. She was 3 years, and 14 days on the voyage, and Francis Drake's ship, the "Golden Hind", was about the same time, performing the voyage, 56 years later. He, however, had over a million pounds worth of Spanish gold in his hold, plundered from Valparaiso, Lima, &c, and after re-fitting his ship at Temati in the Moluccas (visited by the Challenger) reached Plymouth in safety. His voyage so troubled and terrified the Spaniards that they resolved to fortify Magellan's Straits, and thus for ever keep the English & Dutch from the South Seas. For this purpose an expedition left Spain in 1581, under Sarmiento, who built and fortified a city near Sandy Point; but disease, cold and famine soon caused it to be deserted, tho' not before three fourths of the settlers had perished, for when, three years later, an English adventurer, Candlish, came through the straits he found the city in ruins.

In 1615, Schouten,[18] the Dutch navigator found his way round Cape Horn w^h. he named after his native town, Horn, in Holland. I forgot to mention before, that H.M.S. "Challenger" was totally wrecked in 1838, near the Gulf of Penas, on the Patagonian coast, & many of the crew were drowned.

The Archipelago of dreary, barren, mountainous islands to the south of these Straits was called by the Spaniards—Terra-del-Fuego, (land of fire) on account of the volcanoes which appeared at distant intervals to illuminate the perpetual snow which prevails in this region. Narrow Channels, strong currents, and boisterous winds render it unsafe for a sailing vessel to enter this dangerous labyrinth.

The Atlantic side of the straits is more favored by nature than the Pacific and there, a few trees and animals are to be found. Penguins and other Antarctic sea-birds flock to these shores and pursue their prey without molestation. The Terra-del-Fuegan natives are said to be the filthiest and most degraded of all the races under the sun; they are undoubtedly the least favored by nature as regards Country, climate, & subsistence, but shall be able to tell you more about them by and bye, as we shall most likely see them, and also the Patagonians as we proceed towards the Atlantic. We are 200 miles from Sandy Point, where I expect to post this.

January 14th, at Sandy Point

We anchored here this morning and received a Mail. Two of our deserters are here, having been sent on from Valparaiso, by Mail boat, they were captured by the Chilian Police three days after we sailed.

On the 17th we sail for the Falkland Islands. Our Captain has just received a letter from the King of Spain, thanking him for towing into Manilla that derelict schooner which was worth about 10 bob. He calls our vessel "The Werder" which is very much like the Challenger.

We have just missed the Homeward Mail, so it will be some time 'ere this reaches you.

Will tell you about Sandy Point in my next, it appears a miserable looking place. There are two Chilian men of war here.

Our next address is Ascension or S. Vincent.

Yours sincerely,

Joe Matkin[19]

Another three weeks would pass before Matkin took up correspondence again, by which time *Challenger* had arrived at the Falkland Islands. With the ship once again in the Atlantic after an absence of more than two years, spirits were on the rise. Despite the barrenness of the Falklands, the two-week sojourn was a pleasant one—"we have never lived better since we left England than at this place"; pleasant except for the day that drowned seaman Thomas Bush was buried—"one of the gloomiest we have experienced."

H.M.S. "Challenger"
Stanley, Falkland Islands
February 6th, 1876

My dear Mother,

We expect to be at Monte Video in a fortnights time, where I hope to receive some letters, & to post this. Fred would of course send you on my Janry letter from Sandy Point, & you would see that I had also written to Willie.

The time seems to be rolling on wonderfully quick now, & I expect we shall be home within six weeks of your receiving this letter. We remained four days at Sandy Point, & I was on shore several times.

Sandy Point, or Punta Arenas, as the Chilians call it is the most southerly inhabited place on the globe, excepting the opposite coast of Tierra del Fuego: the settlement stands on a flat sandy point of land on the Patagonian side of the straits, & consists of about 80 rough weather board houses, a wooden fort, Catholic chapel &c. The place has sprung up within the last 10 years; it was originally a convict settlement but the discovery of a coal mine in the neighbourhood proved the making of the place & induced other settlers to come. The Liverpool Pacific mail boats call there once a week for coal, & remain six hours; casual steamers passing thro Magellan Straits also call occasionally. The coal mine is six miles inland, but there is a tramway, & a steam engine which drags the coal from the pit right on to the {apology for a} pier {which the Chilian Gov^t has erected, and the coal is shot through holes in the pier into barges underneath[20]}. The mine is owned by a company, most of the shareholders being British, & also most of the workmen, who get enormous wages. The coal is found almost on the surface of the ground & is not to be compared to English coal.

¶There is a Chilian governor there, an English surgeon in charge of the Hospital, & a British Consul who flies our flag from a clothes-prop stuck on top of the roof of his house. There is also a considerable force of miserable looking Chilian soldiers, & a few convicts; but these latter are only exiles & do no heavy labour. There is a gold mine in the vicinity worked by a few Australian diggers but they say they are not doing much. The population is about 500, mostly Chilians of very dilapidated appearance, but there are a good many British & Germans, & a roving population of hunters, mostly British, who go on horse back excessive distances inland, to hunt on the great pampas, or plains of South America; they also trade with the Indians for skins with fire-water & tobacco. These skins fetch a pound apiece at Sandy Point, & when they have gathered sufficient skins to load three or four horses, they start for Sandy Point to dispose of them, this they call coming to town. Vicuna, Guanaco, Emu, & other skins are the most plentiful. The Patagonians themselves make periodical trading visits to Sandy Point, & a tribe was expected there every day, but had not arrived when we left.

¶There were several grog-shops & stores there & we used to get fresh meat & vegetables every day. Meat was 7d per lb, other things dear in proportion. There are any amount of cattle near Sandy Point, but sheep will not thrive. There are white settlers more than 50 miles inland, & they can grow potatoes, cabbages, & turnips, but no grain. The country at the back of Sandy Point is tolerably well wooded, & abounding with lots of wild animals & birds, particularly ducks, geese, & ibis. Sandy Point is in nearly 53° S lat. The climate as at the Falkland Islands, is colder in summer than England in Spring; but still, altho the winters last nearly six months, the cold is not so severe as in England, the ice on the lakes being seldom strong enough to bear a man. The grass there, & also at these islands is very large & coarse, but suitable for cattle & horses: there are plenty of wild horses near Sandy Point; all the Patagonians have them. Mushrooms were very plentiful near the settlement, & also wild cranberries. Lots of shooting parties landed there every day with great success, & the Governor presented our Naturalists with a young Puma or American lion which he had captured, but it died the other day, & is stuffed.

¶There were two Fuegan girls & a woman at Sandy Point, picked up in the Straits by a steamer & left under the care of the Hospital Surgeon, who was trying to civilize them before sending them back to their island. I saw them once or twice, they were in a hut crouching over a smouldering fire. They were washed & partly clothed & I must say they are not half so ugly or degraded in appearance as they have been represented by various travellers— particularly by Commodore Wilkes, the American discoverer of the mythical "Termination Land."[21] They were over 5 feet high & their limbs as well formed as those of other races, they were very broad across the shoulders & had wonderfully large flat faces, & stiff coarse locks of hair; their eyes are something like those of the Chinese, & altogether they look something between Malays, Chinese & Indians. The youngest girl, about 4 years old had quite an intelligent look, & they were all queens compared to the aboriginal women of Australia, or the Papuan or Admiralty Island ladies.

¶We took in a little coal, & left Sandy Point on the 18th, anchoring again the same day at Elizabeth Island, 25 miles further east. We stayed there two days shooting geese, ducks, &c., which were very plentiful there. I was on shore there, with a shipmate, & we shot nine geese. They build their nests in the grass in the interior of the island; the young ones were nearly ready to fly, & could run like partriges. {The English Mail to Valparaiso went past while I was on shore, but our ship could not communicate with her, as she was a long way off; the straits being 18 miles wide just there.[22]} The birds called Tern lay their eggs on the bare ground close to the sea; they are something between a Kingfisher & a swallow. On that part of the island where they were breeding it was impossible to walk without treading on their eggs or young, & the noise they made when any one intruded upon them was exactly like the roar of the surf on a coral reef.

¶We had another concert on the 18th & sailed on the 20th, for these islands, distant 230 miles East of Cape Virgin the Atlantic entrance to Magellan Straits.

The depth across here was only 50 fms; we sighted one of these islands on the 22nd & anchored on Sunday 23rd off this settlement. This group was discovᵈ by Davis in 1592, & named in the reign of James I after Viscount Falkland: it comprises two large islands— East & West Falkland & over 200 small islets & rocks, the whole area being not less than 7000 sq miles in extent, of which East Falkland comprises 4000 & West Falkland 2000 sq miles. There is not a tree on the whole group, but abundance of coarse grass upon which feed enormous herds of half wild cattle & horses, & flocks of sheep. Of course these have been introduced recently. Albatrosses, wild ducks, geese, snipe, penguins &c are wonderfully numerous, & rabbits, which have been introduced quite recently are as plentiful as in England.

¶The coast round these islands is very rough, but there are several good harbours. The land is of moderate elevation, with an undulating surface, & the soil is boggy. Of course grain will not grow here, & there is not a tree to be seen anywhere; but potatoes, cabbages, turnips, & the hardier sorts of vegetables are grown. There are no rivers but numerous lakes & ponds, for in the summer it rains here every day. The principal exports are wool, hides, & tallow; & the islands are frequently visited by American Whaling & Sealing vessels, as well as ships bound round the Horn, for fresh provisions.

¶The French were the first to colonize these islands after losing Canada in 1763; they abandoned them a year or two after, & the British formed a settlement here. The French turned over their colony to the Spaniards & in 1770 these islands were nearly the cause of a war between Great Britain & Spain. Dr Johnson wrote a pamphlet against pushing the matter to extremity, on the ground of the utter worthlessness of the disputed territory. Both countries abandoned the islands as useless in 1774, & in the early part of the present century, the Govt of Buenos Ayres occupied them, & formed a settlement at Port Louis on this island, but it was destroyed in 1830 by the Americans. In 1833 the Govt of Buenos Ayres surrendered their claim to the British, when the whole of the islands were annexed to Great Britn, & this settlement of Stanley established.

¶Stanley harbour would be one of the finest in the world if the hills which encircle it were only higher; it is completely landlocked & has a lighthouse at the entrance. The settlement is built on the slope of the hills; the houses are two storys high, & generally whitewashed; some are of stone & some of wood; they stand in blocks & the streets are tolerably regular. It looks far more clean & pleasant than Sandy Point, but the surrounding country is even more dismal than Patagonia; there is not a tree to be seen nor a sign of vegetation:—it is a second Kerguelen's Land.

¶There is a small detachment of Marines here, & a Barracks; the present Governor is Colonel D'Arcy, whose salary is £1000 per annum. There is also a Bishop here, & two or three missionaries who make periodical visits to the Fuegan Islands to try and civilize

& convert the poor outcasts of that desolate region. They have brought some of the Fuegan men over here, taught them English, & are now learning them farming &c: bye & bye they are to be sent back with presents of a few cattle, horses, sheep, garden seeds & implements, to see what they can accomplish in the way of social improvement in their own country & among their own people. We had one of the men on board here—quite a well made man, could speak English, & laughed at everything; so after this I shall consider the Fuegans as nothing inferior to their antipodean fellow creatures—the Samoiedes and Laplanders of the northern hemisphere, although they are by Nature the least favoured of all her children as regards country & the means of subsistence.

¶There is a dissenting Chapel here as well as a Church of England, & also a Roman Catholic Chapel, for the Governor's wife is a Catholic, & insisted upon having one.

I have been on shore here several times & if there were but nice gardens & fields, could fancy Stanley was an English village. The people here wear real English working dress; the children are very healthy looking & more numerous than in any English place of the same size. There are horses & carts, any amount of stray poultry, cows, cats & dogs, & for the first time since we have left England New Zealand we have found English money & coppers. Copper money we have been strangers to for more than a year. Outside of every house is a peat stack instead of a coal shed the same as in the south of Ireland; & this peat burns much better than wood, & is wonderfully abundant. There are several stores & public houses here, & here as in an English village, everybody knows everyone else & his business. The best of beef, pork, & mutton may be bought here for 3d a pound, & wild geese, ducks, snipe & rabbits in any quantity may be shot just at the back of the settlement; we have never lived better since we left England than at this place; still every thing is dear except meat; butter is 2/- a pound & eggs 4/- a dozen. Many of the houses have vegetable gardens attached, & there are tolerable roads here. Our Concert party gave a very successful entertainment on shore one evening, the proceeds being given to the School fund.

¶We have taken in 15 tons of coal, & given leave to all hands, when of course the public houses were extensively patronised; a month's money was also paid us in English silver. The mail for the Falklands is brought by small schooners every six weeks from Monte Video: these schooners do all the trading to & fro. Our band played on shore once or twice, & the Captain & Officers have given several parties: one afternoon there was a dance party on board, but the ladies here are a very sleepy lot compared to those in Australia & other places we have visited.

¶Last Sunday Dr. Stirling, Bishop of the Falklands, conducted the service on board, & preached an excellent sermon. He said that we had all done our duty to our country on this remarkably interesting voyage, & he had heard that the crew of this ship was one of the most united ship's company's ever afloat.—("especially the deserters" JM). He did not doubt but that there were scores of people in England who were counting the weeks & days that would elapse before the "Challenger" could arrive; he hoped we should find our friends well on our return, & that the money which we shall have so hardly & honourably earned, will not be squandered foolishly in the first English port we touch at, but spent pleasantly among our friends & people. Dr. Stirling comes of a naval family, being the son of Admiral Stirling; one of his brothers is a Commodore, another was Commander with me in the "Audacious" & was married at Hull; and another brother, a Captain Ret. went down in the "City of Boston", with his wife, on the way out to join his ship in North America.

¶On the 31st Janry we took in a lot of wooden piles brought from South America, to form a fence round the cemetery of Port Louis, where we were to arrive the following day. The same evening the Officers gave a large dinner party, & the visitors left the ship at 11 PM in the s[te]am[23] pinnace. When the boat returned [sev]eral of its crew were the worse for [l]iquor. Just as it got close to the ship one of the men—Thomas Bush—able seaman, fell overboard, & being intoxicated, was perfectly helpless in the water. Lieutenant Carpenter at once jumped overboard, & succeeded in the darkness in finding the drowning man, after great exertion he brought him alongside the boat & they were both rescued & brought on board. The water was very cold, & Lieut. Carpenter was quite exhausted but recovered after a time.

Altho' the promptest measures were taken by the Surgeons & numerous vol[unteers] to restore [an]imation to the unfortunate seaman, they were of no avail, & after persevering for upwards of an hour in the usual method for the resuscitation of the apparently drowned, they were compelled to decide that life was extinct. The next morning we steamed round to Port Louis 25 miles further north on the same island & anchored off the small settlement of about half a dozen farm houses. We landed the timber for the cemeterys fence & sent a party to dig the grave of him who only 12 hours before had helped to get this timber on board. At 11 AM the next day the bell tolled for his funeral, & nearly all hands assembled to attend it. The sad procession left the ship in big boats towed one behind the other by the steam pinnace, & as soon as all the party had landed they proceeded slowly over the boggy ground towards the roughly enclosed cemetery. The coffin was borne by the messmates of the deceased, & the Burial Service was read by Commander Maclear. There were about half a dozen graves in the enclosure, & like those of the whalefishermen down at Kerguelen's Land, they seemed to add to the general solitude & desolation of the surrounding country.

It was a wet miserable day, & altogether, in its incidents, one of the gloomiest we have experienced in the "Challenger." The deceased was a remarkably fine, well made young man, & joined the ship at Sheerness in 1872; he was looking forward to being married on his return from this voyage, & as a seaman will be much missed in the ship. We erected a wooden monument over his grave, & after doing some surveying, returned to this port on the 4th.

¶Yesterday we had all the school children of Stanley on board, about 80 of them, & after being shewn everything of interest in the ship, they were regaled with plum cake & tea, & sent back. They were most pleased with the Galvanic battery, the Galapagos Tortoises, & the playful & amusing little goat we brought from Juan Fernandez; they gave three cheers for the "Challenger," & sang the "National Anthem["] before leaving. It was quite a novelty for us to see so many English children around us, & I think we enjoyed it as much as they.

To day as we were to sail we flew the long "Homeward bound pennant" & at 4 PM the anchor was got up & we steamed away for Monte Video, distant 1100 miles north. The Band played "Homeward bound" & the "Girl I left behind me" as we steamed past the settlement, & a couple of nursery maids responded by waving their handkerchiefs. The population of the Falklands is about 1000. We are now nearly out of sight of the most southerly, as well as one of the quietest & pleasantest of Her Majesty's foreign possessions. For a ship in want of fresh provisions, after a long sea voyage, there are few better places to touch at than the Falkland Islands, & altho' the country is naturally sterile & unproductive, being 3 degrees further south than Kerguelen's Land, British settlers & industry have managed to utilize it to the best advantage, & to turn what was only intended at first as a naval and military centre, into quite a commercial & pastoral little colony.

February 16th, at Monte Video

We arrived here yesterday after a rough, but quick passage, of 10 days from Stanley sounding &c across—average depth 1200 fms. We received our letters last night, up to the 10th Dec^r, & we shall get two other English mails before we leave. I had a letter from Charlie, enclosing yours, one from Fred, & one from J.T.S.; also a Mercury from Miss W[ildman] & an Oakham Magazine from Walter Thornton; so I am in good spirits, as the news is pretty good. I hope you had a happier X.mas this year, & that you will not have such a hard winter as you seemed to think by Decr weather.

¶The difference in the climate here, & at the Falklands is something wonderful, & the sudden change has caused a good deal of sickness—Rheumatic fever &c &c. This place is in Lat 35° S & as this is the middle of summer, of course it is much hotter than England in summer. All kinds of fruit are in season, but I have not been on shore yet.

¶We leave here on [the] 29th & shall reach Ascension by the end of March, & St. Vincent, Cape Verdes, by the middle of April. You can write to St. Vincent up to the first week in April as they reach there in 10 days from England. We shall be home by the end of May, & I hope to see you all by the first week in June. You must send this on to Fred & Willie after you have all read it at home, & also let Walter Thornton read it, for I have not time to answer his from here, but will write from Ascension. I will answer Fred's letter next mail; he says he is learning [Animal] Physiology—the young Scamp. This will do for an answer to Charlie's letter. The mail leaves to day so I must conclude with best love to you all, hoping you are all well.

From Your affec^{te} Son,
Joseph Matkin

P.S. Remember me to all enquiring friends.

I hope to hear next week, whether you received my Journal all serene.[24]

The concurrent letter to Swann continues through 24 February at Montevideo. Here we learn that since the Polynesian visits Matkin had been burdened with most of the work of his superior, the ship's steward, because of the latter's illness.

February 24th

I have been too busy since we came in here to get my regular descriptive letter ready, but shall have lots of time on the way across to Ascension, and as everything is written down in my Journal, shall have no trouble about it; the reason of my extra work is owing to the long sickness of the Ship's Steward—my opposite number. He contracted a dangerous illness at Hilo and Tahiti, and has been confined to his hammock almost ever since, & is no better yet, so have his duty to do as well as my own, which as my Paymaster tells me is a great honour, but of no material benefit to me; if he were to go to Hospital, I should receive four times as much pay as I am now doing, and he would also receive his; but as long as he is borne on the Challenger books he alone takes it.

We invalided home in the last Mail boat, a Sub Lieut. dangerously ill from a sickness also contracted at that Southern Paradise —Tahiti. His father is, I believe, domestic Chaplain at Windsor Castle, and dont know what he will think when he learns the nature of his disease. He, Sub. Lieut. Swire, was the tallest, strongest, and finest looking man we had in the ship, but now, thro' bad surgical treatment, he is physically ruined. We sent home in the same vessel a seaman, who was with me in the "Invincible" and "Audacious". He seemed a strong man then but the peculiar voyage of this ship as regards climate has brought on a galloping Consumption, and am afraid the poor fellow will never see England again.

¶I have been on shore, but shall tell you all about that in my next, the weather was intensely hot, much more so than where we are lying, about 3 miles out in the Rio-de-la-Plata: it is about the same as in Melbourne and Sydney at this season. We have received two other mails here.

A Rebellion is expected to break out next week, but as this is the normal state of Uruguay, it creates very little anxiety among the people unconnected with the Gov^t. The English papers seem strangely unsettled in their tone as regards the general peace of Europe. The Buenos Ayres Standard of yesterday (English Paper) contains a Telegram from Lisbon, asserting that enormous naval preparations, were going on in the English Dockyards.

We sail to morrow morning for Tristan d'Acunha, & Ascension, and shall reach the latter in about six weeks, when I shall write again.[25]

A fuller account of the Montevideo port call appears in Matkin's next letter to his brother Fred, begun three weeks after *Challenger* left South America and concluded more than a month later at St. Vincent.

H.M.S. "Challenger"

South Atlantic,

Lat. 34° S, Long. 12° W

March 16th, 1876

Dear Fred,

Your letter reached me at Monte Video on the 15th Febry, but I have only just found time to answer it. We expect to reach Portsmouth by the 24th May, but to pay off at Chatham or Sheerness about the 7th June. The Paymaster is allowed 10 days from the date of paying off to close all Books & Accounts, & I have no doubt he will want me to help him. I need not tell you how busy we shall be & what confusion & excitement will prevail everywhere; the ship will be full of visitors & relatives of the crew, Custom House Officers, Outfitters, Jews, Washerwomen, & bad characters; & altogether it will be such a mixture of slobbering & drunkeness that I hope none of my relations or friends will come near to it. At any rate I should not like any female friend to come, you can come if you think you should like to, for of course I should be very glad to see you, but [I] should have such a little time to be with you, & there is such a very little to see of interest on board the "Challenger" now, that I scarcely think it is worth the trouble & expense in coming to Chatham; however we shall be better able to decide about it when we have reached England.

¶I was so glad to hear of your joining the Evening Science & Art Classes at the Institution; still I should have thought Languages, or Drawing or Music would have been more interesting & useful than the branch you have chosen. I manage to study myself, a little, at times, but I dont go in for anything so abstruse & ambitious as "Animal Physiology": when I get settled down in England I intend to learn Music & French if possible, & after that Spanish & German, if I survive & find myself making reasonable progress. I think I should join St. John's Choir if I were you. I like Choral Services but think Ritualism tomfoolery. I should have liked to have seen our redoubtable Royal Dragoon[26] at the great "Assault at Arms" at York.

¶This is the only sort of Foreign Note Paper I could get in Monte Video, which city I will now begin to describe. We sighted the main land of South America near the entrance to the river Plate, on the morning of 15th Febry & proceeded under steam full speed for Monte Video which is 90 miles up the river. The land was low & similar to the Brazilian coast near Bahia & Pernambuca. In steaming up the river Plate you would almost imagine you were in the open sea if it were not for the freshness & turbid colour of the water: the immense & turbid flood which the La Plata pours into the ocean is perceptible for more than 100 miles out at sea. The La Plata is formed by the union of the rivers Parana, Paraguay, & Uruguay, as well as many tributary streams, & ranks second among the rivers of South America. The Parana has its source among the Bolivian Andes, & is navigable for vessels of light draught for more than 1800 miles. The La Plata is 2,350 miles in length, the area of its basin being 1,200,000 sq miles; it is over 150 miles wide at its mouth, & at Monte Video over 90 miles, while at Buenos Ayres 190 miles up the river it is over 15 miles broad. It was discov'd in 1514 by Juan de Salis who was murdered by the natives in making a descent into the country. It was for some time called the River Salis, but in 1526 Sebastian Cabot, then in the Spanish Service, visited it, & having procured much Silver from the natives, he called it "Rio de la Plata" ("River of Silver"). It is a comparatively shallow river, & we passed lots of vessels at anchor as we proceeded.

¶At 10 AM we passed the lighthouse of Maldonado, a small seaport 50 miles E. of Monte Video. It was near this port that the British Line of Battle ship "Bombay" was burnt in 1863, with over 70 of her crew. The "Bombay" was Flagship of Admiral Eliott, who remained behind at Monte Video whilst the ship steamed down the river for "Target Practice.["] The vessel caught fire whilst the crew were at dinner—it was supposed in the "Spirit Room"—& was burnt to the water's edge in a very short time. The Captain of the vessel—so I have been told by men who were in the ship—called his Galley away as soon as the fire was discov'd & lay off from the ship. There were not sufficient boats to save all hands, & when some of the drowning men swam alongside Captain————[27] galley, he ordered his Coxswain to chop off their hands if they attempted to get into his sacred boat. Commander Stirling who was with me in the "Audacious" was a Lieut. in the "Bombay" & was promoted to Comm'dr for his efficiency on that occasion.

¶From Maldonado down to the anchorage we raced with a French Man of War & beat her by a mile. As we drew over 20 feet of water we were obliged to anchor about 2-½ miles out from Monte Video but it is possible to approach nearer. There were nearly as many merchant vessels & mailboats as at Valparaiso; & 16 Gunboats of all nations including two British, lying close under the town.

¶Monte Video derives its name from a small mountain in the vicinity of the town, upon which is a lighthouse; seen from the sea it looks very compact & imposing—something like Manilla, being built on a level plain—but like all foreign places it disappoints on a closer inspection. There is a wall round the older portion of the city for it has stood more sieges tha[n] any other place in the New World. It is the only goo[d] port in the La Plata, for vessels drawing more than 12 feet of water cannot approach within 12 miles of Buenos Ayres, which is the parent city & greatest enemy of Monte Video, but fortunately on the other side of the river.

¶Buenos Ayres was founded in 1535 by Pedro Mendoza & Monte Video was founded by a Colony from Buenos Ayres. For a long time it was a bone of contention between Spain & Portugual, as the Portuguese claimed the whole of S. America north of the La Plata to the western boundary of Brazil. Soon after Brazil was erected into an independent Kingdom the people of Buenos Ayres rebelled against Spain, & declared themselves Independent, when the Brazilians siezed on this event to take forcible possession of Monte Video. It was recov'd by Buenos Ayres in 1814, & again retaken by the Brazilians in 1821, in which year it became by Treaty what it is now—the capital of the Republic of Uruguay. Buenos Ayres is the capital of the Republic by that name which is the head of the Argentine Confederation of States. Our Officers who have been there, say that it is a remarkably well built & laid out city, standing on an immense level plain facing the La Plata. It has over 200,000 inhabitants; Monte Video has 120,000, & altho' they are of Spanish descent the same as the Buenos Ayreans they hate each other almost as much as they hate the Spaniards. Monte Video was besieged by Rosas the despotic ruler of Buenos Ayres for 9 years between 1850 & 60, when the city suffered severely & its commerce was almost ruined. Since the Restoration of peace however, it has increased rapidly in size & importance, & the country has made great progress; the city is not well fortified & the Republic has a miniature Army & Navy so that the people consider it quite a formidable & respectable Power. Buenos Ayres was taken by the British in 1806, but was retaken by the citizens soon after & our countrymen driven away with considerable loss, owing chiefly to the impossibility of Naval cooperation on account of the shallowness of the river {there being no gun-boats at that time[28]}, & also to the over-anxiety shewn by our brave soldiers to commence pillaging & drinking before the city was conquered. {Buenos Ayres and Monte Video were visited a few years ago by that scourge of South America, Yellow Fever, which almost put a stop to all business & foreign intercourse.[29]}

¶There will proba[bly] be a [r]ow before long between Chili & Buenos Ayres concerning Patagonia, as both Republics claim the East coast of that country.

¶The area of Uruguay is about one sixth larger than that of England, & its present population is under 500,000. It is purely a grazing country at present, its chief wealth lying in the immense herds of cattle, horses, & sheep which graze on the boundless plains in the interior. The chief exports are wool, hides & tallow, & the product of the famous Liebig's Extractum Carnis[30] Works near Monte Video. {Uruguay has a National Debt of over £9,000,000. There used to be a British Minister residing at Monte Video, but at present there is only a Consul.[31]}

¶Our Steam Pinnace used to do all the running between the ship & the shore, on account of the long distance. Meat was only 2d per lb at Monte Video, but everything else was dear; fruit of all kinds was very plentiful but dear. {The weather was to us, coming from the Falkland Islds in 10 days, intensely hot in the day-time, as much so as Sydney & Melbourne.[32]}

¶On 17th Febry the German Frigate "Gazelle"[33] arrived & remained two days before sailing for Kiel. She went to Kerguelen's Land in 1874 to observe the Transit of Venus, & since that time has been on a various cruise taking occasional Soundings & Dredgings at moderate depths only. She was fitted up after the style of the "Challenger", & obtained her scientific apparatus from Plymouth Dockyard. Most of her Officers visited our ship before she sailed.

¶Uruguayan Paper Money is at present worthless in its capital City, & as there is no native coined gold or silver, all kinds of silver & gold coins are taken; but Spanish & Brazilian Dollars fetch the best value; you lose considerably on British, American, French, & Dutch Money: an English sovereign will only pass for 4 dollars 70 cents, & a shilling as 20 cents.

¶A month's money was paid to us in English gold, & the first leave given on Sunday 20th Febry, when most of the men went on shore, & attended the Bull Fight which takes place every Sunday afternoon outside the city—entrance 2 dollars to the seats in the shade & 1 dollar to those in the sun. Had it taken place on a week day, I should have certainly gone, for the say of the thing; as it was I had to get a description from some of my mess-mates. Eight bulls

& 6 horses were killed, & 2 of the "Matadores" or bull-fighters, were within an ace of being killed. Our men describe it as cruel & brutal sport; some of the horses were literally ripped open, & the poor maddened bulls are pricked with spears & knives; irritated by coloured flags, & terrified by combustibles made fast to the handles of spears, whic[h] exploded as soon as the spear was thrust into the animal's body. The combatants consisted of eight men on foot, armed with sharp prods & knives & carrying coloured flags to irritate the beast; & three men on horseback armed with long spears which they try to thrust between the animal's shoulders so as to reach the heart, avoiding at the same time the headlong charges which he makes. One bull, which was very much cheered by the audience—especially the numerous ladies—succeeded in killing three horses one after another, tossing one, with its rider, completely into the air & taking away on its horns, the whole side of the horse: this bull before he received his death wound exhibited a horrible appearance—bleeding from scores of wounds. The Matadores are gaily dressed, & are loudly cheered when they make a particularly successful thrust; if they bungle at it, they are hissed, & if they are unlucky enough to get gored by the bull, the audience, with strict impartiality applaud the bull—singing out "Bravo Toro." My informants assert that when a Matadore is killed, he is buried with disgrace on the spot where he falls:—but this statement must be received with caution. Only one bull was let into the arena at once, & if he didn't exhibit sufficient ferocity then & there, he was lassoed & dragged ignominiously out, to make room for a "gamer" animal. When the bull was killed 4 white mules came in & dragged him out, as well as any horses he had killed. At the conclusion a young bull was brought in & killed by Amateurs, many respectably drest gentlemen presenting themselves as Candidates for the sanguinary honor. This very edifying national Sabbath amusement lasted nearly four hours.

¶On the 20th we Invalided Home in the Mailboat an Officer & one of the crew, suffering from dangerous sickness, & on the 22nd we received from the Gunboats for passage to England, six seamen whose period of foreign service (3 years) had expired. I went on shore that afternoon & spent the day rambling about the city & neighbourhood. When you get close in to Monte Video, it has a much meaner & dirtier appearance than Valparaiso, but is in reality, much cleaner, & of course better laid out, being built on tolerably level ground. There are fewer idlers & loafers too to be seen on the wharves, but still Valparaiso is the finer city in the business portion of the town. No two buildings are alike in M.V. but the streets are very regular, & there are more tramways laid down than in any other place I have visited: the cars were not so comfortable or so showy as those in Valparaiso; they were drawn by four mules & the prices for riding were very moderate. There were one or two fine spacious squares in Monte Video, in which Military Bands played every Evening, & crowds of remarkably well dressed ladies & gentlemen promenaded until II PM There were several large Catholic Churches of the regular stereotyped Spanish pattern; most of them were undergoing repairs, & their interiors exhibited more tawdriness as regards images, ornaments, paintings &c., than those of any other Catholic Churches I have seen. The waxen figures of various male & female saints were so outrageously hideous & un-life-like that an audience at a penny-peep show would have scoffed at them. One figure, gaudily decorated, of a female Saint (apparently the Virgin Mary) was represented as being a daughter of Ham,—she was as black as a crow, & so were many of the surrounding worshippers, for coloured people & half-castes were very numerous in Uruguay. There are no slaves in this country now, & you see lots of pretty, well dressed coloured ladies; the coloured people of Uruguay seemed far more respectable & civilized than those in Brazil. The generality of the shops were not so large & fine as those in Valparaiso, but the Public buildings are finer, & there are many splendid private residences in the suburbs.

¶Monte Video is not so well fortified as Valparaiso, for the country is very poor on account of the frequent political disturbances, & the generally turbulent & unstable character of the people. There was a Rebellion in Uruguay only a few weeks ago, which was with difficulty suppressed, & the Foreign residents report that another is brewing. There is a very numerous & influential Foreign population in Monte Video, British, German, American, French, Spanish, & Italian, but there are more Americans & fewer Germans than in Valp'so. There is an English Daily Paper printed at Buenos Ayres & obtainable at Monte V'do three hours after publication, but it is very feebly conducted & its foreign information is seldom reliable—the European Telegrams, being copied apparently from the native Papers. The Railway which runs parallel with the La Plata for 80 miles, until opposite Buenos Ayres, is owned by a Company, most of the Shareholders being in England.

¶The Soldiers & Vigilantes of Monte Video are about on a footing with those in Brazil & Chili, their clothing, accoutrements &c, being of the commonest description. Their manner of marching &

going thro' their evolutions is peculiarly their own, being totally unlike anything I ever saw before: when they are on duty their drums are always beating, & the buglers play the most "original calls"—each performer appearing to have his own particular tootle.

¶There were 3 fine spacious Markets in M.V. in different parts of the town, the Meat & Fruit supply being splendid; a quarter of Mutton weighing from 12 to 15 lb cost half a dollar. The weather on shore was as hot as Melbourne or Sydney at the same season. Of course Monte Video, Bahia, Valparaiso, Manilla &c are not to be compared to Melbourne or Sydney for general business & convenience, neither are they laid out so well, nor do they possess the comforts & luxuries that are obtainable in our large Colonial Cities. In these Spanish settled countries the great Majority of the people are apparently wretchedly poor, ignorant, & isolated from those above them; there seems no prosperous lower trading class in these cities, & the Catholic religion & style of education does not appear to conduce anything toward drawing the different social classes together. Most of the Officers—Naval & Military, serving in these South American States are Americans or British, & the children of the upper classes are frequently sent to Europe or to the United States to be educated or "finished off."

¶The Public Officials of these little States are invariably incompetent men & notoriously venal & corrupt. Only a day or two before we sailed the Uruguayan Chancellor of the Exchequer was kicked unceremoniously out of office; still I should think in the present state of Uruguayan Finances, that is not a very satisfactory situation. The windows of all the houses in Monte Video, & other Spanish cities, have thick iron or wooden bars to protect them outside; the houses in M.V. are all of stone & brick. There is an English Church & a Clergyman attached to the Consulate.

¶The week after we sailed was "Carnival" week, & great preparations were being made for it especially by the sellers of Dominos, Masks, & Fancy Dresses—some of which were wonderfully quaint & ridiculous. Fightings & even assassinations are very frequent during Carnival Season, & the predicted Rebellion is expected to shew itself during that delectable period. Foreign residents & people unconnected with the Govern't treat these little political Explosions very calmly.

¶On the 25th Febry we steamed down the La Plata, & had seen the last of the Spanish settled cities we shall visit in the Challenger. Our course was to Tristan d'Acunha, thence to Ascension, a distance by our necessary route of nearly 3,800 miles. We sounded nearly every day, & Trawled occasionally; we have lost a good deal of line, & have discov'd an under-current of icy cold water travelling from the South East, towards the Equator. The average depth thus far is about 2,500 fms; on the 2nd Mch we caught a Shark nearly 11 ft long.

¶On the 13th we sounded over the same spot where we sounded 2-½ years ago, on the voyage from Bahia to Tristan d'Acunha & the Cape. It is not often that a ship sails round the world & reaches the same latitude & longitude again, having ploughed a nearly straight furrow right round; we are only 40 minutes behind English time now. We were within 200 miles of Tristan yesterday when the wind suddenly shifted {dead ahead, and as we have not much coal in the ship[34]} the captain decided on not going there again, but proceeding straight for Ascension.[35]

Here there are pages missing, but the narrative continues in the concurrent letter to Swann. *Challenger's* final stop in the South Atlantic was Ascension Island, a replenishment site for the British navy but little else, remarkable to Matkin for its desolation—both natural and civilized—and also for its turtle population.

April 3rd, at Ascension

We arrived here on March 26th, and received a few letters and papers. We have provisioned &c to last us to England and coaled to supply us to S. Vincent; have also received several seamen for passage to England making our total number borne 254.

¶This island was discov'd by a Portuguese Navigator on Ascension Day, 1511, and taken possession of by England as soon as Bonaparte was sent to S. Helena. It is only 34 sq. miles in extent and one of the most desolate, rugged, and inhospitable looking spots on the globe; is nothing but mountain, cinders & sand, and except on the top of Green Mountain where the Hospital is situated, there is not a tree or sign of vegetation visible.

The island produces nothing, & every one on it is on an allowance of provisions and water, the same as on board ship. It is used as a Naval victualling and Re-fitting Depot for the Fleet on the African Station. There are about 200 people on it all under Gov't; the Governor of the island is a Naval Captain, and the discipline, regulations, &c. are something like a Convict Prison, only the people need remain but 2 years. Being so close to the Equator, the Climate is fearfully hot, and it only rains hard about once in three years. There are no springs or wells of fresh water on the island, that for drinking &c. has to be condensed from the sea. There are no shops or hotels, and there are Sentries and Boards of regulations stuck everywhere, so that altogether the general routine is so irksome to all but Officers, that the place may be said to merit the name it bears, of the "Sailor's Hell." Turtle are more plentiful here than at any other place in the world: they land at certain seasons to lay their eggs in the sand, and are then caught and placed in ponds for consumption or export. The Mail Boat generally takes some to Engnd.

We had four given us averaging 300 lbs in weight, two of which we have eaten. They often weigh 600 lbs, and when first hatched are no larger than frogs; the flesh is as good as beef and makes most excellent soup. The eggs are quite white and round, as large as a hens, the shell is as supple as parchment: one of our Turtles contained 300 eggs, which were all eaten before it was dead, for they are several hours dying. They had over 100 Turtles in the ponds when I was there.

On one part of the island the sea bird called "Wide-Awakes"[36] have their Rookery: it covers several acres of ground, over which it is impossible to walk without treading on the eggs or young; the place is called "Wide Awake Fair", and is one of the sights of Ascension.

¶We found H.M.S. "Supply" at anchor here; she took stores down to Kerguelen's Land, in connection with the "Transit of Venus", the year after we were there. Her Officers report that they experienced the worst weather they have ever known on the voyage to that island. Such a fearful surf breaks all round the shores of Ascension, that it is impossible to land except in one particular place, and the rollers are something awful. Seen from the sea the island looks like a great red and black cinder, and after you have landed find it nothing but a mass of lava rocks and clinkers, thrown up without shape or form, and the whole surface is broken up more strangely than S. Vincent, for which island we are just starting; you will most likely remember my account of that rugged spot; it was where our Schoolmaster was accidentally killed.[37]

This segment of the voyage then resumes in the letter to Fred. The Ascension stopover had lasted eight days; by mid-April *Challenger* was again at the Cape Verde Islands.

P.S. April 19th, St Vincent

We arrived here last night from St Jago, the most fertile of the Cape Verde Islands, where we stayed 24 hours for fresh provisions. Our mail soon arrived & I was gladdened with 3 letters—one from Bar^{dn}, one from Swadlincote, & one from yourself, containing the exciting information that we have another niece. I fully intend spending a day or two with you at Stamford, but remember—"you don't get Joseph at any Animal Physiology Classes", I don't mind a Spelling Bee or two. As you are all talking of coming to meet me when I come home, I think I shall come packed in a Hamper—this side up—; then I can be delivered at the door like any other curiosity. I expect a letter from home to day or to morrow, by the mail which left England 10th April, & that this will be the last Foreign mail we shall receive.

¶I think I described St Jago, as well as St Vincent, when we were at this group in August 1873. St Iago was sacked by good old Francis Drake in 1579, & also by Witherington, an English buccaneering Captain of the "bad old times." This dismal Island is certainly the most sterile & rugged spot within the Tropics; it is even worse than Ascension & far more mountainous. No rain has fallen on the island since we were here before, but the town of Mindello has rapidly increased owing to the frequent call here of several lines of mail boats, for coal; & I believe the Portuguese have ceased sending Convicts here now.

¶We crossed the Equator for the sixth time in the "Challenger" on the 7th April in Long. 14° W, & steamed parallel with the African coast until near Cape Verde. The average depth from Ascension until we connected our '1873' line of soundings was about 2,200 fms, & as the course from this island to England has already been sounded once, very little scientific work remains to be done.

¶On the morning of the 10th we witnessed one of the most wonderful sights I ever saw in my life,—an enormous Waterspout, said by our old hands to be the largest they had ever seen. It was a cloudy morning, but no rain fell near us. The waterspout was first seen at 6.30 AM, travelling slowly toward the ship, about 10 miles away. It approached us within about 6 miles, & disappeared finally soon after 7 AM. It looked like an enormous semi-transparent factory chimney—its base resting in the sea, & its apex hidden in a black cloud out of which the lightning frequently flashed. The sea at the base of the column was in such violent commotion that it had the appearance of a ship on fire without any flame. Dense volumes of water could be discerned ascending the column spirally, &

the column itself remained almost perpendicular until the wind caught it, when it was bent in the centre until it looked like the trunk of an old tree, very much bent—breaking soon afterward. It disappeared in the centre first, & the upper part seemed to vanish slowly into the clouds, & the lower portion to sink gradually into the sea. A smaller one was also formed close to it which travelled parallel with it, but with much greater rapidity, & disappeared in about 5 minutes, in a similar manner. I believe some scientific men have asserted that these water spouts are formed up in the clouds owing to their being densely overcrowded with moisture, & descend to the sea; but the general & more reasonable opinion is that they are columns of water raised from the sea to the clouds by an ordinary whirlwind. Still the books say that the amount of water so raised is very small, being nothing but spray from the agitated sea. My own observation leads me to believe that immense volumes of water are carried up from the sea; but I should like to know more about it. You will doubtless read accounts of it in the papers, from some of our Scientific Staff.

¶A Telegram was sent to England last night to announce our arrival, & to stop any other mails from being sent on here. We sail hence for Portsm[th] on the 26th, calling probably at Madeira or St Michaels by the way. We fully expect to reach England by the 24th May, & to pay off by the 7th June.

I have not time to write to Mother this mail but if another should leave in time to reach England before us, I shall certainly write. So send this on to Mother as soon as you have read it & Uncle has read it, and tell Mother to get me some nightshirts made (night-caps I don't wear).

p.p.s. April 20th

The mail from England April 10th has arrived & brought me one letter from Swadlincote, so this is the last I can expect before reaching England.

Goodbye.[38]

Matkin wrote the last two surviving *Challenger* letters after arrival in England. Both are to his cousin. They convey the sense of celebration that all the crew must have enjoyed, compounded as it was by the numerous dignitaries and other visitors who swarmed the ship, but also the eagerness Matkin felt to be rid of the navy. In the closing quotation he selected from Samuel Johnson, we also glimpse a bit of the anxiety that may have lurked in the back of Matkin's mind about the uncertainties of his own future. However tedious the three-and-a-half year expedition often seemed, it had provided the structure and discipline of his life for those years. He had not only survived the voyage when so many others had fallen or deserted; he had thrived, performed honorably, and been given increasing responsibilities. But he had resolved that enough of the world had crossed his bow and that his days at sea must be terminated. So he now faced, at the age of twenty-two, the prospect of finding a new, respectably middle-class vocation ashore, with only his shipboard duties to claim as experience. These uncertainties must have tempered the jubilation of returning home to Rutland.

The final leg of *Challenger's* journey, from St. Vincent to Spithead, included an unplanned one-day stopover in Vigo, Spain, for coaling and victualing. After visiting so many ports, and with England just days away, we might understand, even expect, that Matkin might forgo his detailed narrative at this point. Yet he persevered, as though this was not the last port, but the first.

H.M.S. "Challenger"
Spithead
May 25th, 1876

Dear Tom,

We arrived here at 9 PM last evening, and must commence my description from leaving S. Vincent.

On April 26th the anchor was got up at 5 AM and proceeded out from Porte Grande Harbour, under sail, for England, having no further places to call at according to the projected route. The distance is 2,300 miles; there was scarcely a breath of wind, and the ship was in sight of S. Vincent all day, sailing round the southern point of the opposite Island of S. Antonio, on account of the wind.

May 20th

We have had nothing but head winds since we left S. Vincent; we were blown to the westward of the Azores, and then right on to the coast of Spain, when, being short of provisions and coal, the Captain decided upon putting in to Vigo. The Spanish coast was sighted at about 10 AM 10 miles ahead; several steamers and small sailing vessels were passed during the morning, and the entrance to Vigo Bay was passed at 2 PM when the British Channel Fleet were discovered lying at anchor off the Town. As we steamed past the Fleet the Band of the "Defence" played "Home, Sweet Home". The Fleet consisted of the "Resistance", "Monarch", "Iron Duke", and "Lively" Despatch Boat belonging to the Channel Fleet, and the "Hector" & "Defence", part of the Coast guard Fleet; under the command of Captain Charles Webley Hope, late of the "Invincible" and "Audacious". The other division of the Channel Fleet are lying at Lisbon.

¶The anchor was let go at 2-30 PM about ½ a mile from the town, and some English Papers were sent on board. Lots of Contractors, bum-boats, &c. came alongside as soon as we anchored, and did a good trade as everyone was hungry. Eggs were only 6d per doz, and the other things were cheap in proportion. Vigo is a favorite rendezvous of the Channel Fleet, and when here trade is pretty brisk; with only half of it, as at present, the shop-keepers have the victualling of nearly 4000 additional people.

¶Vigo is a quaint, old-fashioned, but pleasant looking town of the regular Spanish pattern. It is very irregularly built on the slope of a range of hills, and looks like a poorer edition of Lisbon. Judging from the size, the population is about 30,000; there is a very delapidated ancient looking Moorish Castle on the summit of a high hill overlooking the town. The houses are very lofty and ugly, and the streets seem narrow and crooked. The country round the shores of the Bay looks beautiful, and is apparently fertile & well cultivated; there are white houses visible through the trees in all directions, and in its general aspect the country looks very similar to that in the neighbourhood of Lisbon, only there are not so many windmills for crushing the grapes. An English Merchant Ship, and several small trading vessels are at anchor in the bay, but Vigo is not a place of any considerable export trade: during the summer months great numbers of cattle and eggs are sent to Falmouth and Plymouth from this port. It was just outside this port that H.M.S. "Captain" foundered in 1870.[39] After getting close to the land very little wind could be felt, & the sea was quite smooth; the sun seemed to strike warmer, and by the time we came to an anchor, we appeared to be in a different climate; in the evening the seamen of the other ships were bathing overboard, and this morning our men were wearing winter clothing.

Sunday, May 21st

Took in 100 tons of coal to last us to England, the hands working all night, and finishing cleaning ship by 10 AM. The five master Ironclad "Minotaur" arrived at 8 AM and anchored close to us; she is a splendid looking vessel bearing the Flag of Rear Admiral Beauchamp Seymour, commanding the Channel Fleet. Fresh Meat, Vegetables, and bread sufficient to last us four days were taken in during the morning, and an Officer & one seaman joined us from the "Monarch" for passage to England. At 2 PM the anchor was got up once more for another attempt at reaching England. Our Band played "Homeward Bound" and "Auld Lang Syne", as we steamed through the Fleet, and the "Monarch's" band responded, but we could not hear what tune they played. After getting two or three miles down the bay, the splendid iron-clad "Black Prince" was met, on the way to join the rest of the Fleet at Vigo. It was a beautiful warm day at the anchorage, but as soon as we got outside the Bay, the cold wind began to be felt. The distance from Vigo to Plymouth is 520 miles—to Portsmouth about 650; when the "Captain" was lost, H.M.S. "Inconstant" took the news to England, and was 29 hours steaming and sailing to Plymouth.

May 24th

Sighted old England once more at 7 AM when Start Point hove in sight about 15 miles north; stood away up Channel under all possible sail and steam; the wind was very light, and weather beautifully cool and clear; there were more vessels passed during the day than we have seen at sea since we left the Channel in Decr 1872. The Bill of Portland was passed at 4 PM within signalling distance; the distance from Portland to Portsmouth is 60 miles so it will be late before we can arrive at Spithead. The Needles were passed at 7 PM, and soon after 9 the anchor was let go, and the mail came on board as well as lots of Officers' friends &c.

May 25th

We find it very cold to day, almost like winter to us. The ship is full of people; friends and relatives of the Officers, Crew, &c— among others—M^{rs}. Captain Nares and family. In addition to these there were scores of outfitters, bum-boatmen, Jews, and Sharpers, selling sham jewelry, boots, and every conceivable article. Most of the Scientific Staff landed and went on leave, and the Crew and Officers had leave to go on shore, until 10 PM. Close to us are lying the "Valorous" and "Sultan"—the latter—a splendid ship of war, under the command of the Duke of Edinburgh. The supernumary seamen, invalids, prisoners, &c. were discharged to the "Duke of Wellington" to day, and orders were rec^d for the "Challenger" to proceed to Sheerness to morrow to pay off. She is ordered to be paid off all standing, that is without stripping, ready for recommissioning.

In the evening the officers of the "Pandora" dined on board. She is about to sail again for the Arctic regions, and will take mails, &c, for the "Alert" and "Discovery", to be left at the entrance to Smith's Sound, should the ships not be met with. At 6 PM 36 supernumary seamen came on board for passage to Sheerness, and the ship was ordered to proceed to that port as soon as possible. The anchor was got up at 10-30 PM and we steamed from Spithead up Channel; about a dozen of the ship's company who were not able to get on board on time were left, and will have to join the ship by rail.

Yours Sincerely,
Joe Matkin[40]

H.M.S. "Challenger"
Chatham
June 11th 1876

Dear Tom,

This is the last letter you will receive from this famous discovery ship, and shall conclude the description of our long voyage by extracts from my "Journal" since leaving Spithead.

On May 26th we were under full steam all day; the weather wet, cold and miserable completely shutting out all view of the land. A Swedish Gun-boat was passed, and any amount of merchant shipping. Dover was passed at 2 PM and at 6 PM the anchor was let go at the great Nore about 8 miles from Sheerness, the fog being too dense to allow us to proceed with safety.

May 27th

Got up anchor at 5 AM and proceeded into the Medway, anchoring about 4 miles below Sheerness. Thirty supernumeraries were discharged to the "Duncan" flag ship of Vice Admiral Chads, one of the finest line of battle ships in the service. This being kept as the Queen's birthday was a general holiday in Sheerness, all the men of war were dressed with flags, and the "Duncan" fired a royal salute at 12 o'clock. The ship was full of visitors, and we discharged Powder and Shell.

On the 28th, Sunday, all hands went on shore to the Dockyard Church. Leave was given to the ship's company. Ship crowded with visitors of all classes. Lieut. Lord G. Campbell, late of the "Challenger" being among the number.

On the 29th the ship was inspected by Vice Admiral Chadd, Commander in Chief at the Nore, who expressed himself highly satisfied with the appearance of the ship and her company. All the

Deep Sea sounding, Dredging and Trawling apparatus was rigged for his inspection, and will remain in position, as the ship is to be thrown open to the public for the next few days. Crowds of visitors came on board during the day from London & elsewhere, but there is very little for them to see as all the Marine Curiosities, &c, have been sent away. The next day a Grand Luncheon Party was given by the Officers to some of the Lords of the Admiralty, Members of the Royal and Geographical Societies, &c. The ship was crowded with visitors all day, special trains running from London for their convenience; the ship was moved yesterday from her moorings of Queensboro', and towed close alongside the Sheerness Town Pier, which is near the Railway Station. H.M.S. "Duncan" lies close to us.

June 6th

Since Wednesday crowds of visitors have daily visited the ship, and to day orders were received from the Admiralty for the ship to be taken down to Chatham ready for paying off. We left Sheerness at 7 PM and entered the repairing basin at Chatham Dockyard at 11 AM the Ironclad "Rupert" coming out to make room for us. In the afternoon other orders were received for the ship to be taken out of the basin, further down the Medway alongside the Dockyard Jetty, to be ready for paying off on Monday at 9 AM. Nothing but the provisions and scientific stores have to be taken out, as she is to be recommissioned again shortly for coast guard service at Harwich, the crew of the "Penelope" turning out to her from that ship.

Another very distressing occurrence took place at about half-past six this evening, when Sub. Lieut. Henry C. E. Harston committed suicide by poisoning himself. He was subject to fits of despondency, and on this particular evening refused to come to dinner, and retired to his Cabin, where he was heard to make a peculiar snoring noise, & found on inspection to be black in the face. The Surgeon was fetched as soon as possible, but he was already dead. The body was removed from the ship, and an inquest held, when after hearing the evidence the Jury decided that the deceased poisoned himself by drinking Chloral Hydrate, whilst in a state of Temporary Insanity.

Mʳ Harston was a fine looking young man, and a son of retired Captain Harston, R.N. who was dreadfully grieved when the news reached him. The deceased left a letter to the Surgeon apologizing for the trouble he had given, and giving a few weak reasons for his act.

Sunday, June 11th

All hands are busy cleaning out store-rooms, &c, ready for pay-ing off; on the 8th we were returning stores, & closing Books and accounts—Professor Wyville Thomson presented several of the leading hands of the scientific and surveying party among the crew with silver cups, as a souvenir of the Voyage.

June 12th

Ship inspected by the Dockyard authorities; the Savings Bank money was paid at 11 AM and the wages due to the Ship's com-pany at 2 PM when the ship was paid off, and the crew granted 8 weeks leave; several of those who were entitled took their dis-charges from the Navy—myself among the number—finding Sea-life nought but vanity, and vexation of Spirit, especially the latter —my opinion of it coinciding with that of Dr. Samuel Johnson's AD

Challenger decommissioned at Chatham, from T. H. Tizard, H. N. Moseley, J. Y. Buchanan, and John Murray, *Narrative of the Cruise of H.M.S. Challenger with a General Account of the Scientific Results of the Expedition,* 2 vols. in 3 (London: HMSO, 1885), vol. 1, pt. 2, p. 947. Photo courtesy of Scripps Institution of Oceano-graphy.

1776—with which quotation I will conclude my long series of letters from H.M.S. "Challenger":

> A ship is worse than a Jail. There is, in a Jail, better air, better company, better conveniency of every kind: & a ship has the additional disadvantage of being in danger. When men come to like a sea-life, they are not fit to live on land.
>
> Men go to sea, before they know the unhappiness of that way of life; & when they have come to know it, they cannot escape from it, because it is then too late to choose another profession; as indeed is generally the case with men when they have once engaged in any particular way of life.[41]

Hoping to see you to-morrow,
Believe me,
Sincerely Yours,
Joe Matkin.[42]

Conclusion

> When we started the commission, nearly every man on
> the lower deck began to keep a diary, but very few carried
> on the practice for any length of time. Before the ship was
> a year out, the craze had dropped till only one man, so far
> as I can remember, made anything like systematic busi-
> ness of diary-keeping. This was Jack Durran, in No. 1
> Mess. Jack filled five or six volumes—big, thick tomes
> they were, like grocers' day-books, with red and blue
> mottled edges and glossy covers—which would make
> particularly interesting reading now, I'll be bound, could
> they be got hold of.
> —Sam Noble (1926)[1]

There may well have been more than one "Jack Durran" aboard
H.M.S. *Challenger.* To date, however, only the letters of Joseph Matkin
have emerged from the lower decks.[2] Their uniqueness arises not only
from the sailor's perspective they embody, but from the very fact that
they are *letters.* Because they contain a first-hand, on-the-scene report,
one that was not embellished, edited, or published for the general public, the letters engender a special fascination. They bring to the reader a
sense of immediacy, of getting very close to, even intimate with, both
the voyage and the Matkin family. From their informality and first-per-
son directness, we have the feeling of actually being where Joseph Matkin is.

Of course there were serious limits to Victorian informality, even in
family letters. What strikes the modern reader as much as the immediacy are the many paragraphs of geographic and historical information
in the stiff, proper style of Victorian scientific prose. What twentieth-
century son would write his mother, even from near the equator north
of New Guinea, that "evaporation, owing to the great heat of the sun,
is going on night & day to an incredible extent, & the vapour ascending
into the cloud region is met by an upper current of cold air from the
Pole, which condenses it again, & causes the constant rains that are
here met with"? Or describe a sprouting coconut: "The young tree
holds the nut suspended by long white fibres which envelope the inner
shell. It is the province of these fibres to supply the young tree with
nourishment until it no longer requires it"?[3]

336

The careful, impersonal style of these and other descriptive passages derives from Matkin's practice of recording factual material first in a journal—with the formality typical of a nineteenth-century log—then copying portions of it into his letters. But the style is also a reflection of Matkin's desire to emulate the increasingly impersonal prose of Victorian science and government, institutions in which the middle classes were becoming more and more visible. In these passages Matkin was unwittingly preparing himself—and demonstrating his aptitude—for the civil service career to come.

But the precise, dispassionate tone of his formal descriptive passages is only one of a number of styles detectable in Matkin's correspondence. These styles are generally associated with the subjects he addressed. A typology of his subjects and styles, in order of increasing personal involvement, would include the following:

1. Historical travelogues of places visited; taken from printed sources (mostly unspecified), often in the phraseology of scholarly publications.
2. Reports of *Challenger*'s oceanographic activities and results, probably extracted from the ship's log from time to time; the prose is colorless, matter-of-fact.
3. Reports of the ship's movements, projected ports of call, encounters with other ships in and out of port; here we see Matkin at his most "naval," betraying his pride in belonging to the Royal Navy, and to the expedition.
4. Commentary on experiences ashore; personal opinions do surface when Matkin observes foreign cultures, both primitive and civilized (in nineteenth-century terms); but he demonstrates more open-mindedness and less bigotry than we might expect from a representative of the navy of the world's leading empire.
5. References to current happenings in the newspapers, especially the Tichborne affair. Nearly all of these references involve events of a scandalous or spectacular nature. Routine political matters receive no attention; neither of Britain's prime ministers during this period, William Gladstone and Benjamin Disraeli, is ever mentioned.
6. Responses to and requests of his family, directions for addressing of letters, etc.—the important but

unexciting domestic chores of correspondence, often in language hurried by an imminent mail call.

7. Shipboard living conditions and unusual events, both humorous & tragic—the theft of officers' victuals, the loss of a man overboard; the tone is one of emotions engaged but controlled.

8. Personal family affairs—rarely discussed but ultra serious when they do appear.

Overall, we must be impressed not only by the literacy of Matkin's letters—rarely a misspelled or misused word—but equally by the emotional restraint they exhibit. He is at his most exuberant in the occasional joke with his brothers or cousin. Never does he express hatred or outrage. If the "Jack Tar" of British naval tradition was a rough and ready, hard-drinking, hard-fighting, barely literate sailor, Joe Matkin was at odds with that stereotype. The British navy was changing in the late nineteenth century, however; as the "Jack Tars" gave way to the "bluejackets," literacy was on the increase. Of the published *Challenger* accounts, Matkin's letters approach Sublieutenant Herbert Swire's *Voyage* most closely, in their tone and content.

Shifting our focus from the author and his style, what do the letters tell us about the expedition itself? With fifty volumes of *Reports* published and widely distributed, the Matkin letters can hardly be looked to for fresh oceanographic data. Nor should we be surprised that they evince little concern for oceanographic theory or scientific controversy. We might have expected some reference to Darwinian themes, at least when Matkin came into contact with Pacific Island natives, considering that the *Origin of Species* had appeared when Matkin was a schoolboy and the *Descent of Man* had been published the year before *Challenger* sailed. But perhaps Matkin, whose religious views were clearly toward the liberal side, found nothing controversial about Darwinism by the 1870s.

As we should expect, Matkin's contributions are those of emphasis and values, rather than new facts. The letters give us the reactions of one nonscientific crew member to the process of conducting oceanographic research on an unprecedented scale. They demonstrate that early in the voyage curiosity about the deep-sea findings existed at all levels and that this curiosity tended to be replaced by indifference over time. And they express surprise at the costs, in line and equipment, of doing "Big Science" at sea.

The letters also hint at strained relations between the crew and the officers and scientists.[4] They document the physical difficulties of this particular voyage for those below decks, especially from the frequent, radical changes of climate on the one hand, and the ship's habit of lingering in uncomfortable climes on the other. And they make no secret of the sailors' response: jumping ship. Matkin summed up these conditions quite handily in six words: "we are not like other ships."[5]

Some of Matkin's most vivid language relates to maritime rather than oceanographic matters. At a time when Britain ruled the seas and could mount a national expedition for purely oceanographic purposes, the oceans could still be dangerous, vicious, uncivilized. We are advised that there were, even then, still no English charts for parts of the Pacific. And we are reminded of piracy and murder in the Philippine and South China seas, of escaped prisoners and exiles, and of the frequency of ships being lost at sea from storm, grounding, fire, and the like—events we are more likely to associate with the seventeenth century than with the late Victorian era.

We come finally to the ultimate question, one that can only be appreciated after one has read the entire Matkin letter collection and, further, has recognized that Matkin also produced a three-volume journal and maintained a substantial correspondence, now lost, with other relatives and friends during the voyage: what motivations drove him to expend so much time and effort in this cause? It can hardly be argued that Matkin simply had a great deal of time on his hands and chose to write letters to avoid boredom; for surely there must have been great boredom in repeatedly copying out his journal entries to several addressees. Nor is there so much as a hint of anticipated monetary rewards or widespread recognition—from the publication of the journal for example—although several of *Challenger's* officers followed this course.

The answer must lie in his family and his schooling. Both the family business and the Billesdon school placed the foremost premium on the written word. And there is ample evidence that Joseph Matkin absorbed and exemplified that outlook. Books and newspapers played a role in nearly every one of his letters. And recall the comment to his mother about the scholarly progress of the youngest brother: Fred "has improved wonderfully in his writing & will make the best penman in the family."[6]

The parents had high expectations for their sons and educated them accordingly. Matkin's voluminous letter writing might be seen as merely his means of maintaining contact with the family and the familiar, in the face of all that was unfamiliar, exotic, threatening, or just tedious during the expedition. But it was equally a means of reaffirming family values.

The letters should also be seen, however, as a means of validating Matkin's decision to embark upon a career at sea. The family was understandably much upset when brother Will abandoned a promising beginning in business to join the army. Were they also upset when Joseph joined the merchant marine and later the Royal Navy? If so, there is no evidence of it. The Victorian navy had a substantially more positive public image than the army; thus it seems unlikely that the family would have objected as strongly to Joseph's decision to put to sea. It is clear, nevertheless, that he was anxious that his family approve of and be proud of all this travelling so far from home. Joseph was indeed the wanderer of the family, its foreign correspondent, seeing more of the world than most sailors, then leaving the provinces to make his home in London as a civil servant. A bountiful correspondence was an instrument for reassuring the family that so much travelling was still a credit to the Matkin name—that a long and heavy investment was paying off. Joseph used the letters to maintain an image, a persona, among his audience. The letters were the script for the most memorable act of his life, a script that he expected his family—the printing family of Oakham—to applaud, value, and retain both as evidence of the family's worth and reputation and as a reward for sacrificing for his education.

Joseph Matkin was born and raised in what has been called the Age of Equipoise, at the height of Lockean liberalism, laissez faire, and the Smilesian philosophy of self-help and self-reliance.[7] But by the late 1870s and 1880s, liberalism was in decline, the collectivist ideals of the welfare state on the rise. As a clerk for the Local Government Board, Matkin would have seen firsthand the growing influence of parliament and government in the local affairs of towns and cities. It is not too much to imagine that the self-reliant boy of the provinces became a government-reliant man of London. If so, then we may view the life of Joseph Matkin as a microcosm of the evolving social values of late Victorian Britain.

But where does the *Challenger* expedition fit into this scheme? In Matkin's life, *Challenger* was the watershed between the eras of self-reliance and government-reliance, and it combined strong elements of each. The British seaman had traditionally been a self-reliant sort, rising from the rank of "boy" to able seaman, or beyond, by virtue of his own pluck and exertion. But life and work aboard a naval vessel are fundamentally collectivist; very little is accomplished by the energies of a single individual. The seaman works alongside his shipmates, dines alongside his messmates.

In the case of *Challenger*, collectivism was the order of the day in a further sense. The *Challenger* expedition was one of the Victorian era's most salient cases of science by team effort. The long tradition of the individual scientific genius pondering nature's secrets in the privacy of the study was giving way to the modern practice of collaboration among specialists, one of the leading hallmarks of "Big Science." Joseph Matkin was, thus, a beneficiary of and a party to government-sponsored, team-executed scientific research in the grand manner.

He was an observant fellow, as these letters amply demonstrate, and he cannot have been entirely unaware of the power of "improvement" his government was beginning to wield. The move to Whitehall may in fact have been quite calculated; in any case, it was fortunate. For *Challenger* had prepared Matkin to work carefully and conscientiously at the minutiae of a large organization—precisely those qualities sought by the expanding bureaucracies of Whitehall. The transition from ship to shore may have been much smoother for Matkin than for many of his shipmates "below decks."

Appendixes

Chronology of the Life of Joseph Matkin

18 Dec 1851	Marriage of Charles Matkin (1817–1874), printer, Uppingham, formerly bookseller in Lincoln, and Sarah Craxford (c. 1824–1903) of Barrowden
5 Nov 1852	Birth of older brother, Charles Matkin (d. 1924), who inherited family printing business, 1897; eldest son (Charles, 1889–1972) of his ten children took over print shop
2 Dec 1853	Birth of Joseph Matkin (JM), Uppingham, Rutland
1855	Birth of brother William Matkin, (d. 1883), who began a trade, got into some trouble, joined the army, was posted to Edinburgh, became fencing instructor for First Royal Dragoons
ca. 1855	Matkin Printing and Stationery Store established, High St., Oakham
1857	Birth of youngest brother, Fred Matkin
ca. 1858	JM entered Oakham School
ca. 1865	Entered Billesdon Parish Free School
1867	Entered merchant marine
23 Dec 1867	Sailed for Australia aboard *Sussex*
15 Mar 1868	Arrived Melbourne
7 Aug 1868	Discharged upon return from Australia
1868–1870	Made second voyage to Australia aboard *Essex*; remained in Melbourne one year, working in furniture and upholstery shops; returned aboard *Agamemnon*
12 Aug 1870	Entered Royal Navy; assigned to H.M.S. *Invincible*
22 Nov 1871	Transferred to H.M.S. *Audacious* as ship's steward's boy
12 Nov 1872	Transferred to H.M.S. *Challenger*
21 Dec 1872	Sailed with *Challenger* expedition as ship's steward's boy
31 Dec 1872?	Promoted to ship's steward's assistant
16 Dec 1874	Death of Charles Matkin (father)

12 Jun 1876	JM paid off; departed *Challenger* and the navy; returned to Oakham
1878	Moved to London; resided at 40 Upper Berkeley St., Portman Square
1878–1894	Employed as clerk, Lower Division, Local Government Board, Whitehall
9 Sep 1880	Married Mary (Pollie) Swift (1858–1939) of Oakham
1881	Birth of first son, Charles (1881–1965), who subsequently worked in London
1883	Birth of second son, Joseph William (1883–1969)
1886	Resided at 10 Saxon Rd., Selhurst, Croydon; birth of third son, Frederick George (1886–1962), who worked in post office; emigrated to Canada, 1907; worked at various commercial and clerical trades; visited England, 1911, 1920, 1931
1889	Birth of fourth son, Robert Swift (1889–1973), who was subsequently civil servant for Air Ministry and Board of Inland Revenue
1891	Birth of fifth son, Francis Richmond (Frank) (1891–1916), who emigrated to Canada
March 1894	Retired from civil service; returned to Oakham resided at 37 Penn St.
1896	Birth of sixth son, Joseph Hugh (1896–1969)
1903	Death of Sarah Craxford Matkin (mother)
1913	Last year at 37 Penn St., Oakham; moved to London?
1915/16	Separated from wife, who went to live with son Will in Bedford
1924	Wife Mary Matkin visited Canada, remaining one year
27 Oct 1927	JM struck by motorcycle in London, died of complications in St. Pancras Hospital, age 73; resided at 6 Old Compton St., Holborn, at the time; identified at morgue by son Charles
1931	JM's *Challenger* letters passed by Mary Matkin to son Charles
28 Oct 1939	Death of Mary Matkin

Calendar of *Challenger* Letters of Joseph Matkin

Number	Date(s)	Location(s)	Repository	Remarks
1	22 Nov. 1872	Sheerness	JTS	
2	17 Dec. 1872	Portsmouth	JTS	
3	29 Dec. 1872 to 3 Jan. 1873	Off Cape Finisterre, Lisbon Roads	NHM	To mother
4	7 Jan. 1873	Lisbon Roads	JTS	
5	8 Jan. 1873 to 17 Jan. 1873	Lisbon Roads	SIO	To mother
6	20 Jan. 1873	Gibraltar	JTS	
7	2 Feb. 1873	Madeira	SIO	To mother. Includes: broadside (song dedicated to crew of *Challenger*, origin unknown), 2 pp. (pp. 4 and 5), dated 14 Jan. 1873
8	7–9 Feb. 1873	Santa Cruz	JTS	
9	27 Feb. 1873 to 16 Mar. 1873	At sea Off St. Thomas	JTS	
10	3 Mar. 1873 to 16 Mar. 1873	At sea Off St. Thomas	SIO	To mother. Includes "substance" of Prof. Wyville Thomson's lecture
11	7–8 Apr. 1873	Bermuda	JTS	
12	7 May 1873 to 19 May 1873	At sea Halifax	JTS	
13	26 May 1873 to 2 June 1873	At sea Bermuda	SIO	To Charlie
14	27 May 1873 to 5 June 1873	At sea Bermuda	JTS	

continued

Number	Date(s)	Location(s)	Repository	Remarks
15	19 June 1873 to 6 July 1873	Mid-Atlantic, Anchored off Funchal, Madeira	SIO	To mother
16	21 June 1873 to 16 July 1873	Mid-Atlantic Anchoring off Funchal	JTS	
17	16 July 1873	Funchal Roads	SIO	To mother Original held by JM descendants; photocopy at SIO
18	21 July 1873 to 4 Aug. 1873	Canary Islands Anchored at Mindelli, Cape Verdes	JTS	
19	4 Aug. 1873	[St. Vincente]	SIO	To brother. Pages missing
20	16 Aug. 1873 to 21 Sep. 1873	Sierra Leone Bahia, Brazil	JTS	
21	no date to 4 Nov. 1873	[Tristan da Cunha] [Simon's Bay, South Africa]	SIO	[To mother]. Pages missing
22	15 Oct 1873 to 3 Nov. 1873	Tristan d'Acunha [Simon's Bay]	JTS	
23	5–15 Nov. 1873	Simon's Bay	SIO	To Charlie
24	4 Dec. 1873 to 15 Dec. 1873	In Table Bay Simon's Bay	JTS	
25	10 Dec. 1873 to 15 Dec. 1873	In Table Bay Simon's Bay	SIO	To mother
26	15 Dec. 1873	Simon's Bay	SIO	To Charlie
27	25 Dec. 1873 to 18 Mar. 1874	Prince Edward's Island, Melbourne	SIO	To mother
28	25 Dec. 1873 to 16 Mar. 1874	Off Prince Edward's Island, Off King's Island	JTS	
29	23–25 Mar. 1874	Melbourne	SIO	To mother
30	25 Mar. 1874	Hobson's Bay	JTS	
31	10 Apr. 1874	Sydney	JTS	
32	13–15 Apr. 1874	Sydney	NHM	To Fred
33	7–9 May 1874	Sydney	SIO	To mother
34	15 May 1874	Sydney	SIO	To Charlie
35	15 May 1874	Sydney	JTS	

Number	Date(s)	Location(s)	Repository	Remarks
36	21 June 1874 to 1 July 1874	South Pacific Wellington	JTS	
37	23 July 1874 to 6 Aug. 1874	Tongatabu, Friendly Islands Kandavu, Fiji Islands	SIO	To Will
38	23 July 1874 to 7 Aug. 1874	Tongatabu Kandavu	JTS	
39	18 Aug. 1874 to 7 Sep. 1874	New Hebrides Somerset, Cape York	JTS	
40	15 Sep. 1874 to 17 Nov. 1874	Arra Islands, Sea of Arafura, Hong Kong	JTS	
41	25 Nov. 1874	Hong Kong	NHM	To mother
42	16–23 Dec. 1874	Hong Kong	JTS	
43	19–23 Dec. 1874	Hong Kong	NHM	To mother
44	22 Dec. 1874	Hong Kong	NHM	To Charlie
45	28 Dec. 1874– 1 Jan.1875	Hong Kong	JTS	
46	31 Dec. 1874– 5 Jan. 1875	Hong Kong	SIO	To Fred
47	24 Feb. 1875 to 8 Apr. 1875	Humboldt Bay, New Guinea Lat. 30° N, Long. 137° E	SIO	To mother
48	24 Feb. 1875 to 8 Apr. 1875	Humboldt Bay, New Guinea Lat. 30° N, Long. 137° E.	JTS	
49	12 Apr. 1875	Gulf of Yedo	SIO	To mother
50	12 Apr. 1875	Gulf of Yedo	NHM	To Willie
51	30 Apr. 1875	Yokos[u]ka	JTS	
52	4–10 May 1875	Yokohama	JTS	
53	7–10 May 1875	Yokohama	SIO	To mother. Pages missing.
54	1 June 1875 to 16 June 1875	Kobi Yokohama	JTS	
55	13 July 1875 to 4 Aug. 1875	North Pacific Honolulu	SIO	To mother
56	13 July 1875 to 10 Aug. 1875	North Pacific Honolulu	JTS	
57	19 Aug. 1875 to 23 Nov. 1875	Hilo, Hawaii Valparaiso	SIO	To mother

continued

Number	Date(s)	Location(s)	Repository	Remarks
58	19 Aug. 1875 to 21 Nov. 1875	Hilo, Hawaii Valparaiso	JTS	
59	7 Dec. 1875	Valparaiso	JTS	
60	5 Jan. 1876 to 14 Jan. 1876	Messier Channel, Patagonia Sandy Point	JTS	
61	Jan. 1876	Magellan Straits	NMH	To ? Pages missing.
62	13 Jan. 1876	Magellan Straits	NHM	To ? Pages missing.
63	6 Feb. 1876 to 16 Feb. 1876	Stanley, Falkland Islands Montevideo	NHM	To mother
64	6 Feb. 1876 to 24 Feb. 1876	Stanley, Falkland Islands Montevideo	JTS	
65	7 Mar. 1876 to 19 Apr. 1876	South Atlantic St. Vincent	JTS	
66	16 Mar. 1876 to 20 Apr. 1876	South Atlantic St. Vincent	SIO	To Fred. Pages missing.
67	19 Apr. 1876	St. Vincent	SIO	To mother? Fragment.
68	25 May 1876	Spithead	JTS	
69	11–12 June 1876	Chatham	JTS	

Challenger Expedition

Article from the *Hampshire Telegraph*, December 1872
Copied from Joseph Matkin's Journal by John Thomas
Swann, 12 July 1876
(Swann letterbook, pp. 769–776)

An early day of the ensuing week will witness the departure of the most important Naval surveying expedition which has ever been sent forth by any country. We have from time to time recorded the results obtained by Dr. Carpenter, Professor Wyville Thompson, & others, in the brief voyages of the "Lightning" & "Porcupine," results which shewed previously unsuspected variations in the deep sea temperature; the existence of a general oceanic circulation, the presence of life at the greatest depths and the active progress of submarine chalk formation even in the present day. The great scientific and practical importance of the facts revealed by these short and imperfect inquiries was such as to render their continuance a matter of national concern, & Dr. Carpenter who from the first had been most active in their prosecution exerted himself to bring the whole subject under the consideration of the Government. Mr. Lowe, however strict in protecting the public purse against demand for which private enterprise may be legitimately called upon, fully recognizes that matter which it is fitting for the country to undertake should be carried thro' with every advantage that money can secure, and he gave early intimation of his approval of the proposed expedition. In like manner the Admiralty asked only the Official approval of the Royal Society, and as soon as this was given, Mr. Goschen lent himself heartily to the scheme. The suggestion made to the Royal Society was that a ship should be fitted out for an expedition of three or four years duration, during which soundings, thermometric observations, dredging, and Chymical examination of sea water should be carried on continuously with a view to a more perfect knowledge of the physical and biological conditions of the great ocean basins, and in order to ascertain their depth, temperature, specific gravity, and chymic character; at the same time it was recommended that observations should be made on the directions & velocity of the great drifts and currents—especially those of the Gulf Stream, the Equatorial, and Japanese; both at the surface and intermediate strata, as well as the fauna of the deep sea water, and on the zoology & botany of those portions of the globe which are at present comparatively unknown.

351

In order to carry out these recommendations it was necessary to employ a vessel with a main-deck, as no flushed decked vessel could be fitted with the various rooms and appliances required.

H.M.S. "Challenger" was ultimately selected for the purpose and has been lying at Sheerness while the final preparations were being made for her departure.

She belongs to the Class of spar-decked corvettes, and is a modern wooden frigate built ship of about 1500 tons old measurement, but with a displacement of over 2000 tons. She has auxiliary screw power of 400 horses, and carries coal for about 24 days consumption at a speed of 5 knots an hour, together with full sail power under which she will be ordinarily navigated. But as the service on which she is engaged will compel her to keep to sea for unusually long period, and hence to carry very large store of provisions, it has been necessary to reduce her complement of men to 250, and also to reduce her sail power in some degree.

With the exception of two 64 pounders all the guns on the main-deck have been removed in order to obtain the extra accommodation required. On the deck are the Cabins of the Captain and Professor Thompson, the scientific Chief of the expedition, as well as those of the Commander, the Navigating Officer, and some of the scientific staff. There is also a spacious Chart house for surveying purposes, an Analyzing room, a Photographic Studio, and a Laboratory, all fitted with every appliance which science and experience could suggest. On the upper deck are the boats consisting of a 35 ft steam life pinnace, two cutters, one fitted for steam, gigs, whalers, and smaller boats. Here are also three independent steam engines, one of which is an 18 horse power, double cylinder, for heaving in the dredging and sounding line, this engine is fitted with a boiler distinct from the ship boilers, and may be used for distilling water without getting up steam in the latter. The two other engines are for use in the boats, or in the ship as may be necessary. On the after part of the deck beside the usual standard and other compasses is a handsomely mounted Fox & Clip circle, with which it is intended to make an extensive daily series of observations of the Magnetic elements, for which purpose some of the Naval Officers have been specially instructed. On the lower deck are the bunks and mess-places of the crew, and on the after-part a spacious & handsome general mess-room for the Officers, with Cabins.

Every Part of the ship is well ventilated and the greatest possible care has been exercised to secure the comfort and well-being of the officers and crew. This was especially necessary, inasmuch as they will experience many long & dreary sea voyages, & interesting as the work will be to those whose hearts are in it, yet it cannot fail to be often tedious & monotonous. The seamen & marines are all voluntary for the particular service. It is hoped that one of the most interesting physical questions now on debate—namely the circulation of the lower oceanic strata will be finally set at rest by the expedition.

The daily routine throughout the voyage, unless positively prevented by bad weather, will be ascertained, the depth of the ocean, & to observe the other phenomena which have been already alluded to, with a haul of the dredge when circumstances permit. If all goes well she may be looked for in the spring of 1876, after circumnavigating the globe, & traversing the three great oceans from North to South, and from East to West.

APPENDIX D

Challenger at Sheerness, May 27th, 1876

Article from the *Sheerness Guardian*, 3 June 1876
Copied from Joseph Matkin's Journal by
John Thomas Swann
(Swann letterbook, pp. 777–794)

The hazy weather and the state of the Tide which together delayed the arrival of the "Challenger" at her moorings in Sheerness until twenty minutes past seven o'clock on Saturday morning caused much disappointment among the inhabitants of that seaport.

Hoping against hope they stared gloomily thro' the damp and drizzle of the wretched afternoon of Friday. The great question being would the "Challenger" arrive that afternoon, & would she be paid off there or Chatham. The local mind seemed thoroughly engrossed with the paying off part of the matter, & equally well-informed persons ranged themselves on either side. Later in the evening it was known she was at the Nore, and would be in early in the morning, whereupon after its custom Sheerness went to bed, for it shuts up, hotels and all at 11 o'clock—& woe betide the belated wight who is abroad after regulation hours—for he shall find no spot whereon to rest his weary head, and with the great danger of having a night in the lock-up.

On Saturday morning the weather was endurable, & the Challenger was in Sheerness harbour gaily dressed in bunting in honor of the Queen's birthday; the "Duncan" 91 guns, the flag-ship of Admiral Chads, also making a brave show. On board of the ocean wanderer much hearty & general satisfaction was displayed at the happy conclusion of the voyage, for she sailed from Sheerness on a voyage round, & about the world on a certain Saturday just three years, five months, and twenty days before her return to her old moorings. During Monday, Tuesday, & Wednesday she remained at Sheerness open to the inspection of visitors, among whom were a strong contingent from the Royal Society. In the course of her three & half years voyage, she has not merely been round the world, but has zig-zagged up & down in it, covering altogether a distance from Sheerness out & home again of 68,930 miles, or nearly sufficient to put a girdle thrice round the earth. She had crossed the Atlantic several times, & traversed the Pacific to & fro. Starting from Sheerness on the 7th of Decr 1872, she, after touching at Portsmouth, Lisbon & Gibraltar, made a cruise among the Canary Islands, & thence across the Atlantic to Bermuda & Halifax. From Bermuda her course lay again across the ocean to S. Michael & Madeira, then down to the Equator to visit the rocks of St. Paul, which rear their lonely head in mid-Atlantic, & so on to Bahia. Again she crossed to the Cape of Good Hope, and went on to Melbourne, Sydney, & Wellington, & Tongatabu & Levuka, & to Cape York, Banda & Amboina, among the spice islands of the Indian seas, to Manilla & Hong Kong, & back to Manilla again; thence to Camiguin with its volcano in eruption, to Samboanga, rich in botanical specimens, to that abode of the primitive savage, Humboldt Bay, in New Guinea, where the natives armed with bows and arrows surrounded the ship on her arrival, blowing powerful notes on a kind of trumpet. Their hostility and astonishment rose to a tremendous pitch when the ship got under weigh; every bow being bent, and every arrow drawn to the head, & pointed at her screw as it slowly revolved.

At the Admiralty Islands less hostility was encountered, the anxiety of the inhabitants being less the ship should pass them by without staying to barter hoop-iron and hatchets. From these magnificent Islands the Challenger made her way to Yokohama, & Honolulu, to Hilo, Tahiti, and to Juan Fernandez, the ancient hiding place of buccaneers and the enforced residence first of William Sawkniss, & found by Dampier, & afterwards of Alexander Selkirk—and now leased by the Chilian Govt to a breeder of cattle.

From there to Valparaiso, through the Straits of Magellan to the Falkland Islands & thence to Monte Video, Ascension—S. Vincent, and Vigo. The good ship sped on her course encountering as she neared home the most disagreeable weather of the entire voyage. The work of the "Challenger" can hardly be estimated by that portion of her collection which has been exhibited to the public. It will be possible to unpack many of the boxes of preparations; but all that could be was done to give an idea of the variety & extent of the investigations made during her voyage. Out of 680 boxes of specimens of various kinds at least 300 have been sent home from convenient places at which the ship has touched, the botanical specimens, especially having been got rid of as quickly as possible, as well as the natural history specimens preserved in spirits. But there was no want of matter to interest the scientific mind, whether directed towards anthropology, or plunged into those tremendous depths of ocean whence creatures once supposed to be extinct such as the "Pterocrinus Aldrichonea" have been raised & restored to their rank as extant beings. Rare "Umbelularia" from the Phillipine Archipelago also attracted the attention of the learned in natural history, while physical geographers noted the extraordinary ocean depths measured between Manilla & Japan.

Very odd fish, if fish they may be called, are the "gorgonia" or seafan, trawled off the coast of Patagonia, the "asterobphyton" or plant star, captured off Cape Virgin, at the entrance of the Straits of Magellan.

From Manilla comes curious sponges, from Sebu rare specimens of "Euplectella"; from the Galapagos Island Tortoises of the race which it is feared is doomed to extinction; but probably all these will yield in interest to the tiny foraminifera dredged from the ocean depths more easily counted by miles than fms. Preserved in glass tubes are many specimens of the minute globerigina [sic], minute creatures who live on the surface of the sea, but whose relics sinking through thousands of fms. to the bottom of the sea undergo during that incalculable period various mutations from the action of the salt water until at last they are deposited on submarine hills, valleys & plateaux in the form of red mud. From a spot far to the south-east of the inhospitable Crozet, Kerguelen & McDonald Islands come a specimen of diatomaceon mud, raised from a depth of 1,950 fms. This is not only interesting from its scientific value, but as marking the nearest point to the South Pole made by the "Challenger" in her already famous voyage. It is possible that to a large majority of those who visited the discovery ship the means by which these results have been obtained will be at least as interesting as the specimens themselves. In the bow of the ship is the sounding & dredging apparatus employed in compelling [the] ocean to reveal his secrets. There are mighty reels of cord 2 & 3 nautical miles in length made of the finest Italian hemp. These lines have done their work remarkably well, & have shewn powers of resistance such as their slender proportions hardly indicate. The actual deep sea sounding is not from the bows of the ship; the line is carried from the reel amidships, and is then rove over the fore-yard, the steam being eased by accumulators of indiarubber. The brass tube which brings up samples of the sea bottom is fitted with a "butterfly valve" which allows it to take up about a pint of mud & then closes. Every one of the deep sea soundings has cost three hundred weight of iron, for when the tube is filled the great "sinkers" which surround it are detached, as the strain of hauling them up would probably cause the line to part. For mere shallow dredgings "valve lead," as they are called, are used—heavy sinkers with a tube & butterfly valve attached to them.

Leaving submarine depths for the upper air the visitor found enough and to spare to interest him, whether he cared for animate or inanimate nature. The collection of canoes, weapons, & articles of savage dress & ornaments was supplemented by several fine volumes of photographs— forming an admirable record of the voyage. Almost immediately the eye marked the distinction between the two great races of the South Sea Islands; the dark skinned woolly-haired Papuans, magnificent in form, & rugged of features, & the lighter colored Tahitians, slender in build, and in some cases almost beautiful. For good looks there is no comparison between the two races, but an impression prevails among those who have seen South Sea character, that the woolly haired Papuan is the man to do the work of holding his own against the exterminating influence brought to bear upon the natives of the Pacific Islands. Many curious relics of Humboldt Bay, New Guinea are preserved in the museum of the "Challenger," axes headed with green stone precisely like the remains of the stone age found in the burial places of primitive Europe. Spears pointed with splinters of obsidian, little hatchets armed with fragments of shells, bows of great size and power. The specimens from the Admiralty Islands are yet more primitive, the bow & arrow being absent, the javelin, the most artificial weapon known, and a full-dress suit consisting of three conch shells. By dint of models, specimens and photographs, a very good idea may be formed of the domestic & marine architecture of Oceania. There is the canoe with its outrigger connected with the main hull in which the rowers sit by a fighting stage, on which the warriors take up their position, towering over their humbler assistants like the ancient Greek Chieftain over his charioteer. Sometimes the stage extends quite across the canoe, springing up to windward to afford additional space and opportunity for trimming the boat. Equally curious are the low round primitive huts of the Admiralty Islanders, the elaborate structure of the more advanced lake dwellers of New Guinea, and the pile built or pulpit dwellings of the Philippines & other islands of the Indian Sea. Turning to the curious artificial incubator or hatching mound, the visitor encounters the work of birds in place of that of man, formed of leaves & twigs by the Australian Jungle Fowl, who lays her eggs in the mound & leaves them to be hatched by the heat generated by the moisture of the Wet Season. The Photographs are full of interest, and altho' often taken under very adverse circumstances, are exceedingly well executed. Strange places, and stranger people are preserved for ever in the photographic log of the "Challenger."

Shock headed groups from the Aroo Islands, Philippine villages precisely resembling those of the relics of which are found in Swiss lakes, views of streets in Manila which vividly recall the well known features of Venice, portraits of Japanese and Tahitian beauties side by side with the broad features of Queen Charlotte of Tongatabu, and great landscapes and seascapes, the great crater of Kilauea in the Sandwich Islands, the desolate rocks of St. Paul, the "stone rivers" of the Falkland Islands, the great glacier at Port Churrucea, beautiful scenes in the Straits of Magellan, a delicious bathing place at Fiji, scenes from the nutmeg, pepper, & cocoa groves of the Indian Archipelago, and from the savage solitudes of Tristan d'Acunha.

It seemed at first sight rather hard upon the Officers & Crew of the

"Challenger" that they should be detained on board after they had returned from their successful voyage, but they took the infliction in very good part, and received the many visitors who intruded on them during the week with all imaginable courtesy. One of their number, Lieutenant Carpenter, has already received the medal of the Royal Humane Society for saving the life of a sailor in the Falkland Islands. On Monday morning Vice Admiral Chads, Commander of the Nore, paid a visit to the vessel, & the men were put through the customary drill. He remained on the ship about 2 hours, & expressed his pleasure at again seeing the Officers & Sailors. He also complimented the men upon their smart appearance, and the expertness with which they went through the various exercises.

The collection of Photographs comprising nearly 500, is highly interesting, views were taken of every place visited by the vessel. Mr. J. J. Wild, the Artist who accompanied the expedition, has a great number of beautiful paintings of scenes passed in the cruise. These, we believe, will be engraved and published shortly. At S. Vincent a Photograph was taken of an Albino negress; the features of this woman are exactly those of the black negress, but her skin and hair are perfectly white. Two large living Tortoises, one 40 years old, and the other 100 have been brought home. The largest one will walk quite as fast with two men standing on its back, as when only carrying its own weight.

Lord Clarence Paget, Mr E. J. Reed, M.P., Sir H. Cole, Professors Allman, Crum-Brown, Octingen, and Ericker, Baron von Wrangell, Dr. Burdiermann, Secretary of the General Loan Committee of the Scientific apparatus exhibition, Kensington, and the following members of the Science & Art department—Norman Macleod, Majors Testing & Donelly paid a visit to the Challenger on Tuesday.

Invitations to visit the ship were sent by the Admiralty to all the English and Foreign Members of the Kensington Loom Apparatus Collection, many of whom accepted them.

On Thursday & Friday the Challenger swung for the adjustment of Compasses, and to make Magnetic Observations.

It is thought that 10 or 12 days must elapse before the stores can be got out to enable her to pay off.

About 600 persons visited the ship on Wednesday.

APPENDIX E

"List of Books for
H.M.S. 'Challenger' "

The books listed below were taken aboard H.M.S. *Challenger* prior to sailing for the use of the scientific staff (Tizard Papers, MS86/072, Envelope 7, National Maritime Museum Library, Greenwich). This list is reproduced with the kind permission of the Trustees of the National Maritime Museum.

According to the published *Narrative* (vol. I, pt. I, p. 45):

> The Library consisted of several hundred volumes, including Voyages, Travels, standard works on Zoology, Botany, Chemistry, Transactions and Proceedings of Societies, &c. These were either supplied by the Admiralty, or were the property of the Scientific Staff. It does not appear that any useful purpose would be served by giving a list of these books.

The list is reproduced here for the first time as a representation of the state of the literature in oceanography circa 1870, an indication of the texts regarded as essential by the *Challenger* "scientifics," and a clue to the sources that might have been available to Matkin. Annotations indicating the full title and publication details of the work (first edition, unless otherwise noted) have been inserted when they could be established with reasonable certainty.

20 Decr. 1872

List of Books for
H.M.S. "Challenger"

Adams' Notes of a Naturalist in Japan
> Arthur Adams. *Travels of a Naturalist in Japan and Manchuria.* London: Hurst & Blackett, 1870.

Adams' Genera of Mollusca. 3 vols
> Henry Adams and Arthur Adams. *The Genera of Recent Mollusca; Arranged According to their Organization.* 3 vols. London: John Van Voorst, 1853–1858.

359

"Adventure & Beagle"—Voyage of. 4 vols
> Robert Fitzroy. *Narrative of the Surveying Voyages of His Majesty's Ships Adventure and Beagle, between the Years 1826 and 1836.* . . . London, 1839.

Agassis' Travels in Brazil
> Louis Agassiz and Elizabeth Cary Agassiz. *A Journey in Brazil.* Boston: Tichnor & Fields, 1868.

Airey's Treatise on Magnetism
> Sir George Biddel Airy. *A Treatise on Magnetism for the Use of Students in the University.* London: Macmillan & Co., 1870.

Alc. d'Orbigny's (Ferussac et D'Orbigny) Mollusques vivants et fossiles ou description de toute les espèces de coquilles et mollusques—Tome I Les Cephalopodes—Paris—1845 8 vo with 8 tables
> Alcide Dessalines D'Orbigny. *Mollusques vivants et fossiles ou description de toutes les espèces de coquilles et mollusques classées suivant leur distribution géologique et géographique.* . . . Paris, 1845.

Alcock's Japan. 2 vols
> Sir Rutherford Alcock. *The Capital of the Tycoon: A Narrative of a Three Year's Residence in Japan.* 2 vols. London: Longman, Green, Longman, Roberts & Green, 1863.

Allmans Hydroids—2 parts (Ray Soc.)
> George James Allman. *A Monograph of the Gymnoblastic or Tubularian Hydroids.* 2 pts. London: Ray Society, 1871–1872.

Ansons Voyage round the World
> George Anson. *Voyage Round the World in the Years MDCCXL, I, II, III, IV.* London: J. & P. Knapton, 1748.

Antarctic Regions—Reports of Observations to be made
> [unidentified]

"Astrolabe". Voyage of. Capt Durville
> Jules Dumont D'Urville. *Voyage de la Corvette d'"Astrolabe."* 16 vols. Paris: J. Tastu, 1830–1835.

Astronomical Observations made in Capt Cook's voyages—3 vols
> William Wales and William Bayly. *The Original Observations, Made in the Course of a Voyage Towards the South Pole, and Round the World, in His Majesty's Ships the Resolution and Adventure.* . . . London, 1777; William Bayly. *The Original Observations Made in the Course of a Voyage to the Northern Pacific Ocean.* . . . London, 1782.

[page 2]

Baches Tides of the Gulf Stream
> Alexander Dallas Bache. *On the Tides of the Atlantic and Pacific Coasts of the United States; the Gulf Stream; and the Earthquake Waves of December 1854.* New Haven: E. Hayes, 1856.

Bligh's Voyage to the South Seas
> William Bligh. *A Voyage to the South Sea, Undertaken by Command of His Majesty, For the Purpose of Conveying the Breadfruit Tree to the West Indies. . . .* London: George Nicol, 1792.

Boid's Description of the Azores
> Edward Boid. *A Description of the Azores, or Western Islands, from Personal Observation.* London: Edward Churton, 1835.

Bowring's Phillippine Islands
> Sir John Bowring. *A Visit to the Philippine Islands.* London: Smith, Elder & Co., 1859.

British Association—Catalogue of Stars
> *The Catalogue of Stars of the British Association. . . .* London, 1845.

British Museum Catalogues
> Grays Seals & Whales with Suppt.
> > John Edward Gray. *Catalogue of Seals and Whales in the British Museum.* 2d ed. London, 1866; Supplement 1871.

Günther's Fishes—8 vols
> A. Günther. *Catalogue of the Fishes in the Collection of the British Museum.* 8 vols. London, 1859–1870.

Lizards of Australia & New Zealand
> John Edward Gray. *The Lizards of Australia and New Zealand in the Collection of the British Museum.* London, 1867.

Bronn—Thier. Reichs—vols 1.2.3.5.6
> H. G. Bronn. *Die Klassen und Ordnungen des Thier-Reichs wissenschaftlich dargastellt in Wort und Bild.* Leipzig and Heidelberg, 1859–1866.

Bunsen—Gasometrische Methoden
> R. W. E. Bunsen. *Gasometrische Methoden. . . .* Braunschweig, 1857.

Burneys Collection of Voyages—5 vols
> James Burney. *A Chronological History of the Discoveries in the South Sea or Pacific Ocean. . . .* 5 vols. London: G. & W. Nicol, 1803–1817.

Chanticleer—Voyage of—2 vols
> William Henry Bayley Webster. *Narrative of a Voyage to the Southern Atlantic Ocean in the Years 1828, 29, 30, Performed in H.M. Sloop Chanticleer.* . . . 2 vols. London: R. Bentley, 1834.

Claparède—Annelides du Golfe de Naples—3 vols
> Édouard Claparède. *Les Annélides Chétopodes du Golfe de Naples.* . . . Geneva, 1868.

" —Glaneurs Zootomiques
> Édouard Claparède. "Glanures Zootomiques parmi les Annélides de Port-Vendres (Pyrénées orientales)." *Mémoires de la Société de Physique et d'Histoire Naturelle de Genève.* Vol. 17 (1864). Pp. 463–600.

" —Les Annelides de Port Vendres
> [unidentified]

[page 3]

Claus—Gründzuge der Zoologie
> Carl Claus. *Grundzüge der Zoologie, Gebrauche an Universitäten und Höhern Lehranstalten.* Marburg/Leipzig: N. G. Elwert'sche Universitäts-Buchhandlung, 1868.

Cook's Voyages—8 vols 4to
> James Cook. *Cook's Voyages.* 8 vols. London: W. Strahan & T. Cadell, 1773–1784.

Dana's Corals & Coral Islands
> James Dwight Dana. *Corals and Coral Islands.* London: Sampson Low, Marston, Low & Searle, 1872.

Darwin's Cirripedia—2 vols (Ray Soc.)
> Charles Darwin. *A Monograph on the Sub-Class Cirripedia, with Figures of All the Species.* . . . 2 vols. London: Ray Society, 1851.

" —Geological Observations—2 vols
> Charles Darwin. *Geological Observations on the Volcanic Islands, Visited during the Voyage of H.M.S. Beagle;* and *Geological Observations on South America.* 2 vols. London: Smith, Elder & Co., 1844–1846.

Diesing—Systema Helminthum—2 vols
> Karl Moritz Diesing. *System Helminthum.* 2 vols. Vindobonae [Vienna]: W. Braumüller, 1850–1851.

"Dolphin"—Voyage of 8vo
> *A Voyage Round the World, in His Majesty's Ship the Dolphin, commanded by the Honourable Commodore Byron.* . . . London: J. Newbery & F. Newbery, 1767.

Dove's Law of Storms
> Heinrich Wilhelm Dove. *The Law of Storms Considered in Connection with the Ordinary Movements of the Atmosphere.* . . . 2d ed. London: R. H. Scott, 1862; 1st ed. Berlin, 1857.

Dumeril's—Histoire naturelle des Poissons—II vols (Suites à Buffon)
> Auguste Henri Andre Dumeril. *Histoire Naturelle des Poissons, ou, Ichthyologie générale.* 2 vols. in 3. Paris: Roret, 1865–1870.

Earls Eastern Seas 8vo
> George Windsor Earl. *The Eastern Seas, or, Voyages and Adventures in the Indian Archipelago, in 1832-33-34.* . . . London: W. H. Allen, 1837.

Edwards—Milne—Leçons sur la Physiologie—9 vols
> Henri Milne-Edwards. *Leçons sur la Physiologie et l'Anatomie Comparée de l'Homme et des Animaux.* 14 vols. Paris: V. Masson, 1857–1881.

Ellis's Polynesian Researches. 4 vols
> William Ellis. *Polynesian Researches, During a Residence of Nearly Six Years in the South Sea Islands.* . . . 2d ed. enlarged and improved. 4 vols. London: Fisher, Son & Jackson, 1831.

Encyclopedie l'Histoire Naturelle II vols
> *Encyclopédie Méthodique.* . . . Tome 116-125: Histoire Naturelle. 10 vols. Paris: Chez Panckoucke, 1782-1825.

Enderby's Account of the Auckland Islands
> Charles Enderby. *The Auckland Islands: A Short Account of Their Climate, Soil, and Productions; and the Advantages of Establishing There a Settlement at Port Rous, for Carrying on the Southern Whale Fisheries.* London: P. Richardson, 1849.

[page 4]

Endlisher—Genera Plantarum—2 vols
> István László Endlicher. *Genera Plantarum Secundum, Ordines Natureles Disposito.* Vindobonae [Vienna]: Fr. Beck, 1836–1840.

Erskines—Pacific Islands
> John Elphinstone Erskine. *Journal of a Cruise Among the Islands of the Western Pacific.* London: John Murray, 1853.

Frey—Das Mikroskop
> Heinrich Frey. *Das Mikroskop und die Mikroskopishe Technik. Ein Handbuch.* . . . Leipzig: W. Engelmann, 1863.

Forrest's Voyage to New Guinea
> Thomas Forrest. *A Voyage to New Guinea, and the Moluc-cas, from Balambangan: Including an Account of the Magindano, Sooloo, and Other Islands; . . . Performed in the Tartar Gallery . . . During the Years 1774, 1775 and 1776. . . .* London: J. Robson, 1779.

Gegenbaur—Grundzüge der Verg. Anatomie
> Carl Gegenbaur. *Grundzüge der Vergleichenden Anato-mie. . . .* Leipzig: W. Engelmann, 1859.

Grisebach. Die Vegetation der Erde 2 vols
> August Heinrich Rudolph Grisebach. *Die Vegetation der Erde nach irher klimatischen Anordnung. Ein Abriss der vergleichenden Geographie der Pflanzen.* 2 vols. Leipzig: W. Engelmann, 1872.

Gulf Stream—A Collection of Papers in 1 vol—4to
> [unidentified]

Handbuch der Zoologie—2 vols
> Jules Victor Carus and C. E. A. Gerstaecker. *Handbuch der Zoologie.* 2 vols. in 3. Leipzig: W. Engelmann, 1863–1875.

Hartung—Geologische Beischreibung der Inseln der Madeira und Porto Santo
> Georg Hartung. *Geologische Beischreibung der Inseln der Madeira und Porto Santo.* Leipzig: W. Engelmann, 1864.

" —Die Azoren—2 vols
> Georg Hartung. *Die Azoren in ihrere äusseren Erscheinung und nach ihrer geognostischen Natur geschildert. . . .* Leipzig: W. Engelmann, 1860.

"Herald"—Voyage of—2 vols—8vo
> Berthold Carl Seemann. *Narrative of the Voyage of H.M.S. Herald during the Years 1845-51, under the Command of Captain Henry Kellett. . . .* 2 vols. London: Reeve & Co., 1853.

Hochstetter—Neu Seeland
> Ferdinand Christian von Hochstetter. *Neu-Seeland. . . .* Stuttgart: Cotta, 1863.

Hooker's Flora Antarctica—2 vols
  " —Flora of Tasmania and New Zealand—4 vols
> Joseph Dalton Hooker. *The Botany of the Antarctic Voy-age of H.M. Discovery Ships Erebus and Terror, in . . . 1839-1843, Under the Command of Captain Sir James Clark Ross.* 3 parts in 6 vols.; I. Flora Antarctica; II. Flora Novae Zelandicae; III. Flora Tasmaniae. London: Reeve Brothers, 1844-1860.

Hutton's Mathematical Tables
> Charles Hutton. *Mathematical Tables: Containing Common, Hyperbolic, and Logistic Logarithms.* . . . London: G. G. J. and J. Robinson, 1785.

[page 5]

Jahrbuch—Mineralogie—Geologie &c 2 vols
>   ″ —No 7—1872
> *Neues Jahrbuch für Mineralogie, Geologie, und Paläontologie.*

Jeffrey's Conchology—5 vols
> John Gwyn Jeffreys. *British Conchology, or An Account of the Mollusca Which Now Inhabit the British Isles and Surrounding Seas.* 5 vols. London: J. Van Voorst, 1862–1869.

Johnston's Physical Atlas
> Alexander Keith Johnston. *The Physical Atlas, A Series of Maps & Illustrations of the Geographical Distribution of Natural Phenomena.* . . . Edinburgh: W. & A. K. Johnston and Cowan & Co., 1848.

Jones on the Rochon Micrometer
> [unidentified]

Jukes' Narrative of the Voyages of the "Fly"—2 vols—8vo
> Joseph Beete Jukes. *Narrative of the Surveying Voyage of H.M.S. Fly, Commanded by Capt. F. P. B. Blackwood in Torres Strait, New Guinea and Other Islands of the Eastern Archipelago, 1842–1846.* . . . 2 vols. London: T. & W. Boone, 1847.

King's Survey of Australia—2 vols
> Philip Parker King. *Narrative of a Survey of the Intertropical and Western Coasts of Australia, Performed Between the Years 1818 and 1822.* . . . 2 vols. London: J. Murray, 1827.

Kolf's Voyages along the Coast of New Guinea
> Dirk Hendrik Kolff. *Voyages of the Dutch Brig of War Dourga, Through the Southern and Little-known Parts of the Moluccan Archipelago and Along the Previously Unknown Southern Coast of New Guinea, Performed During the Years 1825 & 1826.* . . . London: J. Madden & Co., 1840.

Kolliker's Lehrbuch der Entwicklungsgeschichte—2 vols
> Albert Kölliker. *Entwicklungsgeschichte des Menschen und der höheren Thiere.* Leipzig: W. Engelmann, 1861.

Lamont's Wild Life among the Pacific Islands
> E. H. Lamont. *Wild Life among the Pacific Islanders.* London: Hurst & Blackett, 1867.

Leydig—Ueber den Bau und die systematische Stellung der Raderthiere—
Abstract from Siebolds und Kollikers Zeitschrift—3 parts
> Franz Leydig. "Ueber den Bau und die systematische Stel-
> lung der Räderthiere." *Zeitschrift für wissenschaftliche
> Zoologie*. Vol. 6. 1855. Pp. 1–120.

Loomis's Astronomy
> Elias Loomis. *An Introduction to Practical Astronomy,
> with a Collection of Astronomical Tables*. New York: Har-
> per & Bros, 1865.

[page 6]

Lyells Principles of Geology
> Charles Lyell. *Principles of Geology, Being an Attempt to
> Explain the Former Changes of the Earth's Surface by Ref-
> erence to Causes Now in Operation*. 3 vols. London:
> J. Murray, 1830–1833.

" —Elements of Geology
> Charles Lyell. *Elements of Geology*. London: J. Murray,
> 1838.

Maurys Sailing Directions—2 vols
> Matthew Fontaine Maury. *Explanations and Sailing Direc-
> tions to Accompany the Wind and Current Charts*. . . .
> 8th ed. 2 vols. Washington: W. A. Harris, 1858–1859.

Magnetical Observations at Kew. Results of
> [unidentified]

Masters Botany for Beginners
> Maxwell Tylden Masters. *Botany for Beginners; an Intro-
> duction to the Study of Plants*. London, 1872.

Mivart on the Genesis of Species
> St. George Jackson Mivart. *On the Genesis of Species*.
> London: Macmillan & Co., 1871.

Muller, S.—Ueber den glatten Hai des Aristotles
> Johannes Müller. *Über den glatten Hai des Aristotles, und
> über die Verschiedenheiten unter den Haifischen und
> Rochen in der Entwicklung des Eis*. . . . Berlin: König-
> lichen Akademie der Wissenschaften, 1842.

" —Physik und Meteorologie—3 vols + Atlas
> Johann Heinrich Jakob Müller. *Lehrbuch der Physik und
> Meteorologie*. 3 vols. plus atlas. Braunschweig: Vieweg &
> Sohn, 1856, 1858.

Murray's Missions in Western Polynesia
> Archibald Wright Murray. *Missions in Western Polynesia:
> Being Historical Sketches of These Missions, From Their
> Commencement in 1839 to the Present Time*. London:
> J. Snow, 1863.

"Novara"—Voyage of 3 vols 8vo
>Karl von Scherzer. *Narrative of the Circumnavigation of the Globe by the Austrian Frigate Novara. . . .* 3 vols. London: Saunders, Otley & Co., 1861–1863.

Palmers Kidnapping in the South Seas
>George Palmer. *Kidnapping in the South Seas. Being a Narrative of a Three Months' Cruise of H.M. Ship Rosario.* Edinburgh: Edmonston & Douglas, 1871.

Perry's Expedition to the China Seas and Japan
>Matthew Calbraith Perry. *Narrative of the Expedition of an American Squadron to the China Seas and Japan: Performed in the Years 1852, 1853 and 1854. . . .* 3 vols. Washington: A. O. P. Nicholson, 1856.

Petermanns Mittheilungen—1865 to '72 6 vols—67 + 68 missing
  ″ —No XI. 1872
>*Mittheilungen aus Justus Perthes' Geographischer Anstalt über wichtige neue Erforschungen auf dem Gesammtgebiete der Geographie.* A. Petermann, ed.

Piddingtons Law of Storms
>Henry Piddington. *The Sailor's Horn-Book for the Law of Storms: Being a Practical Exposition of the Theory of the Law of Storms, and its Uses to Mariners of All Classes in All Parts of the World, Shewn by Transparent Storm Cards and Useful Lessons.* New York/London: J. Wiley, 1848.

 ″ —Horn Book of Storms in India and China Seas
>Henry Piddington. *The Horn-Book of Storms for the Indian and China Seas.* Calcutta: Bishop's College Press, 1844.

[page 7]

Poggendorff. Annalen der Physik und Chemie—1871 and '72—6 vols
>*Annalen der Physik und Chemie.* J. C. Poggendorf, ed.

Powells New Homes for the Old Country
>George Smyth Baden-Powell. *New Homes for the Old Country. A Personal Experience of the Political and Domestic Life, the Industries, and the Natural History of Australia and New Zealand.* London: R. Bentley & Son, 1872.

Rang et Souleyet—Hist. Nat. des Mollusq: Pteropodes
>Paul Charles Alexandre Léonard Rang and [Louis François August] Souleyet. *Histoire Naturelle des Mollusques Ptéropodes, Monographie Comprenant la Description de Toutes les Espèces de ce Groupe de Mollusques. . . .* Paris: J.-B. Baillière, 1852.

"Rattlesnake"—Voyage of 2 vols
John Macgillivray. *Narrative of the Voyage of H.M.S. Rattlesnake, Commanded by the Late Captain Owen Stanley . . . During the Years 1846-50. . . .* 2vols. London: T. & W. Boone, 1852.

Redfield's Observations in relation to the Cyclone of the West Pacific
William C. Redfield. *Observations in Relation to the Cyclones of the Western Pacific; Embraced in a Communication to Commodore Perry.* 1850.

Reports by Secretaries of Embassy and Legation in Parliamentary Papers of 1872. Presented by Command—Nos 503, 543, 549, 563, 567, 595, 597

Ross's Antarctic Voyage—2 vols.
James Clark Ross. *A Voyage of Discovery and Research in the Southern and Antarctic Regions: During the Years 1839-43.* 2 vols. London: J. Murray, 1847.

Rosser's Notes on the Physical Geography of the North and South Atlantic
William Henry Rosser. *Notes on the Physical Geography and Meteorology of the South Atlantic; Together with the Sailing Directions for the Principal Ports of Call, and for the Islands.* London: J. Imray, 1862; *North Atlantic Directory: The Physical Geography and Meteorology of the North Atlantic.* London, 1864.

Sachs—Handbuch der physiologischen Botanik 4 vols.
Julius von Sachs. *Handbuch der Experimental-Physiologie der Pflanzen. . . .* 4 vols. Leipzig: W. Engelmann, 1865.

"Samarang"—Voyage of 2 vols 8vo
Sir Edward Belcher. *Narrative of the Voyage of H.M.S. Samarang During the Years 1843-46; Employed Surveying the Islands of the Eastern Archipelago. . . .* 2 vols. in 1. London: Reeve, Benham & Reeve, 1848.

Schnizlein—Iconographie Fam Nat. Regne Vegetablis—4 vols
Adelbert Schnizlein. *Iconographia Familiarum Naturalium Regni Vegetabilis . . . Abbildungen der Natürlichen Familien des Gewächsreiches. . . .* 4 vols. Bonn, 1843-1870.

Seemann's Fijian Islands—8vo
Berthold Carl Seemann. *Viti: An Account of a Government Mission to the Vitian or Fijian Islands, in the Years 1860-61.* Cambridge: Macmillan & Co., 1862.

v Siebold—Vergleichen Anatomie der wirbellosen Thiere und Stannius—
Vergleichende Anatomie der Wirbelthiere—1 La. auflage und 1 Fische und
Amphibien, 2 en. aufl.
>Karl Theodor Ernst von Siebold and Hermann Stannius.
>*Lehrbuch der Vergleichenden Anatomie.* 1. Wirbellose
>Thiere—Siebold; 2. Wirbelthiere—Stannius. Berlin: Veit,
>1848, 1846.

[page 8]

Snow's Two Years Cruize off Tierra del Fuego, the Falkland Islands, etc—
2 vol.s
>William Parker Snow. *Two Years' Cruise off Tierra del
>Fuego, the Falkland Islands, Patagonia, and the River
>Plate.* . . . 2 vols. London: Longman, Brown, Green,
>Longmans, & Roberts, 1857.

Stokes's Discoveries in Australia 2 vols
>John Lort Stokes. *Discoveries in Australia; With an
>Account of the Coasts and Rivers Explored and Surveyed
>During the Voyage of H.M.S. Beagle in the Years 1837-38-
>39-40-41-42-43.* . . . 2 vols. London: T. & W. Boone,
>1846.

Storer's Dictionary of Chemical Solubilities
>Frank H. Storer. *First Outlines of a Dictionary of Solubili-
>ties of Chemical Substances.* Cambridge, Mass.: Sever &
>Francis, 1864.

"Sulphur"—Voyage of 2 vols—8vo
>Sir Edward Belcher. *Narrative of a Voyage Round the
>World, Performed in Her Majesty's Ship Sulphur, During
>the Years 1836-42, Including Details of the Naval Opera-
>tions in China, from Dec. 1840, to Nov. 1841.* . . . 2 vols.
>London: H. Colburn, 1843.

Tandon's World of the Sea
>Christian Horace Bénédict Alfred Moquin-Tandon. *The
>World of the Sea.* Translated and enlarged by the Reverend
>H. Martyn Hart. . . . London: Cassell, Petter and Galpin,
>1869.

Thomson's New Zealand—2 vols
>Arthur Saunders Thomson. *The Story of New Zealand:
>Past and Present—Savage and Civilized.* 2 vols. London:
>J. Murray, 1859.

Tidal Observations—Report of British Association 1870
>Committee for the Purpose of Promoting the Extension,
>Improvement and Harmonic Analysis of Tidal Observa-
>tions, Sir William Thomson, et al. *Report of the Fortieth
>Meeting of the British Association* . . . *1870.* Pp. 120-151.

Vogelsang—Philosophie der Geologie
> Hermann Peter Joseph Vogelsang. *Philosophie der Geologie und Mikroskopische Gesteins-studien.* Bonn: M. Cohen & Sohn, 1867.

Walker's Terrestrial and Cosmical Magnetism
> Edward Walker. *Terrestrial and Cosmical Magnetism, The Adams Prize Essay for 1865.* Cambridge: Deighton, Bell & Co., 1866.

Wallace's Malay Archipelago—2 vols
> Alfred Russel Wallace. *The Malay Archipelago, The Land of the Orang-utan and the Bird of Paradise; a Narrative of Travel, with Studies of Man and Nature.* 2 vols. London: Macmillan, 1869.

Wallich's North Atlantic Sea Bed
> George Charles Wallich. *The North-Atlantic Sea-bed: Comprising a Diary of the Voyage on Board H.M.S. Bulldog, in 1860; and Observations on the Presence of Animal Life, and the Formation and Nature of Organic Deposits, at Great Depths in the Ocean.* London: J. Van Voorst, 1862.

Watt's Dictionary of Chemistry with Supplement—6 vols
> Henry Watts. *A Dictionary of Chemistry and the Allied Branches of Other Sciences.* 5 vols. London: Longmans, Green & Co., 1866–1868.

Wilke's Voyage round the World 5 vols
> Charles Wilkes. *Narrative of the United States Exploring Expedition. During the Years 1838, 1839, 1840, 1841, 1842.* 5 vols. Philadelphia: C. Sherman, 1844.

Wundt—Handbuch der Physiologie neue Aufl.
> Wilhelm Max Wundt. *Lehrbuch der Physiologie des Menschen.* Erlangen: F. Enke, 1865; rev. ed., 1868.

"Abstract of Voyage of H.M.S. 'Challenger' "

H.M.S. 'Challenger',
Sailing from Sheerness, Decr. 7th, 1872,
& arriving there May 27th, 1876,
after her cruise round the World
(Swann letterbook, pp. 796–798)*

Grand Totals

Distance made good	68930 miles
Coals expended	4994 tons 8 cwt
No. of days at sea	725
No. of days in harbor at place left	564
No. of Deep sea soundings obtained	375
No. of Serial Temperatures	255
No. of Successful Dredgings	112
No. of Unsuccessful Do.	19
No. of Successful Trawlings	131
No. of Unsuccessful Do.	16

Ship's Company numbered 243,
November 15th, 1872, the day of commissioning,
and 226, June 12th, 1876, the day of paying off.

Died—Nov. 72 J. Brooker, Leading Stoker, Heart Disease
———— J. Tubbs, Marine—Drowned, Sheerness
Mar. 73 W. H. Stokes, accident, concussion of brain
April 73 A. Ebbels, Naval Schoolmaster, apoplexy
———— J. Long, or^d seaman, Yellow Fever, Bahia
June 74 D. Wurton, A.B., Drowned, Cook's Straits
75 W. Pembre—Caffre Servt, Decline, Hong Kong
June 75 J. McDonald—Sick Berth Steward, Delerium Tremens
Sep. 75 Dr. Von Suhm (Naturalist), Erysepelas
Jan. 76 T. Bush. A.B. Drowned, Falkland Isl.
April 76 J. May—Carp. Crew (Consumption)

Deserted at various places	60	of original
Invalided, Discharged,		ship's
Imprisoned, Exchanged	29	company

*Compare with statistics assembled in Daniel Merriman, "Challengers of Neptune: the 'Philosophers,' " *Proceedings of the Royal Society of Edinburgh*, Section B, 72 (1972): 15–45.

Total of original ship's company left in
the ship on the day of paying off   144

Total number borne on the ship's books
during the commission   400

Died at Hong Kong after the ship had sailed
  George Crutchley, Marine, Insanity

Committed suicide whilst paying off at Chatham, June 6th
  Sub. Lieut. H. C. E. Harston, who drank a portion of Hydrate of Chloral

Notes

Introduction

1. John Laffin, *Jack Tar: The Story of the British Sailor* (London: Cassell, 1969).

2. I thank Oskar Spate for pointing out the two lower-deck accounts of Cook's third voyage, neither of which, as he indicated, is entirely satisfactory: *John Ledyard's Journal of Captain Cook's Last Voyage*, ed. J. K. Munford (Corvallis, Ore.: Oregon State University Press, 1963), relies heavily on John Rickman; and Heinrich Zimmermann's *Account of the Third Voyage of Captain Cook* (Wellington: Alexander Turnbull Library, 1926), which is too brief.

3. Henry Baynham, *From the Lower Deck: The Old Navy, 1780–1840* (London: Hutchinson, 1969); idem, *Before the Mast: Naval Ratings of the Nineteenth Century* (London: Hutchinson, 1971).

4. John Bechervaise had sailed with Beechey in H.M.S. *Blossom* in 1825; William Simpson was aboard H.M.S. *Plover* in 1848 during the search for the ill-fated expedition of Sir John Franklin; and Daniel Hartley was a petty officer in H.M.S. *Alert* on the Arctic expedition of 1875 (Baynham, *Before the Mast*, chaps. 3–5).

5. See Roy M. MacLeod, "Science and the Treasury: Principles, Personalities, and Policies, 1870–1885," in G. L'E. Turner, ed., *The Patronage of Science in the Nineteenth Century* (Leyden: Noordhoff International Publishing, 1978).

6. A. L. Rice, *British Oceanographic Vessels 1800–1950* (London: Ray Society, 1986), pp. 30–39.

7. "The 'Challenger,' Her Challenge," *Punch*, 1872. Reproduced in the Swann Letterbook, pp. 7–10.

8. On this procedure see especially JM to Sarah Craxford Matkin, 25 December 1873, SIO, M27. For a calendar of the letters, see Appendix B.

9. JM to Sarah Craxford Matkin, Yokohama, 7 May 1875, SIO, M53.

10. See JM to Sarah Craxford Matkin, 2 February 1873, SIO, M7.

11. On Charlie: JM to Charles Matkin, 5–15 November 1873, SIO, M23. On Fred: JM to Sarah Craxford Matkin, 10–15 December 1873, SIO, M25.

12. "You need never send any of my letters to Swadlincote & Barrowden as I always write there." JM to Sarah Craxford Matkin, Melbourne, 23–25 March 1874, SIO, M29.

13. Bryan Waites, "The Spirit of Rutland," *Rutland Record* 1 (1980), pp. 3–4. The county of Rutland ceased to have a separate administrative identity in 1974 as a result of the reorganization of local government throughout the United Kingdom. It is now part of Leicestershire.

14. The City Directory of Oakham for 1855 indicates that the Matkin Printing and Stationery Store was established in High St. in that year. In the 1858 directory, "Hawthorn & Matkin" are described as "Booksellers, Stationers and Printers." (Hawthorn's identity is not known, other than that he also operated a printing shop in Uppingham at the time.) Sarah Matkin ran the business from the time of her husband's death in 1874 until 1897, when control passed to the eldest son, Charles. The Rutland County Museum in Oakham has a printing press used by the Matkin print shop in the nineteenth century.

Judging from the fact that the elder Charles Matkin left behind "Effects under £450" in his will, it would appear that the Matkins were well established amongst the shopkeeping middle class of Rutland (will of Charles Matkin, 12 October 1875, Somerset House).

15. For an account of one student's experiences at Oakham School in the early 1800s, see John L. Barber, "Oakham School 140 Years Ago," *Rutland Record* 3 (1982/83): 118–120. Also useful is Roger Blackmore, "English Schools: Oakham," *This England,* Winter 1991, pp. 48–50.

16. Commemorative plaque, The Old School, Billesdon, Leicestershire. In 1837 the school's enrollment consisted of twenty-two free pupils and eighteen others who paid between 7s. 6d. and 10s. per year. "The costs of stationery were defrayed by the parents of the children" (Janet D. Martin, "Billesdon," vol. 5, pp. 6–15, of *The Victoria History of the Counties of England: A History of Leicestershire,* ed. J. M. Lee and R. A. McKinley [London: University of London, 1964], 5 vols).

17. JM to Sarah Craxford Matkin, Gulf of Yedo, 12 April 1875, SIO, M49.

18. Matkin apparently wrote occasionally to the Creatons. One letter to Matkin from the headmaster's wife survives, proclaiming "how charmed we *all* were to read your most kind & *deeply interesting* letter—it *was* good of you to take so much trouble & to spend such a lot of yr time in writing to me" (Ellen Creaton to Joseph Matkin, Dovercourt nr Harwich, 11 July 1876, SIO).

19. Invaluable in this context is David Layton's *Science for the People: The Origins of the School Science Curriculum in England* (New York: Science History Publications, 1973). On science in the secondary schools, see Gordon W. Roderick and Michael D. Stephens, *Scientific and Technical Education in Nineteenth-Century England* (Newston Abbot: David & Charles, 1972); and for the universities, D. S. L. Cardwell, *The Organisation of Science in England,* rev. ed. (London: Heinemann 1972).

20. Layton, *Science for the People,* p. 43.

21. Practicing scientists feared that science taught with only its practical utility in view would be a science unappreciated for its own sake and,

eventually, a science that no longer moved forward. Metropolitan scientists at the Government School of Mines were especially insistent, moreover, that observational sciences like botany or geology be emphasized, not the experimental sciences of mechanics and agricultural chemistry advocated by the early educators. See Layton, chap. 6.

22. Ibid., p. 35.

23. H.M.S. *Agamemnon*, with the U.S.S. *Niagara*, had laid the first trans-Atlantic telegraph cable in 1858 (Rice, *British Oceanographic Vessels*, p. 28).

24. Letter, T. R. Padfield, PRO, to D. C. Day, 5 July 1984, SIO; ADM 188/46, no. 64786, PRO.

25. Four letters from H.M.S. *Invincible*, 21 March–11 November 1871; seven letters from H.M.S. *Audacious*, 26 November 1871–30 May 1872, SIO.

26. Eugene L. Rasor, *Reform in the Royal Navy: A Social History of the Lower Deck 1850–1880* (Hamden, Conn.: Archon Books, 1976), p. 16. See also Robert D. Foulke, "Life in the Dying World of Sail, 1870–1910," *Journal of British Studies* 3 (1963): 105–136.

27. "Continuous Service" involved the signing of a ten-year agreement. Matkin was probably recruited by the older hire-and-discharge method: a seaman signed on at the beginning of a particular ship's commission (three to five years) and was "paid off" at its conclusion. See Rasor, *Reform*, pp. 26–28.

28. These reforms were effected, in large measure, through the efforts of several enlightened admirals who had experienced directly the problems of recruiting and discipline before serving in Whitehall. Most notable among them was Admiral Sir William Fanshawe Martin (1801–1895), commander in chief of the Mediterranean squadron in 1860 and a Naval Lord later in the decade. See Rasor, *Reform*, pp. 120–121. In the foregoing section I have drawn heavily on Rasor's work, which is invaluable for this period of British naval history.

29. There is considerable confusion about the date of promotion. Although the paymaster of his previous command (H.M.S. *Audacious*) had expressed a desire to see Matkin advanced in rank from ship's steward's boy to ship's steward's assistant as soon as he reached his nineteenth birthday (2 December 1872), *Challenger's* surviving ledger indicates that Matkin was not promoted until December 1873—his twentieth birthday (JM to Sarah Craxford Matkin, Hull, 30 May 1871, SIO; H.M.S. *Challenger*, Quarterly Ledger, 15 November 1872–31 December 1873, ADM 117/196, List 5, PRO). From 15 December 1872 until 31 December 1873, Matkin was rated as "Ship's stewards Boy" on List 10 ("Boys of all classes") of the ledger; his wages were those of a boy 1c: 7 pence per day. From 2 December 1873, his rating was "Ship's Steward's Asst." on List 13 ("Permanent Supernumeraries"); his wages increased to 1 shilling 4 pence per day, those of an ordinary seaman (ADM 117/196-7, PRO). But according to family information (and substantiated in M7; see Chapter 1), this promotion took place a year earlier, on 31 December 1872. Matkin later confirmed (M23), however, that the pay raise took place on 1 December 1873 (see Chapter 3).

A possible explanation for this discrepancy is that the promotion and the pay raise were separate events. Matkin's superiors may have felt that, during this expedition, he was filling a billet judged to involve greater responsibilities and requiring more authority than would normally be accorded a "boy," and that he should thus be promoted at the outset of the voyage rather than at his normal time. Other regulations, however, may have prohibited his actually receiving the pay associated with the higher rank until he reached his twentieth birthday.

For reasons unknown, throughout the Quarterly Ledger he was listed variously as "George" (or "Geo.") "Matkin" or "Matkins" but never as "Joseph." It has been suggested that this error may have arisen initially because Matkin preferred to be called "Joe" and ship's record-keepers transcribed this as the similar sounding "Geo." (Letter, Eileen Brunton, NHM, to Deborah Day, SIO, 29 October 1984).

30. JM to Sarah Craxford Matkin, Sydney, 7 May 1874, SIO, M33.

31. List 5 (Petty Officers, Seamen), Quarterly Ledger for *Challenger*, 15 November 1872–31 December 1873, ADM 117/196, PRO.

32. William Henry Smyth, *The Sailor's Word-book: An Alphabetical Digest of Nautical Terms* . . . (London: Blackie & Son, 1867): "*Ship's steward*, the person who manages the victualling of mess departments. In the navy, paymaster's steward." See also the *Oxford English Dictionary*.

According to Matkin family tradition, JM was the ship's writer aboard *Challenger*, and he was selected for the expedition because of his writing ability. The *Challenger* log lists him as ship's steward's boy and ship's steward's assistant, however; and in letter M11, Matkin mentions that his hammock is next to that of the ship's writer, whose name was Richard Wyatt. Of course JM may have assisted the ship's writer when his steward's duties allowed; and he may have written letters for illiterate crew members, as ship's writers often did. I thank A. L. Rice for suggesting the latter possibility.

33. Rasor, *Reform*, p. 101.

34. JM to Sarah Craxford Matkin, 2 February 1873, SIO, M7.

35. On modern oceanographic research vessels the common terms are "scientificos" and "doctors." See H. Russell Bernard and Peter D. Killworth, "Scientists and Crew: A Case Study in Communication at Sea," *Maritime Studies and Management*, 2 (1975): 112–125.

36. To be precise one should probably say "*three* cultures," because the ship's officers constituted a middle group between the civilian scientists, with their desire to execute the scientific goals of the expedition with the greatest possible thoroughness and care, and the naval crew, with its preference to run the ship safely and efficiently and get on to the next port. Socially, the officers would have had more in common with the scientists, including class origin and education, while in terms of shipboard experience and values, they would have had closer affinities to the crew. Bernard and Killworth (ibid., p. 115) have observed that, aboard modern research vessels, the marine technicians tend to have this intermediate status and, as a result, often play a brokering role in reducing the tensions, which the authors see as "a particular example of a very general phenomenon in our society, the conflict between academics and the working class."

37. See Bernard and Killworth, "Scientists and Crew"; idem, "On the Social Structure of an Ocean-Going Research Vessel and Other Important Things," *Social Science Research* 2 (1973): 145-184; Ben Finney, "Scientists and Seamen," in A. A. Harrison, Y. A. Clearwater, and C. P. McKay, eds., *From Antarctica to Outer Space: Life in Isolation and Confinement* (New York: Springer Verlag, 1991), pp. 89-101.

38. Bernard and Killworth, "Scientists and Crew," pp. 119-122. The authors go so far as to claim that "the maintenance of subcultural privacy is essential to the well-being of a R/V [research vessel]. Attempts to reduce the scientist/crew tension by forcing greater interaction will be counterproductive" (p. 122).

39. "The critical variable in determining the hostility of the crew towards scientists is port time and the willingness of scientists to relinquish their sea time for needed rest and recreation by the crew" (ibid., p. 124).

40. Thomson, *Voyage*, vol. 1, p. 114.

41. Moseley, *Notes*, p. 1; see also pp. 578-579.

42. Swann letterbook, pp. 448-449, Humboldt Bay, New Guinea, 24 February 1875, M48. Admittedly, French and Spanish colonials receive stronger criticism than their British counterparts.

43. Seagoing ships were supplied with libraries by the Admiralty beginning in 1828; most books were of a moral, religious, or educational nature. The scope broadened as donations were received from private sources. See J. Winter, *Hurrah for the Life of a Sailor: Life on the Lower Deck of the Victorian Navy* (London: Michael Joseph, 1977), p. 25. I thank Julie Chittock of the National Maritime Museum for this reference.

44. Samuel Smiles (1812-1904), the Scottish author, had been the eldest of eleven children charged with supporting a family of modest means when his father died. He went on to careers in medicine, journalism and finally authorship of the biography of George Stephenson and a series of books espousing devotion by the working classes to strong Victorian values of self-reliance and perseverance. The most popular of these works, *Self-Help* (translated into seventeen languages), was published in 1859 when Joseph Matkin was six years old. See Asa Briggs, "Samuel Smiles: The Gospel of Self-Help," *History Today*, 37 (May 1987): 37-43; M. D. Stephens and G. W. Roderick, *Samuel Smiles and Nineteenth-Century Self-Help in Education* (Nottingham: University of Nottingham, 1983); and T. Travers, *Samuel Smiles and the Victorian Work Ethic* (New York: Garland Publishing, 1987).

45. JM to Sarah Craxford Matkin, Hong Kong, 25 November 1874, NHM, M41.

46. I am indebted to Eileen Brunton and Deborah Day for information on the location of the Matkin family grave.

47. See especially M56 in which he remarked on the English church in Honolulu: "the service was burlesqued by that mixture of millinery, & tom foolery called 'Ritualism'" (Swann letterbook, p. 599, 10 August 1875).

48. I thank Margaret Deacon for this suggestion about Matkin's attitude toward royalty.

49. See especially letter M21: "I have not a single companion in the ship that I call a companion, though of course we are all sociable and friendly" (JM to Sarah Craxford Matkin [Tristan da Cunha, 15 October?]–Simon's Bay, 4 November 1873, SIO).

50. I thank an anonymous referee for reinforcing this point.

51. List 13, Quarterly Ledger for *Challenger*, 1 January 1874–31 December 1875, ADM 117/197; 1 January 1876–12 June 1876, ADM 117/198, PRO.

52. *Imperial Calendar*, 1878–1894. The Local Government Board was formed in 1871 to administer public health and poor relief throughout the country, in accordance with recommendations of the Royal Sanitary Commission. It gradually took over many areas of municipal supervision, including engineering and medical inspections, vaccination, burial grounds, and wartime refugees (PRO Current Guide, pt. 1).

53. Many of the details of Matkin's pre- and post-*Challenger* life are from interviews and correspondence with Matkin descendants Mary Matkin Stone, Phyllis McLaren, and James Bugslag.

54. Sir John Murray was killed in a motor vehicle accident in 1914.

Chapter 1

1. For a listing of equipment, see Tizard et al., *Narrative*, vol. 1, pt. 1, pp. 41–45.

2. Ship's log for *Challenger*, 18 November 1872, ADM 53-10536, PRO.

3. Sublieutenant Lord George Granville Campbell, son of the eighth duke of Argyll and author of *Log-letters from "The Challenger."*

4. See Introduction, p. 5.

5. *Challenger* was the first major British expedition to carry an official photographer billet. Although more than thirty years had elapsed since the first successful photographic process was announced by Daguerre (1839), photography had not yet established a strong professional ethic, judging from the difficulty the expedition had in retaining its photographer. Three photographers "jumped ship" in succession and had to be replaced at later ports. Nevertheless, some eight hundred glass negatives were produced. A catalogue compiled by Eileen Brunton now resides at the Mineralogical Library, NHM.

6. Matkin was presumably referring to the natural history workroom. See Thomson, *Voyage*, vol. 1, p. 12.

7. Had Matkin enclosed a sketch or print of *Challenger*'s crest? The ship's seal, showing a standing armored figure with arm outstretched, is now located at the Natural History Museum (Eileen Brunton, personal communication). The ship's figurehead itself has survived and now stands above the entrance to the Institute of Oceanographic Sciences, Wormley, Godalming, Surrey. See D. R. C. Kempe and H. A. Buckley, "Fifty Years of Oceanography in the Department of Mineralogy, British Museum (Natural History)," *Bulletin of the British Museum Natural History* (Historical Series), 15 (1987): 59–97, esp. pp. 76–77.

8. Swann letterbook, pp. 13–17, Sheerness, 22 November 1872, MI.

9. The *Illustrated London News* for 14 December 1872 carried an article and a half-page illustration of *Challenger*. *The Graphic, An Illustrated Weekly Newspaper* of 28 December displayed "The Staff and Promoters of the *Challenger* Deep-Sea Expedition" (see p. 24 above).

10. Petropavlovsk, Kamchatka Peninsula, Russia.

11. The Marquess of Lorne, John Douglas Sutherland Campbell (1845–1914), was elder brother of Sublieutenant Lord George Campbell.

Admiral Sir Francis Leopold McClintock (1819–1907), Arctic explorer, discovered the diaries and log book of the lost expedition of Sir John Franklin; published *The Voyage of the "Fox"* in 1859 (Peter Kemp, ed., *Oxford Companion to Ships and the Sea* [London/New York/Melbourne: Cambridge University Press, 1976], p. 506).

12. Matkin had been at Hull frequently in his previous ships.

13. Swann letterbook, pp. 18–22, Portsmouth, 17 December 1872, M2.

14. Sam Noble, *Sam Noble, Able Seaman: 'Tween Decks in the 'Seventies* (New York: Frederick A. Stokes Co., 1926), pp. 9–10.

15. Miss Emma Wildman, an employee of the printing firm and family friend, frequently mentioned by Matkin.

16. For the control of scurvy. Lemon and lime juice were issued regularly to British seamen beginning in the 1790s. See Christopher Lloyd and Jack L. S. Coulter, *Medicine and the Navy, 1200–1900*, 4 vols. (Edinburgh/London: E. and S. Livingstone, 1957–63), vol. 4, chap. 7; and Margaret Deacon and Ann Savours, "Nutritional Aspects of the British Arctic (Nares) Expedition of 1875–76 and Its Predecessors," in James Watt, E. J. Freeman, and W. F. Bynam, eds., *Starving Sailors: The Influence of Nutrition upon Naval and Maritime History* (London: National Maritime Museum, 1981), pp. 131–162.

17. The rum ration was instituted in 1655 and abolished in 1970. From 1850 the daily ration was a half-gill (one-eighth of a pint). See Eugene L. Rasor, *Reform in the Royal Navy: A Social History of the Lower Deck 1850–1880* (Hamden, Conn.: Archon Books, 1976), pp. 82–83.

18. H.M.S. *Captain*, a double-screwed, six-gun turret ship of a new and controversial design, was lost with most of its crew in a storm in the Bay of Biscay shortly after commissioning in 1870 (*Annual Register*, 1870, pt. 2, pp. 107–110).

19. See Thomson, *Voyage*, vol. 1, pp. 107–108.

20. See Appendix E for a listing of books taken aboard for the scientific staff. Matkin is probably referring to the crew's own library, however.

21. The organlike keyboard instrument popular in the late nineteenth century. Metal reeds were activated by a bellows driven by the player's feet. Matkin had probably learned to play at Billesdon school, which had purchased a harmonium in 1856 (Janet D. Martin, "Billesdon," vol. 5, p. 14, of *The Victoria History of the Counties of England: A History of Leicestershire*, ed. J. M. Lee and R. A. McKinley [London: University of London, 1964], 5 vols). He played for church services aboard *Audacious* on at least two occasions (JM to Sarah Craxford Matkin, Hull, 30 May 1871, SIO).

22. The merchant vessel aboard which Matkin had sailed to Australia, 1867–68.

23. For comparison, Thomson's account of the day's dredging says simply that "the weather was boisterous and the attempt was unsatisfactory" (*Voyage*, vol. 1, p. 108). The official *Narrative* devotes just over a page to the entire Portsmouth-Gibraltar leg (Tizard et al., *Narrative*, vol. 1, pt. 1, pp. 46–47).

24. Thomson reports "dredging at a depth of 1,975 fathoms a little to the N.W. of the Burlings, the dredge fouled something at the bottom, an unusual occurrence in such deep water, and carried away" (*Voyage*, vol. 1, p. 108). Spry reported 3,000 fathoms of line lost (*Cruise*, p. 13). The official *Narrative* (vol. 1, pt. 1, p. 46) speculated the dredge had caught on a telegraph cable.

25. JM to Sarah Craxford Matkin, 29 December 1872, NHM, M3.

26. Matkin consistently misspelled Wyville Thomson's surname.

27. Swann letterbook, pp. 23–27, Lisbon Roads, 7 January 1873, M4.

28. The channel squadron served as a backup to the Royal Navy's principal home squadrons based at Portsmouth, Plymouth, and the Nore. See William Laird Clowes, *The Royal Navy: A History from the Earliest Times to the Death of Queen Victoria*, 7 vols. (London: Sampson Low, 1903), vol. 7. I thank Margaret Deacon for this reference. See also, Rasor, *Reform*, pp. 70, 156n24.

29. King Liuz I and Queen Marie Pia of Savoy.

30. Sir Charles Murray.

31. I.e., a strong head wind.

32. *Challenger*'s crew included a schoolmaster, Adam Ebbels, to assist with shipboard education during the long voyage, including that of Captain Nares' nine-year-old son. The position of naval schoolmaster had been accorded commissioned status in 1861, at which time a university degree became a requirement (Rasor, *Reform*, p. 105.) Both Ebbels and his replacement would meet with tragic ends, as Matkin later reported.

33. I.e., marine invertebrates, possibly crustacea—barnacles, crabs, shrimps, and lobsters—which belong taxonomically to the same phylum (Arthropoda) as insects. "Insect" was often used by scientific laymen to refer to any articulated or segmented animal, even a coral polyp, anemone, or jellyfish.

34. JM to Sarah Craxford Matkin, 8 January 1873, SIO, M5.

35. Matkin refers presumably to the church and monastery of Santa Maria at Belem. The latter was also known as the monastery of St. Jeronimo; a monument to Portuguese navigation, it was built in 1499 on the site from which Vasco da Gama had departed on his first voyage of discovery to the East Indies (Spry, *Cruise*, pp. 14–15; Wild, *At Anchor*, p. 10).

36. Campbell, *Log-letters*, pp. 4–5.

37. The species of fish acquired on that occasion (16 January 1873), according to Thomson (*Voyage*, vol. 1, pp. 117–118), were *Mora mediterranea* and *Coryphaenoides serratus*. Both were bloated and distorted by the internal expansion of gas as they were brought up from "a considerable

depth," perhaps explaining Matkin's description. The latter species was described by A. Günther in the *Challenger Reports, Zoology*, vol. 22 (1887), p. 134, plate XXXII under the name *Macrurus aequalis* (see p. 41 above). The depth was 600, not 1,900, fathoms.

38. A family friend.

39. Swann letterbook, pp. 28–41, Gibraltar, 20 January 1873, M6.

40. Perhaps *Halosaurus macrochir* from 1,090 fathoms on 28 January (*Challenger Reports, Zoology*, vol. 22, p. 237, plate LIX).

41. Provincial newspapers, probably the *Grantham* [Lincs.] *Journal* and the *Lincoln, Rutland and Stamford Mercury*.

42. Napoleon III, i.e., Louis Napoleon Bonaparte (1808–1873), Emperor of France, 1852–1870.

43. See Introduction, n. 29.

44. Walter Thornton, a chum in Oakham.

45. Correctly, the Lord Provost of Edinburgh and his lady.

46. JM to Sarah Craxford Matkin, Madeira, 2–4 February 1873, SIO, M7.

47. Matkin refers to Admiral Horatio Nelson's attempted capture of Santa Cruz in 1797, during the Napoleonic Wars. Nelson lost his right arm, not his eye, in this battle; the eye had been damaged three years earlier in the Corsican campaign. Matkin's history was competent but not flawless.

48. All of *Challenger*'s original cannon, except two, had been removed for the expedition.

49. Presumably Matkin was referring to unborn sharks, in which case "the old gentleman" should have been called "the old lady."

50. Swann letterbook, pp. 42–49, Santa Cruz, 7–9 February 1873, M8.

51. Thomson, *Voyage*, vol. 1, pp. 44, 74–75, 242–244; *Summary of Results*, pt. 1, p. 137; Tizard, et al., *Narrative*, vol. 1, pt. 1, pp. 95–97.

52. The "Slip watter-bottle." See Thomson, *Voyage*, vol. 1, pp. 34–37; Tizard et al., *Narrative*, vol. 1, pt. 1, pp. 111–112. That this bottom water was potable is highly unlikely.

53. The current drag; *Narrative*, vol. 1, pt. 1, pp. 79–82.

54. Artemus Ward, pen name of Charles Farrer Browne (1834–1867), American humorist, who became very popular in England during a visit there in 1866. Ward's career began in the printing trade.

55. See below. The promised sequel was never delivered, so far as is known. A copy of the lecture was sent to his mother, who was to circulate it through the family.

56. On these shark experiences, Moseley's *Notes* (pp. 8–11) is the most detailed account.

57. Probably the fish *Stomias affinis*, one of seventy-eight different species from Station 23 (*Challenger Reports, Zoology*, vol. 22, p. 205 & plate LIV; Tizard et al., *Narrative*, vol. 1, pt. 1, pp. 161–165).

58. Swann letterbook, pp. 50–61, at sea, 27 February–16 March 1873, M9.

59. A family friend whose son, Edward, Matkin would visit in Australia. See Chapter 4.

60. Lieutenant C. W. Baillie, inventor of the Baillie sounding machine.

61. Date should be 1869, the year in which H.M.S. *Porcupine* was first utilized for dredging.

62. William B. Carpenter (1813–1885), F.R.S., physiologist and administrator, University of London. A key promoter of the *Challenger* expedition and participant in earlier explorations around the British isles.

63. W. B. Carpenter, J. G. Jeffreys, and Wy. Thomson, "Preliminary Report of the Scientific Explorations of the Deep Sea in H.M.S. *Porcupine* during the Summer of 1869," *Proceedings of the Royal Society* 18 (1870): 397–492.

64. The greatest ocean depths now known are in trenches such as the Challenger Deep, which is nearly 6,000 fathoms. The average depth of the oceans, however, is about 2,000 fathoms (Alastair Couper, ed., *The Times Atlas of the Oceans* [London: Times Books, Ltd., 1983], p. 27).

65. The "clustered sea-polype" *Umbellularia groenlandica* (Thomson, *Voyage*, vol. 1, pp. 149–151; renamed *Umbellula thomsoni* in *Summary of Results*, pt. 1, p. 123).

66. Captain J. F. L. P. Maclear, second in command, examined the phosphorescent light emitted with an electroscope (Thomson, *Voyage*, vol. 1, p. 151).

67. Probably Thomson's *Euplectella suberea*, a hexactinellid sponge (*Voyage*, vol. 1, pp. 138–141).

68. Probably the crustacean *Willemoesia leptodactyle*, named after Willemoes-Suhm and illustrated in Thomson, *Voyage*, vol. 1, p. 189. This would date Thomson's lecture to 4 March 1873 (Thomson, *Voyage*, vol. 1, p. 187).

69. JM to Sarah Craxford Matkin, at sea, 3–16 March 1873, SIO, M10. Another transcription of the Thomson lecture appears in the Swann letter-book, pp. 832–845.

Chapter 2

1. The passenger ship *Northfleet*, carrying about four hundred emigrants bound for Hobart, Tasmania, was anchored two miles off Dungeness, Kent, on the night of 22 January 1873, when it was struck amidships by an unknown steamer going full speed. The steamer left the scene immediately, and the *Northfleet* sank forty-five minutes later in eleven fathoms of water with a loss of more than three hundred lives, despite the proxmity of many other ships at anchor. (*Annual Register* 1873, pt. 2, pp. 9–15).

2. *Good Words, A Weekly Magazine*, Edinburgh and London, 1860–1906. Wyville Thomson composed a total of fifteen "Letters from H.M.S. *Challenger*" for this popular journal. Matkin later wrote that the crew did not always appreciate Thomson's rendering of events (see Chapter 4, n. 17).

These *Challenger* articles have not been enumerated elsewhere; the complete listing is as follows:

Good Words 14 (1873):
"I.   Bermuda, April 7, 1873," pp. 394–397.
"II.   [Lisbon to Gibraltar]," pp. 504–510.

"III.   Gibraltar to the West of Teneriffe," pp. 854–858.
Good Words 15 (1874):
"I.   From the West of Teneriffe to St. Thomas," pp. 45–48.
"II.   Bermudas," pp. 94–103.
"III.   Bermudas (continued)," pp. 157–165.
"IV.   Azores," pp. 381–387.
"V.   The Azores (continued)," pp. 494–499.
"VI.   St. Paul's Rocks," pp. 552–561.
"VII.   Tristan d'Acunha," pp. 618–627.
"VIII. Tristan d'Acunha (concluded)," pp. 671–674.
"IX.   Kerguelen Island," pp. 743–751.
"X.   Kerguelen Island (concluded)," pp. 814–821.
Good Words 16 (1875):
"The Antarctic," pp. 489–493.
Good Words 17 (1876):
"A Morning Ride," [Bahia, Brazil], pp. 452–454.

3. Campbell (*Log-letters*, p. 17) also commented on the conditions of the Negroes of St. Thomas.

4. The *Varuna* of Liverpool, according to Spry (*Cruise*, p. 61) and Tizard et al. (*Narrative*, vol. 1, pt. 1, p. 130).

5. Probably the dolphinfish, *Coryphaena hippurus*.

6. The ship's writer, Richard Wyatt, writer 3c.

7. I.e., stroke.

8. Exeter.

9. Lieutenant Pelham Aldrich.

10. Swann letterbook, pp. 62–76, Bermuda Dockyard, 7–8 April 1873, MII.

11. The steamer *Atlantic*, one of the original five vessels of the White Star Line, ran aground and broke up in a fierce storm near Halifax on 1 April 1873. Of the 931 persons aboard, 560 died, mostly emigrants. (Charles Hocking, *Dictionary of Disasters at Sea During the Age of Steam: Including Sailing Ships and Ships of War Lost in Action 1824–1962*, 2 vols. [London: Lloyd's Register of Shipping, 1969], vol. 1, p. 58).

12. One of the earliest public aquaria, the Brighton aquarium had just opened in August 1872. See W. Saville Kent, "The Brighton Aquarium," *Nature* 8 (1873): 531–533.

13. The first of numerous references to the difficulties the expedition encountered with sailors "jumping ship." Before the voyage was completed, 60 crew members had deserted according to Matkin's count. Eleven crewmen had died, 29 were invalided, discharged, imprisoned or exchanged, leaving only 144 members of the original 243 on board on the day of paying off (Swann letterbook, p. 797). See Appendix F.

14. No indication in Thomson of what this might be. Campbell mentioned a "new worm . . . the size of a small eel" (*Log-letters*, p. 22); and the *Summary of Results* (pt. 1, p. 214) reported two specimens of the annelid *Harmonthoe benthaliana* were obtained (Station 45).

15. See n. 11.

16. Including, according to Spry, numerous members of the Halifax Institute of Natural Science (*Cruise*, p. 74). See Eric L. Mills, "The *Challenger* Expedition: How it Started and What It Did," in Mills, ed., *One Hundred Years of Oceanography: Essays Commemorating the Visit of H.M.S. "Challenger" to Halifax, May 9–16, 1873* pp. 1–23 (Halifax: Dalhouse University in cooperation with The Nova Scotia Museum, 1975); idem, "H.M.S. *Challenger*, Halifax, and the Reverend Dr. Honeyman," *Dalhousie Review* 53 (1973): 529–545.

17. The battleship H.M.S. *Royal Alfred*, the last major wooden ship to be built at Portsmouth, was launched in 1864 and served as flagship on the North American station from 1867 to 1874. See Conrad Dixon, *Ships of the Victorian Navy* (Southampton: Ashford Press Publishing in association with the Society for Nautical Research, 1987), p. 86.

18. The Royal Navy was organized into a series of squadrons, each assigned to a geographic area or "station" of concern to Britain, e.g., the Mediterranean Station, the North American Station, the East Indies Station. See William Laird Clowes, *The Royal Navy: A History from the Earliest Times to the Death of Queen Victoria*, 7 vols. (London: Sampson Low, 1903), vol. 7.

19. "Roger," here and in subsequent letters, refers to the famous fraud trial of Arthur Orton, the so-called Tichborne Claimant. Orton (1834–1898), actually the son of a London butcher and formerly a resident of Wagga Wagga, Australia, surfaced in 1865, claiming to be Sir Roger Charles Tichborne, eldest son of James Francis Doughty-Tichborne, tenth baronet, and heir to the Tichborne estates. Sir Roger was believed by his family to have been lost at sea off the coast of Brazil in 1854, but his mother became convinced that Orton was her lost son. The case came to trial in 1871, and it took a jury more than a hundred days to judge Orton an imposter. Subsequently, in the trial Matkin refers to (lasting from September 1873 until February 1874), Orton was found guilty of perjury and sentenced to fourteen years in prison. This was the "media event" of 1873–1874 and one of the longest and most expensive trials in British history. See James Beresford Atlay, *The Tichborne Case*, Notable British Trials Series (London: W. Hodge, 1912); Douglas Woodruff, *The Tichborne Claimant: A Victorian Mystery* (New York: Farrar, Strauss and Cudahy, 1957).

Curiously, the person believed to be the last to see Sir Roger alive was Captain Oakes of the *Northfleet* (n. 1). Because Oakes had been subpoenaed to testify at the Tichborne trial, he did not sail on the *Northfleet*'s last, disastrous voyage (*Annual Register* 1873, pt. 2, pp. 9–15).

20. The "Yankee professor" may have been Alexander Agassiz (1835–1910), the wealthy American naturalist and oceanographer (Thomson, *Voyage*, vol. 1, pp. 387–388), although there is no indication in G. R. Agassiz' *Letters and Recollections of Alexander Agassiz with a Sketch of His Life and Work* (Boston/New York: Houghton Mifflin Co., 1913) that he was aboard during this leg of the expedition. Contemporary Halifax newspapers give no indication, except that Agassiz "returned to the United States" from Halifax after visiting the ship; I thank Eric Mills for this information.

21. Swann letterbook, pp. 77–93, At sea, 7 May–Halifax, 16 May 1873, M12.

22. Matkin's figures here do not agree with those listed in the *Summary of Results:* 27 May 1873—2,650 fathoms; 28 May 1873—2,500 fathoms.

23. For this day, Thomson says simply, "We took one or two hauls of the dredge during the day with but little success" (*Voyage*, vol. 1, p. 404).

24. "Irresistable" in the concurrent letter to Swann.

25. Captain Charles Clerke (1741–1779), captain of H.M.S. *Discovery*, assumed command of the expedition after Cook's death in 1779; he died of consumption later the same year.

26. JM to Charles Matkin, At sea, 26 May–Bermuda, 2 June 1873, SIO, M13.

27. See above, n. 20.

28. An idiom for inebriation. More commonly "three sheets to the wind."

29. Swann letterbook, pp. 94–108, At sea, 27 May–Bermuda, 5 June 1873, M14.

30. "Liebig's extract of beef [or meat]," a nutritional supplement developed by the renowned German organic chemist Justus von Liebig. It enjoyed an "extraordinary and rapid rise into popularity," some consumers regarding it as a substitute, not just a supplement, for meat. See W. A. Shenstone, *Justus von Liebig: His Life and Work (1808–1873)* (London/Paris/Melbourne: Cassell and Co., Ltd., 1895), pp. 163–164; Christopher Lloyd and Jack L. S. Coulter, *Medicine and the Navy, 1200–1900*, 4 vols. (Edinburgh/London: E. and S. Livingstone, 1957–1963), vol. 4, pp. 101–102.

31. Apparently there was at one point a plan to return to England for repairs to the ship's fresh-water system.

32. No mention in Thomson, Campbell, Spry, Moseley, Linklater. For June 19, Thomson (*Voyage*, vol. 2, pp. 11–12) mentions only grey mud and Foraminifera at 2,750 fathoms.

33. These figures agree closely with those given by Thomson (*Voyage*, vol. 2, p. 8), with the exception of that for 20 June, which Thomson gives as 2,750.

34. The phosphorescent tunicate *Pyrosoma spinosum*, a colony of thousands of individual ascidians (Campbell, *Log-letters*, pp. 23–24; Thomson, *Voyage*, vol. 2, pp. 85–86; *Summary of Results*, pt. 1, p. 258).

35. The schizopod *Gnathophausia gigas* (*Summary of Results*, pt. 1, p. 258).

36. Possibly A. D. Bache's *Tides of the Gulf Stream* or Rosser's *Notes on the Physical Geography of the North and South Atlantic;* both were included in the "scientifics' " library (see Appendix E).

37. No record of a second lecture survives.

38. Perhaps "asteroids" (i.e., starfish).

39. This colorful trek is well described by Campbell (*Log-letters*, pp. 28–32) and by Thomson (*Voyage*, vol. 2, pp. 28–47).

40. *Flabellum* and *Ceratotrochus* (Thomson, *Voyage*, vol. 2, pp. 50–56).

41. JM to Sarah Craxford Matkin, Mid Atlantic, 19 June—Madeira, 16 July 1873, SIO, M15.

42. Swann letterbook, pp. 109–136, Mid Atlantic, 21 June–Madeira, 16 July 1873, M16 (quote from pp. 121–122). Arthur Orton, the Tichborne Claimant, still had some months remaining in the limelight; he would not be sentenced until 28 February 1874.

43. I.e., a travelling salesman.

44. Possibly a greeting to relatives living upstairs.

45. JM to Sarah Craxford Matkin, Funchal Roads, 16 July 1873, SIO, M17.

46. Subsequent letters suggest that Joe Pepper was a relative of Swann and a medical student.

47. The much publicized visit of the Shah of Persia to Britain, June–July, 1873. He was received and knighted by Queen Victoria at Windsor (*Annual Register* 1873, pt. 2, pp. 58–66).

48. The continuing saga of the trial of Arthur Orton, impersonator of Sir Roger Tichborne.

49. List 5, Quarterly Ledger for *Challenger*, 15 November 1872–31 December 1873, ADM 117/196, PRO.

50. Probably H.M.S. *Simoom*.

51. In the Ashanti War of 1872–1873 Kofi Karikari, King of the Ashanti people, was pitted against the British led by Sir Garnet (later Viscount) Wolseley. The conflict arose from a dispute over rights to the Elmina district of the Gold Coast. Further comments appear in Matkin's letter from the Cape (see Chapter 3).

52. Swann letterbook, pp. 137–151, Canary Islands, 21 July–St. Vincent, 4 August 1873, M18.

53. JM to [Will Matkin] (incomplete), 4 August 1873, SIO, M19.

54. Africa was still very much the "dark continent" in the 1870s, in great measure because of the prevalence of diseases lethal to Europeans. Henry M. Stanley's famous encounter with David Livingston had occurred just eighteen months earlier, and his expedition to trace the Congo River to its mouth would not set off for another year.

55. Thomson attributed "this perfect blaze of phosphorescence" to the ascidian *Pyrosoma* of the surrounding Guinea current (*Voyage*, vol. 2, pp. 84–89).

56. Spry expanded upon this point: "The disagreeable practice of shaving, &c., those who for the first time 'cross the line' was not permitted, although there were many who were anxious to join in the usual sport. This old-fashioned custom, which the present age seems inclined to get rid of, is gradually falling into disuse, and but few ships' companies now pay that homage on entering Neptune's dominions as they were wont to," (*Cruise*, p. 87).

57. Though disappointed, Thomson gave a slightly more sympathetic account of the governor's action (*Voyage*, vol. 2, pp. 114–117).

58. Matkin was apparently ignoring such prominent men as Sir John Richardson, Joseph Hooker, and Thomas Henry Huxley, who had served

with distinction as surgeon-naturalists during the nineteenth century. On the quality of Scots naval surgeons, see Lloyd and Coulter, *Medicine and the Navy*, vol. 4, pp. 2–4, 81–82.

59. Thomson was less in doubt about the schoolmaster's fate: "It turned out, however, that the poor fellow had not been murdered or robbed at all. His body was found a week or two after we left, lying, dried up with the scorching heat, on a ledge near Wellington Peak; he had wandered too far and had been overcome by heat and fatigue and unable to return—very probably he had had a sun-stroke. His purse and watch were intact; even the vultures had failed to discover him, he had gone too far beyond the ring around the town where they chiefly find their food." (*Voyage*, vol. 2, p. 73).

60. Swann letterbook, pp. 152–174, Sierra Leone, 16 August–Bahia, Brazil, 21 September 1873, M20.

61. JM to Sarah Craxford Matkin, [Tristan da Cunha, 15 October?]–Simon's Bay, 4 November 1873, SIO, M21.

62. Not the Caribbean Trinidad, but the uninhabited island of Trinidade east of Rio de Janeiro and belonging to Brazil.

63. Probably H.M.S. *Sutlej*.

64. Other islands, including Hawaii and Pitcairn, have a greater claim on this distinction.

65. The distance is actually 2,500 miles.

66. Matkin's account of the Stoltonhoff saga differs in some details from the more extended version given by Spry (*Cruise*, pp. 96–110), which was based on a dictation taken by Paymaster Richard R. A. Richards. See also Thomson, *Voyage*, vol. 2, pp. 169–177. In his history *Tristan da Cunha, 1506–1902* (London: Allen and Unwin, 1940), J. Brander demonstrates that Britain's desire to annex the island was stimulated by *Challenger*'s visit.

67. Swann letterbook, pp. 175–193, Tristan d'Acunha, 15 October–150 miles from the Cape, 27 October 1873, M22. Dickens' works, no doubt from the crew's library. The author had died just three years earlier.

Chapter 3

1. Sir John Edmund Commerell (1829–1901), then commander in chief on the African west coast, survived this injury and went on to become Admiral of the Fleet.

2. This paragraph suggests that Swann was preparing for the ministry.

3. I.e., the trial of the Tichborne Claimant; see Chapter 2.

4. Paray-le Monial, a village in east central France, near the upper reaches of the Loire.

5. Swann letterbook, pp. 175–201, Tristan d'Acunha, 15 October–Simon's Bay, 3 November 1873, M22.

6. Matkin's own twentieth birthday.

7. In 1873 the Dutch had begun a very costly war for supremacy in Sumatra that would continue for the next thirty years.

8. JM to [Sarah Craxford Matkin], [Tristan da Cunha, 15 October?]–Simon's Bay, 4 November 1873 (fragment), SIO, M21.

9. The first mention of the logbook Matkin kept throughout the voyage. Each of the three volumes would be shipped home when full.

10. This was the long-awaited pay raise; see Introduction, n. 29. The estimate of savings turned out to be wishful thinking. Upon return to England he was paid off £4.10.5.

11. I.e., Kaffirs, Bantu-speaking tribes of eastern South Africa.

12. Navigating Sublieutenant Arthur Havergall.

13. JM to Charles Matkin, Simon's Bay, 5–15 November 1873, SIO, M23.

14. The River Welland flows northeast from Leicestershire to the North Sea, passing through Barrowden and Stamford. Historically it formed the border between Rutland and Northamptonshire. See Charles Phythian-Adams, "The Emergence of Rutland and the Making of the Realm," *Rutland Record* 1 (1980): 5–12.

15. The British had established an astronomical observatory for the Southern Hemisphere at Cape Town in 1829. Frank Maclear was among its early directors.

16. Preparations were afoot to observe, from widely dispersed points on earth, the transit of Venus across the face of the sun on 9 December 1874—the same astronomical phenomenon that had first taken Cook to the Pacific and Tahiti a century earlier.

17. Swann letterbook, pp. 202–218, Table Bay, 4 December–Simon's Bay, 15 December 1873, M24.

18. JM to Sarah Craxford Matkin, Table Bay, 10 December–Simon's Bay, 15 December 1873, SIO, M25.

19. I.e., his new sister-in-law.

20. JM to Charles Matkin, Simon's Bay, 15 December 1873, SIO, M26.

21. JM to Sarah Craxford Matkin, Prince Edward's Island, 25 December 1873–Melbourne, 18 Mar 1874, SIO, M27. Matkin also mentions writing a third letter, to an unnamed party in Swadlincote, probably his aunt and uncle.

22. Ibid.

23. James Weddell (1787–1834) aboard the brig *Jane of Leith;* the year was 1823. Captain Sir James Clark Ross (1800–1862).

24. The Southern Elephant Seal, largest member of the seal family.

25. George Anson, *Voyage Round the World in the Years MDCCXL, I, II, III, IV* (London: J. & P. Knapton, 1748).

26. JM to Sarah Craxford Matkin, 25 December 1873–18 March 1874, SIO, M27.

27. Ibid.

28. *Bathydraco antarcticus.* See *Summary of Results*, pt. I, p. 495.

29. Matthew Fontaine Maury, *Physical Geography of the Sea and its Meteorology* (1st ed., New York: Harper and Brothers, 1855).

30. JM to Sarah Craxford Matkin, 25 December 1873–18 March 1874, SIO, M27.

31. *Challenger* was the first steamer to cross the Antarctic Circle.

32. *Challenger* reached no farther south than the 66°40′ S. latitude achieved that day.

33. Lieutenant Charles Wilkes (1798–1877), who led the United States Exploring Expedition to the Pacific and Antarctic, 1838–42, and authored the *Narrative* of the expedition (1844), reported sighting a land mass from a distance of sixty miles. Dubbed "Termination Land," it was not seen again. See Tizard et al., *Narrative*, vol. 1, pt. 1, p. 407; Wild, *At Anchor*, p. 77.

34. The *Astrolabe* and the *Zelée* under Jules Dumont d'Urville.

35. The *Vostok* and the *Mirni* under the command of Fabian von Bellingshausen, 1819–1821, reached 69°52′ S.

36. According to the *Summary of Results* (pt. 1, p. 501), the dredge came up empty on this date; an abundant haul was made two days later.

37. In his popularization of this leg of the voyage for the magazine *Good Words* (16 [1875]: 489–493, at 492), Wyville Thomson abbreviated this incident to a single sentence: "We tried to get under the lee of an iceberg, but while reefing, an eddy caught the ship and dragged her towards the berg, which she fouled, carrying away her jib-boom." Spry (*Cruise*, p. 143) mistakenly dates these events to 26 February.

38. Thomson again condensed this near tragedy to "When the gale was at its height we saw the loom of an iceberg on the lee-bow, and we were drifting directly upon it. As there was no time to steam ahead, Captain Nares went full speed astern with the four boilers, took in the fore try-sail and set the weather after-clue of the reefed main topsail aback, and under this sail the ship fortunately gathered stern-way, keeping broadside to the wind, and we drifted past the berg" (*Good Words* 16 [1875], 492.)

39. The trawl on this date yielded "one of the most interesting hauls made during the cruise." "100 specimens of invertebrates and fishes . . . belonging to about 37 species, of which 32 are new to science," (*Summary of Results*, pt. 1, pp. 502, 504).

40. Since 1 knot = 1 nautical mile per hour, Matkin's "knots per hours" is redundant—a strange mistake for a seasoned sailor. Curiously, Wyville Thomson made the same error ("The Antarctic," *Good Words* 16 [1875]: 493).

41. Matkin meant 334 nautical miles per day.

42. Four species of fish were obtained, three of them new (*Summary of Results*, pt. 1, pp. 510).

43. Four species of fish were again brought up.

44. Swann letterbook, pp. 219–274, off Prince Edward Island, 25 December 1873—off King's Island, 16 Mar 1874, M28.

45. Achille François Bazaine (1811–1888) was marshal of France's Army of the Rhine during the Franco-Prussian War of 1870. Having failed to engage the German army aggressively at Metz, Bazaine was tried before a military court in 1873 and found guilty of negotiating with and prematurely capitu-

lating to the enemy. An initial sentence of degradation and death was com-
muted to twenty years seclusion.

46. JM to Sarah Craxford Matkin, 25 December 1873–18 March 1874, SIO,
M27.

Chapter 4

1. Swann letterbook, pp. 324–325, Tongatabu, 23 July–Kandavu,
7 August 1874, M38.

2. The wedding of his eldest brother, Charles.

3. JM alludes to the fact that Charles, as eldest son, will inherit the family printing business, while the three younger sons must seek their own
fortunes.

4. "Custom," i.e., more business patronage, customers.

5. JM to Sarah Craxford Matkin, Melbourne, 23–25 March 1874, SIO,
M29.

6. Cricket had been played in Australia since the early nineteenth century. The first English team visited in 1861–1862. Australia won the first of
the famous test matches in 1877.

7. Swann letterbook, pp. 275–278, Hobson's Bay, 25 March 1874, M30.

8. Charles Matkin [Sr.] to Joseph Matkin, Oakham, 29 May 1984, SIO.

9. Gold was discovered near Ballarat in 1851, and the city soon became
one of Australia's leading mining towns. The "Eureka stockade" incident,
in which forty men were killed, occurred there in 1854—one of the rare
occasions in Australian mining history when lives were lost. Good quality
ore was mined in the region for decades, making Ballarat among the largest inland cities in Australia.

10. Swann letterbook, pp. 279–285, Sydney, 10 April 1874, M31.

11. New Caledonia was annexed by France in 1853; a penal colony was
maintained there until 1897.

12. This was indeed only a rumor. The protectorate of British New
Guinea was not proclaimed until 1884. Four years later it became a depen-
dency of Queensland and in 1906 was renamed the Territory of Papua.
If there was a plan to transport "a few natives of New Guinea" to
England, it was not carried out.

13. Issues of the *Stamford Mercury* newspaper.

14. JM to Fred Matkin, Sydney, 13–15 April 1874, NHM, M32.

15. This is the only mention of such plans, none of which would come
to pass.

16. See pp. 12–13 above.

17. See Chapter 2, n. 2. It is not clear just what the crew's objections
might have been, unless it was Thomson's emphasis. He devoted nearly
all his "good words" to heavy descriptions of the natural history of the
oceans (often surprisingly technical for a popular weekly) and of places
visited (many of which the crew would not have had the resources to
visit). He rarely alluded to events among the *Challenger* "community";
the crew was all but invisible in his accounts.

18. JM to Sarah Craxford Matkin, Sydney, 7–9 May 1874, SIO, M33.

19. Matkin had apparently received a photograph of his new sister-in-law.

20. Paris Communards. See n. 23.

21. *Challenger* actually departed Sydney ten days later, on 8 June.

22. JM to Charles Matkin, Sydney, 15 May 1874, SIO, M34.

23. Henri Rochefort (Marquis de Rochefort-Lucay, 1830–1913), the radical French politician and journalist. Briefly a member of the government that followed the revolution of 1870, Rochefort sympathized with the Paris Communards and was subsequently court-martialed and banished to New Caledonia. Escaping to San Francisco in 1874, he returned to France under the general amnesty of 1880.

24. Swann letterbook, pp. 286–292, Sydney, 15 May 1874, M35.

25. Probably "Admiralty Bay."

26. The iron steamer *British Admiral*, on its first voyage, Liverpool to Australia, struck a reef off Kings Island, Bass Strait, on 23 May 1874. Nearly all the eighty passengers and crew were lost (*Annual Register* 1874, pt. 2, p. 79).

27. Swann letterbook, pp. 293–309, South Pacific, 21 June–1 July 1874, M36.

28. Reached on 17 July 1874 at Station 171a between Raoul and Tonga (Tizard et al., *Narrative*, vol. 1, pt. 1, p. 476).

Challenger's Lieutenant Pelham Aldrich would return to the Tonga region in 1888–1889 in command of H.M.S. *Egeria*, at which time depths of more than 4,000 fathoms would be reached. I thank Dr. Robert L. Fisher for bringing this to my attention.

29. Now Eua.

30. Spry, *Cruise*, p. 185, mentions "chinam" as a stain for hair, giving it a "reddish tinge." Linklater, *Voyage*, p. 108, refers to the application of "coral lime."

31. The Wesleyan Missionary Society was founded in 1814; the first missionaries arrived in Tonga in 1826.

32. Swann letterbook, p. 325, Tongat[a]bu, Friendly Islands, 23 July 1874, M38.

33. Ibid.

34. Ibid.

35. Ibid., p. 326.

36. Vanua Levu and Viti Levu.

37. Swann letterbook, p. 328, At sea, 2 August 1874, M38.

38. Matkin was confused here: Viti Levu has nearly twice the area of Vanua Levu; the Rewa, the principal river of Viti Levu, was explored by the party of officers and scientifics (Linklater, *Voyage*, p. 115).

39. Fiji did in fact become a colony of Great Britain in September 1874 (*Annual Register*, 1874, pt. 1, p. 129).

40. Swann letterbook, p. 332, At sea, 2 August 1874, M38.

41. Ibid., pp. 333–335.

42. Ibid. pp. 336–337.

43. John Coleridge Patteson (1827–1871) became the first bishop of Melanesia in 1861; he was killed by natives of Nukapu, Santa Cruz Islands, 20 September 1871. See Charlotte Yonge, *Life of John Coleridge Patteson, Missionary, Bishop of Melanesia,* 2 vols. (London: Macmillan and Co., 1874).

44. JM to Will Matkin, Tongatabu, Friendly Islands, 23 July—Kandavu, Fiji Islands, 6 August 1874, SIO, M37.

45. A temporary setback only. See n. 39.

46. Current names/spellings where they differ from Matkin's: Espiritu Santo, Epi, Malakula, Ambrym, Pentecost.

47. On August 24, off the Louisiade Archipelago (Spry, *Cruise,* p. 199).

48. Matkin displays his knowledge of what would later be called the Pacific "Ring of Fire." Subsequent geological research would demonstrate, however, that the Sandwich Islands (Hawaii) are not a part of this "system."

49. The predicted development of the village of Somerset did not take place. Established in the 1860s, the settlement was transferred to Thursday Island off the northwest coast of the mainland in 1877. The island had a population of 2,283 in 1981. See C. G. Austin, "Early History of Somerset and Thursday Island," *Journal of the Historical Society of Queensland"* (1949). I thank Leanne Piggott and Roy MacLeod at the University of Sydney for supplying this information.

50. The Scottish religious sect, originally followers of Richard Cameron (1648?–1680), which became the Reformed Presbyterians in 1743. All but a few congregations united with the Free Church of Scotland in 1876.

51. Matkin had evidently sent a photograph of a Tongan woman, probably the one taken by the ship's photographer (see p. 189).

52. Swann letterbook, pp. 339–355, New Hebrides, 18 August—Somerset, Cape York, 7 September 1874, M39.

53. The Aru Islands, at the western edge of the submarine continental shelf between New Guinea and Australia.

54. Probably Pulau Trangan.

55. Now Dobo.

56. Kai or Kei Islands.

57. The correct date was 3 November; either Matkin or the transcriber, Swann, erred.

58. See Chapter 3, n. 45. Marshal Bazaine had been imprisoned on Ile Sainte Marguérite but escaped to Italy in 1874.

59. Swann letterbook, pp. 356–390, Arra Islands, Sea of Arafura, 15 September—Hong Kong, 17 November 1874, M40.

Chapter 5

1. Swann letterbook, p. 557, Kobi, 1 June–Yokohama, 16 June 1875, M54.

2. Swann letterbook, p. 390, Hong Kong, 17 November 1874, M40.

3. The first grandchild in the family, a daughter born to brother Charlie and his wife Harriett.

4. JM to Sarah Craxford Matkin, Hong Kong, 25 November 1874, NHM, M41.

5. "Your letter of July 27th . . . was the first intimation I had concerning the March affair. I need not say anything about it." JM to Charles Matkin, Hong Kong, 22 December 1874, NHM, M44.

6. In fact, Thomson would publish two more articles in 1875 and 1876. See Chapter 2, n. 2.

7. Matkin is surely confused here; the Ladrones (i.e., the Marianas) are more than 2,000 miles from Hong Kong.

8. Matkin's geography is slightly in error: Macao is on the Chinese coast at the entrance to the Canton River.

9. There was no "war" as such "between 1848 and 1859," only continuing friction between the decadent Manchu government and the imperialistic Western powers. The first Anglo-Chinese "Opium War" was fought 1839-1842; a second offensive was conducted by the British and French against the Chinese, 1858-1860.

10. During the Greek war for independence from the Ottoman Empire, Sir Edward Codrington commanded British, French, and Russian vessels against a Turkish and Egyptian fleet (1827). His flagship was called the *Asia* at that time.

11. JM to Charles Matkin, 22 December 1874, NHM, M44.

12. Ibid.

13. Ibid.

14. Thomson must have issued this threat—not recorded elsewhere—in momentary frustration. The chief scientist stayed with the expedition to the end.

15. Matkin is probably referring to the Austro-Hungarian expedition of Payer and Weyprecht aboard the *Tegethoff*, 1872-1874, during which Franz Josef Land was discovered and named after the Austrian emperor.

16. Captain Nares commanded the Arctic expedition of *Alert* and *Discovery* in 1875-1876. *Alert* reached a higher latitude and wintered farther north than any ship had done before. See G. S. Nares, *Narrative of a Voyage to the Polar Sea During 1875-76* (London: Sampson Low, 1878); Geoffrey Hattersley-Smith, "The British Arctic Expedition, 1875-76," *Polar Record* 18 (1976): 117-127; and Margaret Deacon and Ann Savours, "Sir George Strong Nares (1831-1915)," idem, 127-141.

17. Captain Frank Turle Thompson replaced Nares; Lieutenant A. Carpenter replaced Pelham Aldrich.

18. See Chapter 3, n. 16.

19. Swann letterbook, pp. 418-419, Hong Kong, 16 December 1874, M42.

20. JM to Charles Matkin, 22 December 1874, NHM, M44.

21. Ibid.

22. Swann letterbook, p. 424, 16 December 1874, M42.

23. JM to Sarah Craxford Matkin, Hong Kong, 19–23 December 1874, NHM, M43.

24. I.e., when intoxicated.

25. JM to Fred Matkin, Hong Kong, 31 December 1874–January 1875, SIO, M46. Nearly identical is the Swann letterbook, pp. 427–444, Hong Kong, 28 December 1874–1 January 1875, M45.

26. Alphonso XII (1857–1885), first king of all Spain, proclaimed himself sole representative of the monarchy on 1 December 1874.

27. Swann letterbook, pp. 445–448, Humboldt Bay, New Guinea, 24 February 1875, M48.

28. Probably the common sea snake.

29. Presumably water buffalo.

30. The siliceous sponge, *Euplectella aspergillum*. Matkin was correct in its description but misled as to its origin.

31. The Meangis Islands (Tizard et al., *Narrative*, pt. 2, p. 668); now Kepulauan Talaud.

32. Obsidian. See Spry, *Cruise*, p. 271; and Tizard et al., *Narrative*, pt. 2, p. 705.

33. On this passage, the ship passed Maty Island (now Wuvulu) and the Hermit Islands at the western end of the Bismarck Archipelago; "Tiger" and "Bertrand" could not be identified.

34. The Admiralty Islands were discovered by the Dutch in 1616 and visited in 1767 by Carteret, who named them.

35. Admiralty Island, later named Manus, the largest of about 40 islands.

36. Wild Island, named after the ship's artist, J. J. Wild (see *At Anchor*, p. 136).

37. Swann letterbook, p. 476, Admiralty Islands, 10 March 1875, M48.

38. This sounding, 4,475 fathoms reached on 23 March 1875 in the Mariana Trench (11°24′ N, 143°16′ E), was the greatest depth reached during the expedition (Tizard et al., *Narrative*, pt. 2, p. 734).

39. I.e., the Mariana Islands. Magellan gave the name "Ladrone" to these islands in 1521; they were renamed "Las Marianas" after Maria Anna of Austria, widow of Philip IV of Spain, in 1668. Matkin had corrected his earlier confusion (n. 7).

40. JM to Sarah Craxford Matkin, Humboldt Bay, New Guinea, 24 February–Lat. 30° N, Long. 137° E, 8 April 1875, SIO, M47.

41. JM to Sarah Craxford Matkin, Gulf of Yedo, 12 April 1875, NHM, M49.

42. JM to Will Matkin, Gulf of Yedo, 12 April 1875, NHM, M50. This is the first intimation of serious intentions for a future career. In three years' time Matkin would indeed become a government clerk in Whitehall.

43. "The most frightful railway accident which has yet taken place in England" (*Annual Register* 1874, pt. 2, pp. 128–129) occurred at Shipton, near Oxford, Christmas Eve 1874. A Great Western Railway train en route to Birmingham, travelling at thirty-five miles per hour, broke a wheel on

one of its carriages. Several cars derailed and plunged down a twenty-foot embankment, one coming to rest in a canal. Thirty-one people died.

44. En route from England to South America, *La Plata* foundered in the Bay of Biscay during a gale, 29 November 1874. All but 15 of the crew of 77 were lost. The *Cospatrick* was carrying 429 emigrants from England to Auckland when it caught fire at midnight, 17 November 1874. Only 3 of the 430 aboard survived. (*Annual Register* 1874, pt. 2, pp. 119–121, 130–132).

45. Spry (*Cruise*, p. 281) gave a population of 2.5 million.

46. I.e., Yokosuka. There are frequent errors of spelling and of fact in Matkin's account of Japan, but these errors were common among nineteenth-century English travellers. I heartily thank Professor Lane Earns for assistance in sorting out this section.

47. The date should be either 1854, when Admiral Perry opened Japan, or 1859, when the treaty ports were opened to foreign trade and settlement.

48. Matkin may be referring to Chinese trade in the later seventeenth century. No Chinese ships came to Nagasaki before Westerners arrived; the port was founded by the Portuguese and Jesuits in 1570.

49 I.e., "Tycoon" or "Shogun."

50. Francis Xavier did not die a martyr, but of illness, in China; the number of Japanese he converted is greatly exaggerated.

51. Matkin is referring to Toyotomi Hideyoshi (ruled 1582–1598), who was known by the title of *taiko* (retired chancellor) rather than shogun (Edwin O. Reischauer, *Japan: The Story of a Nation*, 4th ed. [New York: McGraw-Hill, 1990], p. 66–67).

52. I.e., Nagasaki. The banning of Christianity was the act of Toyotomi Hideyoshi.

53. Giovanni Battista Sidotti (1667–1714) was the last Catholic missionary to enter Japan during the Tokugawa's ban on Christianity. He arrived in 1708 and died in confinement.

54. The date should be 1609.

55. I.e., Shikoku, Kyushu and Hokkaido. Nippon is the Japanese name for Japan, not for the largest of the islands, Honshu.

56. The date should be 1868.

57. Only the ports of Nagasaki, Kanagawa, and Hakodate were opened to the West in 1859.

58. I.e., Miyako, "the capital," an early name for Kyoto.

59. Sir Harry Smith Parkes (1828–1885), consul at Canton during the British-Chinese hostilities of 1856–1861, was minister to Japan from 1865 to 1882.

60. Swann letterbook, pp. 479–499, Yokosuka, 30 April 1875, M51.

61. JM to Sarah Craxford Matkin, Yokohama, 7–10 May 1875, SIO, M53.

62. I.e., the Daibutsu (Great Buddha) at Kamakura.

63. Kamakura.

64. Edward Vaughn Hyde Kenealy (1819–1880), Irish barrister and MP, was leading counsel for Arthur Orton, the Tichborne Claimant (see Chapter 2, nn. 19, 42). After the verdict went against his client, Kenealy started

up a paper to plead his case and denounce the judges. He was disbarred in 1874 but elected to parliament in 1875. He went on to publish *The Trial at Bar of Sir Roger C. D. Tichborne, Bart., in the court of Queens Bench* . . . *for Perjury*, 8 vols. (London, 1875–1880).

65. The elimination of Vancouver Island from the itinerary changed this schedule.

66. Swann letterbook, pp. 506–534, Yokohama, 4–10 May 1875, M52.

67. I.e., Kyushu from Shikoku.

68. I.e., Shikoku from Honshu.

69. Probably either William F. Gragg, *A Cruise in the U.S. Steam Frigate Mississippi, Wm. C. Nicholson, Captain, to China and Japan* . . . (Boston: Damrell & Moore, 1860); or James D. Johnston, *China and Japan— Narrative of a Cruise of the U.S. Steam Frigate Powhattan* . . . (Philadelphia: C. Desilver; Baltimore: Cushings and Bailey, 1860). Both ships arrived in Nagasaki in 1858.

70. Probably Nagoya.

71. Swann letterbook, pp. 535–566, Kobi, 1 June 1875, M54.

Chapter 6

1. Spry, *Cruise*, p. 306.

2. Ibid., p. 307.

3. Swann letterbook, pp. 567–568, North Pacific, 13 July 1875, M56.

4. During the American Civil War, the U.S.S. *Kearsarge* sank the Confederate warship *Alabama* off Cherbourg, 19 June 1864. The *Alabama* and other Confederate vessels built in Britain became the subject of international arbitration in 1871–1872, resulting in the adoption of the Treaty of Washington, which specified rules governing the activities of neutral nations toward the navies of belligerent nations. Britain was required to pay the United States $15,500,000 in damages. Matkin would likely have been interested in this case, as the judgment was reached just a few months before *Challenger* left England.

5. The Japan Trench, the second of the Pacific trenches sounded by *Challenger*.

6. Swann letterbook, p. 575, 13 July 1875 M56.

7. He must have meant 1876.

8. Swann letterbook, p. 575, 13 July 1875 M56.

9. But not for another twenty-three years, in 1898, although King David Kalakaua signed a reciprocity treaty with the United States in 1876.

10. The figure 6,000 square miles would have been appropriate for the "Big Island" of Hawaii, which Matkin listed (below) as 4,000 square miles.

11. Possibly John Chawner Williams (1818–?), son of John Williams (1796–1839), a London missionary to the South Seas, see Gavan Daws, *A Dream of Islands: Voyages of Self-discovery in the South Seas* (New York/ London: W. W. Norton & Co., 1980).

12. Matkin's meaning here is unclear, unless he is referring to Mormon polygamy and emphasis on large families. Mormon missionaries had, in fact, arrived in Hawaii by 1850; see R. Lanier Britsch, "The Lanai Colony: A Hawaiian Extension of the Mormon Colonial Idea," *Hawaiian Journal of History* 12 (1978): 68–83.

13. Swann letterbook, p. 584, 24 July 1875, M56.

14. Widow of the admiral and Arctic explorer whose expedition in search of the Northwest Passage was lost in 1847. Jane Lady Franklin (1792–1875) organized the search party that in 1859 finally determined the plight of her husband's expedition. She travelled widely in later years, including the Pacific.

15. Kawaiahao (Congregational) Church.

16. Swann letterbook, p. 587, 4 August 1875, M56.

17. Ibid. 594–595.

18. Ibid., p. 596.

19. King David Kalakaua (1836–1891); r. 1874–1891.

20. Presumably another reference to the imposter of Sir Roger Tichborne, Arthur Orton.

21. Swann letterbook, p. 599, 4 August 1875, M56.

22. JM to Sarah Craxford Matkin, North Pacific, 13 July–Honolulu, 4 August 1875, SIO, M55.

23. Swann letterbook, pp. 599–600, 10 August 1875, M56.

24. "The tortoises were fed a good deal on pine-apples, a number of which were hung up in the Paymaster's office. The animals used to prop themselves up against a board put across the door of the office to keep out dogs, unable to surmount the obstacle, and used to stare and sniff longingly at the fruit. They also learned to know their way along the deck to the Captain's cabin, where there was another store of Pine-apples, and where they were often fed" (Moseley, *Notes,* p. 595).

25. Matkin's reference here and later (Chapter 7) is to Thomas Cochrane (1775–1860), the controversial British admiral who had commanded the naval forces of Chile (1817–1822) in that country's revolt against Spanish rule and of Brazil (1823–1825) against Portuguese. Like his father the 9th earl, Cochrane was an enthusiast for scientific invention; he was an early promoter of steam and screw propulsion for warships and, in that sense, a benefactor of *Challenger*. His son, Sir Arthur Auckland Leopold Pedro Cochrane (1824–1905), commander in chief, Pacific station, 1873–1875, was admiral of the *Repulse*.

26. Disappointing indeed, for, as Lieutenant Spry pointed out, "The principal object of our visit to Hilo was that opportunities might be afforded to those who desire to visit the celebrated Crater of Kilauea" (*Cruise*, p. 322).

27. A streptococcal infection.

28. Caroline Island, one of the southernmost of the Line Islands of Polynesia, not to be confused with the Carolines of Micronesia.

29. Matkin actually did attend a French class in London three years later (JM to Mary Swift, 12 March 1878, SIO).

30. Prince Alfred (1844-1900), Duke of Edinburgh and Duke of Saxe-Coburg and Gotha, was second son of Victoria and Albert. He commanded H.M.S. *Galatea* during a voyage around the world in 1867-1868.

31. Swann letterbook, p. 639, Tahiti, 3 October 1875, M58.

32. Swann letterbook, p. 645, Juan Fernandez, 15 November 1875, M58.

33. JM to Sarah Craxford Matkin, Hilo, Hawaii, 19 August-Valparaiso, 23 November 1875, SIO, M57.

34. Swann letterbook, p. 648, Valparaiso, 21 November 1875, M58.

Chapter 7

1. Charles Darwin, *Journal of Researches into the Natural History and Geology of the countries visited during the Voyage of H.M.S. "Beagle" Round the World, under the Command of Capt. Fitzroy, R.N.*, new edition (New York: D. Appleton, 1898; 1st ed., 1839), p. 501.

2. I.e., Punta Arenas in the Straits of Magellan.

3. H.M.S. *Vanguard* collided with H.M.S. *Iron Duke*, both iron-clad men-of-war, near Kingston, Ireland, on 1 September 1875. The *Vanguard* sank but without loss of life. *Annual Register* 1875, pt. 2, pp. 79-81.

4. *The British Workman and Friend of the Sons of Toil*, published 1855-1921.

5. See Chapter 6, n. 25; and Thomas Cochrane Dundonald, *Narrative of Services in the Liberation of Chili, Peru and Brazil, from Spanish and Portuguese Domination*, 2 vols. (London: J. Ridgway, 1859).

6. Bernardo O'Higgins (1776-1842) served under José de San Martin in the liberation of Chile and ruled the country during its first five years of independence.

7. The Spanish frigate *Esmeralda*, captured 5 November 1820.

8. April 1809.

9. In fact, Cochrane was reinstated in the British navy in 1832 and commanded the North American and West Indian station, 1848-1851. He died in 1860.

10. A strong pronouncement, especially when placed alongside his earlier comment from Hawaii that Honolulu was "more like England than any town we have been to since leaving Wellington" (M55, see Chapter 6).

11. Swann letterbook, pp. 649-667, Valparaiso, 7 December 1875, M59.

12. JM to [Sarah Craxford Matkin?], Magellan Straits, January 1876 (fragment), NHM, M61.

13. Ibid.

14. For Wild's rendition of this adventure, see *At Anchor*, pp. 185-188.

15. Port Grappeler.

16. The *Karnach*.

17. Little was made of this incident in other accounts. According to the official *Narrative* (vol. 1, pt. 2, p. 865), "Towards evening the sky became overcast and gloomy, and shortly after midnight the wind freshened and

blew in squalls from the northward, so the cable was veered to five shackles, when finding the stern close to a rock awash between Centre and One Tree Islands, steam was got up in three boilers." Spry, the engineer, reported simply: "Steam was, however, at command, and no danger resulted" (*Cruise*, p. 353).

18. Jacob Lemaire and Willem Shouten discovered Cape Horn in 1616.

19. Swann letterbook, pp. 668–691, Messier Channel, Patagonia, 5 January–Sandy Point, 14 January 1876, M60.

20. Swann letterbook, p. 693, Stanley, Falkland Islands, 6 February 1876, M64.

21. See Chapter 3, n. 33.

22. Swann letterbook, p. 699, Stanley, Falkland Islands, 6 February 1876, M64.

23. Original letter damaged here and below.

24. JM to Sarah Craxford Matkin, Stanley, Falkland Islands, 6 February —Montevideo, 16 February 1876, NHM, M63.

25. Swann letterbook, pp. 715–718, Montevideo, 24 February 1876, M64.

26. Presumably a reference to brother Will.

27. The line-of-battleship H.M.S. *Bombay*, Captain Colin Campbell commanding, caught fire during target practice, blew up, and sank off Montevideo, 22 December 1864. Two officers and ninety men were lost. Charles Hocking, *Dictionary of Disasters at Sea During the Age of Steam: Including Sailing Ships and Ships of War Lost in Action 1824–1962*, 2 vols. (London: Lloyd's Register of Shipping, 1969), vol. 1, p. 93.

28. Swann letterbook, p. 727, South Atlantic, 7 March 1876 M65.

29. Ibid.

30. See Chapter 2, n. 30.

31. Swann letterbook, p. 728, South Atlantic, 7 March 1876 M65.

32. Ibid.

33. Very little was said about this historic encounter of the two great oceanographic expeditions of the 1870s; but see Wild, *At Anchor*, pp. 194–195. On the *Gazelle* expedition, see Hans Ulrich Roll, "On the Roots of Oceanography in Germany," in Walter Lenz and Margaret Deacon, eds., *Ocean Sciences: Their History and Relation to Man* (Hamburg: Bundesamt für Seeschiffahrt und Hydrographie, 1990), pp. 3–19.

34. Swann letterbook, p. 741, South Atlantic, 15 March 1876, M65.

35. JM to Fred Matkin, South Atlantic, 16 March–St. Vincent, 20 April 1876, SIO, M66.

36. The sooty tern.

37. Swann letterbook, pp. 741–746, Ascension, 3 April 1876, M65.

38. JM to Fred Matkin, 16 March–20 April 1876, SIO, M66.

39. See above, Chapter 1, n. 18.

40. Swann letterbook, pp. 750–759, Spithead, 25 May 1876, M68.

41. Samuel Johnson, in a letter to James Boswell, 16 March 1759 (Robert Debs Heinl, Jr., *Dictionary of Military and Naval Quotations* [Annapolis: U.S. Naval Institute, 1966], p. 294).

42. Swann letterbook, pp. 760–767, Chatham, 11 June 1876, M69.

Conclusion

1. Sam Noble, *Sam Noble, Able Seaman: 'Tween Decks in the Seventies* (New York: Frederick A. Stokes Co., 1926), p. 104.

2. There is one other candidate, but his name and rank are unknown: the State Library of Victoria in Melbourne holds an original journal of 341 pages (MS 10264) kept during the expedition. Internal evidence suggests, however, that the author was an officer (copy of letter, HE/PC70/CIW, State Library of Victoria, to Eileen V. Brunton, NHM, 4 July 1989).

3. JM to Sarah Craxford Matkin, Humboldt Bay, New Guinea, 24 February–10 March 1875, SIO, M47.

4. Drastic punishment of sailors did not occur, however; flogging had been suspended throughout the peacetime navy in 1871.

5. JM to Sarah Craxford Matkin, 24 February–10 March 1875, SIO, M47.

6. JM to Sarah Craxford Matkin, Table Bay, 10 December–Simon's Bay, 15 December 1873, SIO, M25.

7. W. L. Burn, *The Age of Equipoise: A Study of the Mid-Victorian Generation* (London: George Allen and Unwin, 1964); A. H. Thornton, "Introduction—the Smilesian Philosophy," in Michael D. Stephens and Gordon W. Roderick, eds., *Samuel Smiles and Nineteenth Century Self-Help in Education* (Nottingham: University of Nottingham, 1983), pp. 1–15.

Selected Bibliography

Works about the *Challenger* Expedition

Bailey, Herbert S. "The Background of the 'Challenger' Expedition." *American Scientist*, 60 (1972): 550–560.

Burstyn, Harold L. "Pioneering in Large-Scale Scientific Organisation: The *Challenger* Expedition and Its Report. I. Launching the Expedition." *Proceedings of the Royal Society of Edinburgh*, Section B (Biology), 72 (1971–1972): 47–61.

———. "Science and Government in the Nineteenth Century: The *Challenger* Expedition and Its Report." *Bulletin de l'Institut Oceanographique*, no. special 2 (1968): 603–613.

Campbell, Lord George. *Log-letters from the "Challenger."* London: Macmillan and Co., 1876.

Henderson, J. Welles, and Harris B. Stewart. "A Recently Discovered *Challenger* Sketchbook." *Proceedings of the Royal Society of Edinburgh*, Section B, 72 (1971–1972): 223–229.

Leyten, L. "Sir Charles Wyville Thomson (1820–1882) and Sir James Murray (1841–1914)—The *Challenger* Expedition." In R. Harré, *Some Nineteenth-Century British Scientists*, 1–30. Oxford: Pergamon Press, 1969.

Linklater, Eric. *The Voyage of the Challenger.* London: John Murray, 1972.

Maclean, G. "Appendix IV. Report on the Health of the Crew of H.M.S. *Challenger*, during the Years 1873–76." In T. H. Tizard, H. N. Moseley, J. Y. Buchanan, and John Murray, *Narrative of the Cruise of H.M.S. "Challenger"* . . ., vol. 2 (1), pp. 1027–1031. London: HMSO: 1885.

Merriman, Daniel. "Challengers of Neptune: the 'Philosophers.' " *Proceedings of the Royal Society of Edinburgh*, Section B, 72 (1971–1972): 15–45.

Micheli, Elisabeth Ross. "The British Empire Lays Claim to the Undersea World: The *Challenger* Expedition, 1872–1876." M.Phil. thesis, Cambridge University, 1988.

Moseley, Henry Nottidge. *Notes by a Naturalist on the "Challenger."* London: Macmillan and Co., 1879.

Müller, Gerhard H., ed. *Die Challenger-Expedition zum tiefsten Punkt der Weltmeere, 1872–1876.* Rudolf von Willemoes-Suhms Briefe von der Challenger-Expedition mit Auszügen aus dem Reisebericht des Schiffsingenieurs W. J. J. Spry. Stuttgart: Thieneman, 1984.

Rehbock, Philip F. "Huxley, Haeckel and the Oceanographers: the Case of *Bathybius haeckelii.*" *Isis* 66 (1975): 504–533.

Report on the Scientific Results of the Voyage of H.M.S. "Challenger" during the Years 1873–76. 50 vols. London: H.M.S.O., 1880–1895.

Rice, A. L. "Oceanographic Fame—and Fortune: The Salaries of the Sailors and Scientists on HMS *Challenger.*" *Archives of Natural History* 16 (1989): 213–220.

Spry, W. J. J. *The Cruise of Her Majesty's Ship "Challenger".* London: Sampson Low, Marston, Searle, and Rivington, 1876.

Swire, Herbert. *The Voyage of the Challenger.* 2 vols. London: Golden Cockerel Press, 1938.

Thomson, Sir Charles Wyville. *The Voyage of the "Challenger": The Atlantic.* 2 vols. London: Macmillan and Co., 1877.

Wild, John James. *At Anchor: A Narrative of Experiences Afloat and Ashore During the Voyage of H.M.S. "Challenger" from 1872 to 1876.* London: Marcus Ward, 1878.

Willemoes-Suhm, Rudolf von. "Briefe an C. Th.E. von Siebold von R. von Willemoes-Suhm." *Zeitschrift für Wissenschaftliche Zoologie* (Leipzig, 1874–1877).

Yonge, Sir Maurice. "The Inception and Significance of the *Challenger* Expedition." *Proceedings of the Royal Society of Edinburgh,* Section B, 72 (1971–1972): 1–13.

Other Sources

Baynham, Henry. *Before the Mast: Naval Ratings of the Nineteenth Century.* London: Hutchinson, 1971.

Clowes, William Laird. *The Royal Navy: A History from the Earliest Times to the Death of Queen Victoria.* 7 vols. London: Sampson Low, 1903.

Deacon, Margaret. *The History of Oceanography: An Annotated Bibliography.* New York: Garland Publishing, forthcoming.

———. *Scientists and the Sea, 1650–1900. A Study of Marine Science.* London: Academic Press, 1971.

Dixon, Conrad. *Ships of the Victorian Navy.* Southampton: Ashford Press Publishing, 1987.

Halsband, Robert. "Editing the Letters of Letter-Writers." *Studies in Bibliography* 11 (1958): 25–37.

Herdman, William A. *Founders of Oceanography and Their Work.* London: Edward Arnold & Co., 1923.

Hocking, Charles. *Dictionary of Disasters at Sea during the Age of Steam: Including Sailing Ships and Ships of War Lost in Action 1824–1962.* 2 vols. London: Lloyd's Register of Shipping, 1969.

Kemp, Peter, ed. *The Oxford Companion to Ships and the Sea.* London/ New York/Melbourne: Oxford University Press, 1976.

Kline, Mary-Jo. *A Guide to Documentary Editing.* Baltimore/ London: The Johns Hopkins University Press, 1987.

Laffin, John. *Jack Tar: The Story of the British Sailor.* London: Cassell, 1969.

Layton, David. *Science for the People: The Origins of the School Science Curriculum in England.* New York: Science History Publications, 1973.

Lenz, Walter, and Margaret Deacon, eds. *Ocean Sciences: Their History and Relation to Man.* Proceedings of the Fourth International Congress on the History of Oceanography, Hamburg, 23–29 September 1987. Hamburg: Bundesamt für Seeschiffahrt und Hydrographie, 1990.

Lewis, Michael. *The Navy in Transition 1814–1864: A Social History.* London: Hodder and Stoughton, 1965.

Lewis, Wilmarth S. "Editing Familiar Letters," In J. A. Dainard, ed., *Editing Correspondence: Papers Given at the Fourteenth Annual Conference on Editorial Problems, University of Toronto, 3–4 November 1978,* 25–37. New York/London: Garland Publishing, 1979.

Lloyd, Christopher. *The British Seaman 1200–1860. A Social Survey.* London: Collins, 1968.

———. *The Nation and the Navy: A History of Naval Life and Policy.* London: Cresset Press, 1954.

McConnell, Anita, ed. *Directory of Source Materials for the History of Oceanography.* UNESCO Technical Papers in Marine Science 58. Paris: UNESCO, 1990.

Mills, Eric L. *Biological Oceanography: An Early History, 1870–1960.* Ithaca, N.Y.: Cornell University Press, 1989.

Mills, Eric L., ed. *History of Oceanography.* Newsletter of the Commission of Oceanography, Division of History of Science, International Union of the History and Philosophy of Science, no. 1– (1989–).

———. *One Hundred Years of Oceanography.* Halifax: Dalhousie University in cooperation with The Nova Scotia Museum, 1975.

Nicholls, David. *Nineteenth-Century Britain 1815–1914.* Folkestone: Dawson; Hamden, Conn.: Archon Books, 1978.

Rasor, Eugene L. *British Naval History Since 1815: A Guide to the Literature.* New York/London: Garland Publishing, 1990.

Rehbock, Philip F. "The Early Dredgers": 'Naturalizing' in British Seas, 1830–1850." *Journal of the History of Biology* 12 (1979): 293–368.

Rice, A. L. *British Oceanographic Vessels 1800–1950.* London: Ray Society, 1986.

Schlee, Susan. *The Edge of an Unfamiliar World: A History of Oceanography.* New York: E. P. Dutton, 1973.

Sears, Mary, and Daniel Merriman, eds. *Oceanography: The Past.* New York: Springer-Verlag, 1980.

Stephens, W. B. *Sources for English Local History.* Manchester: Manchester University Press; Totowa, N.J.: Rowman and Littlefield, 1973.

Thompson, F. M. L. *The Rise of Respectable Society: A Social History of Victorian Britain 1830-1900.* Cambridge, Mass.: Harvard University Press, 1988.

Wood, Gordon. "Historians and Documentary Editing." *Journal of American History* 67 (1981): 871–877.

Wüst, Georg. "The Major Deep-Sea Expeditions and Research Vessels, 1873-1960." *Progress in Oceanography* 2 (1964): 1–52.

Index

(Illustrations in [brackets])

About the Author

Philip F. Rehbock received degrees in economics and the history of science from Stanford University and The Johns Hopkins University. From 1965 to 1970 he served as a supply corps officer in the U.S. Navy, working with many of Joseph Matkin's modern American naval counterparts. His publications concern the biological and ocean sciences of the nineteenth century and include *The Philosophical Naturalists: Themes in Early Nineteenth-Century British Biology.* He is presently a professor at the University of Hawaii, where he holds a joint appointment in the History and General Science departments.

Production Notes

Composition and paging were done on the
Quadex Composing System and typesetting
on the Compugraphic 8400 by the design
and production staff of University of
Hawaii Press.

The text typefaces are Trump and
Compugraphic Optima, and the display
typeface is Trump.

Offset presswork and binding were done by
The Maple-Vail Book Manufacturing Group.
Text paper is Writers RR Offset, basis 50.

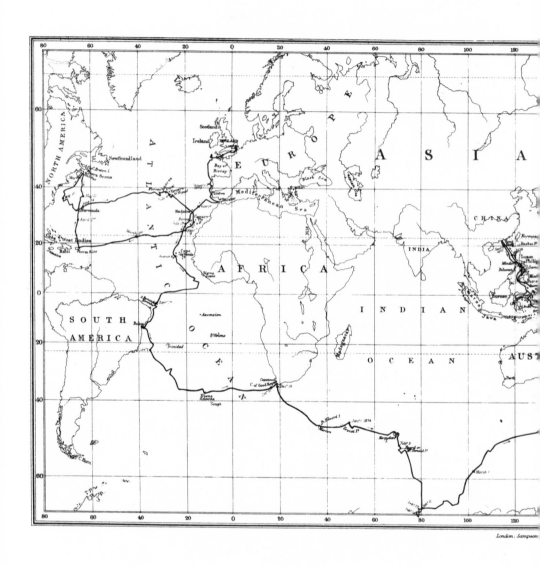

London: Sampson